# Advanced Blasting Technology for Mining

# Advanced Blasting Technology for Mining

Guest Editors

**Krzysztof Skrzypkowski**
**Andrzej Biessikirski**

MDPI

Basel • Beijing • Wuhan • Barcelona • Belgrade • Novi Sad • Cluj • Manchester

*Guest Editors*

Krzysztof Skrzypkowski
Faculty of Civil Engineering
and Resource Management
AGH University of Krakow
Krakow
Poland

Andrzej Biessikirski
Faculty of Civil Engineering
and Resource Management
AGH University of Krakow
Krakow
Poland

*Editorial Office*
MDPI AG
Grosspeteranlage 5
4052 Basel, Switzerland

This is a reprint of the Special Issue, published open access by the journal *Applied Sciences* (ISSN 2076-3417), freely accessible at: https://www.mdpi.com/journal/applsci/special_issues/KEG2N3IPO7.

For citation purposes, cite each article independently as indicated on the article page online and as indicated below:

Lastname, A.A.; Lastname, B.B. Article Title. *Journal Name* **Year**, *Volume Number*, Page Range.

**ISBN 978-3-7258-6768-4 (Hbk)**
**ISBN 978-3-7258-6769-1 (PDF)**
**https://doi.org/10.3390/books978-3-7258-6769-1**

# Contents

# About the Editors

**Krzysztof Skrzypkowski**

Krzysztof Skrzypkowski is an Associate Professor of Mining Engineering at the AGH University of Krakow, Faculty of Civil Engineering and Resource Management, Department of Mining Engineering and Occupational Safety, in Poland, working as a university professor since 2021. His scientific research concerns the stability of underground mine workings, particularly the selection of supports for room and roadway excavations. In addition, his research interests focus on energy consumption in underground mining, computer-aided design, and determining the geotechnical parameters of backfilling materials. As the head of a rock bolting laboratory, he performs model, numerical, and industrial tests on various cooperation mechanisms between mining supports and rock masses. He is the author of over 210 scientific publications, several patents, and a utility model, adressing topics including artificial pillars, especially interrooms, support housing, load absoroption in rock bolt support, support pillars for mine roofs, clamps for mining anchors, anchor plate systems with bolt load indicators, yielding with rock bolt support load indicators, cribs in the mining industry, methods for measuring anchor loading and dynamometric anchor washers, and liable supporting feet in longwall mine linings. As an academic lecturer, he works closely with scientific institutes, industries, and universities involved in exploiting mineral and rock raw materials.

**Andrzej Biessikirski**

Andrzej Biessikirski has, since 2023, held the position of Research Professor at the AGH University of Krakow, Faculty of Civil Engineering and Resource Management, Department of Mining Engineering and Occupational Safety, Poland. His work has a strong interdisciplinary focus, poised at the intersection of explosives engineering, mining science, and applied physico-chemical analysis. He is the author or co-author of 92 scientific publications, making substantial contributions to both fundamental research and industrial applications in blasting engineering and high-energy materials. His research covers the design, characterization, and performance evaluation of explosive materials, with particular emphasis on ammonium nitrate(V)-based explosives, blasting additives and modifiers, detonation properties, and post-blast fume analysis. A major stream of his research addresses rock fragmentation and muck pile characterization in open-pit and block-caving mining systems, using both classical models (e.g., Kuz–Ram) and modern digital approaches. He is recognized for the application of photogrammetry and the Structure-from-Motion (SfM) method, cumulative size distribution analysis, and indirect fragmentation assessment techniques, including tools such as Split Desktop 2.0 and O-Pitblast.

Overall, his research integrates chemistry, materials science, mining engineering, geomechanics, and digital diagnostics to improve blasting efficiency, operational safety, and environmental performance in modern mining and civil engineering.

applied sciences

MDPI

*Editorial*

# Advanced Blasting Technology for Mining

**Krzysztof Skrzypkowski * and Andrzej Biessikirski**

Faculty of Civil Engineering and Resource Management, AGH University of Krakow, 30-059 Kraków, Poland; abiess@agh.edu.pl
* Correspondence: skrzypko@agh.edu.pl; Tel.: +48-126172160

The use of explosives in both open-pit and underground mining is associated with a sudden increase in pressure during the detonation of explosive charges. Explosives are used for destructive impacts on the rock (exploitation), but they also induce vibrations around the detonation site [1–4]. To minimize the impact of blast vibrations on nearby railway structures, it is recommended to optimize drilling and explosive charge arrangements by controlling charge height and regulating explosive quantities per blast and per segment (Contribution 1). Dudek et al. (Contribution 2) state that quantifying the resource efficiency impact of alternative blast designs and demonstrating operational and environmental trade-offs aligns with practical engineering choices in surface mining by integrating blasting and diesel-fueled equipment emissions into a unified assessment of ramp construction strategies. Paraseismic vibrations that propagate through the rock mass can be a source of adverse impacts, such as damage to structures within and outside the mine, as well as the risk of landslides on the slopes. Xue et al. (Contribution 3) used a laser Doppler measurement method to provide novel, technical support for measuring and analyzing flameproof enclosure failure modes. Iacob et al. (Contribution 4) developed specific algorithms to analyze the spatial distribution of perforations generated by metal fragments on panels located around the detonation site. Pyra and Żołądek (Contribution 5) found that the use of photogrammetry based on unmanned aerial vehicles (UAVs) led to improved safety conditions, including a documented reduction in rockfall by an average of 42% near protected structures. Furthermore, the photogrammetric approach allowed for a ten-fold reduction in the number of images captured and an 80% reduction in model processing time, without compromising mapping accuracy. Another area of interest is the transport of explosives, which requires special procedures. Maranda et al. (Contribution 6) found that sensitizing the matrix directly in the blasthole obviates the need for transporting explosive materials via public roads, a factor that is critical for ensuring transport safety. This work determined the classification of sensitizers, their physicochemical properties, and their interactions with the emulsion matrix. A further study by Nachlik and Pytlik (Contribution 7) investigated laboratory testing of explosives. This testing requires the use of test stands that reflect industrial conditions, and, at the same time, indicate areas for optimization. Nachlik and Pytlik revealed that inter-laboratory comparison tests are crucial for determining the overall level of sensitivity of the tested explosives, and applying sound and light detectors during sensitivity to friction tests may partially eliminate the human factor, resulting in more compliant results.

One of the basic parameters determining the effectiveness of mining with explosives is the correct arrangement of holes in rows. Casale et al. (Contribution 8) revealed that the relationship between row spacing, spacing, and detonation sequence was embedded within the powder factor formula, where these parameters influence fragmentation and displacement outcomes. They develop a simplified method for determining the powder factor that can serve as both a design validation tool and a parameter for estimating the cost of interventions on unstable slopes. With regard to underground mining, for both ore and coal, one area of particular interest is the impact of vibrations on the rock bolt and arch yielding support. Zheng et al. (Contribution 9) employed numerical modeling and discovered that there is a quantitative relationship between the explosion equivalence factor and damage evolution, and determined that a dynamic damage model should be developed for underground excavations. Pei et al. (Contribution 10) confirmed that propagation of blast-induced shock waves in a T-shaped tunnel exhibits consistent patterns across four distinct batches, shaped by reflections and diffractions. To achieve adequate excavation progress, special blasting patterns are developed, which are adapted to ensure appropriate fragmentation of the mined material and to avoid roof collapse in the case of roofs composed of high-strength rocks that accumulate the elastic energy of the rock mass. Liu et al. (Contribution 11) stated that to achieve good pre-cracking in complex stress fields, the direction of the local maximum principal stress must be aligned with the direction of the inter-hole crack. Delayed sandblasting can be used to improve the crack root, allowing cracks to expand and merge in adjacent holes.

Finally, Gautam et al. (Contribution 12) found that roof blasting depends on the nature of local roof falls, results of strata monitoring studies, and the hanging span of roof strata near the line of extraction. In this work, the researchers developed blasting design strategies to promote caving and minimize roof overhang for the continuous miner technology used in the bord and pillar mining method.

The use of explosives in both open-pit and underground mining is widespread [5] and continues to evolve thanks to the availability of materials, their safe use, and the monitoring of their impact on structures, mine workings, and mine supports. Further research at universities and research institutes will be developed and verified under real-world conditions to reduce operating costs while achieving the desired crushing of ore and minimizing the impact on the surrounding environment and rock mass.

**Author Contributions:** Conceptualization, K.S. and A.B.; writing—review and editing, K.S. and A.B.; supervision, K.S. and A.B.; project administration, K.S. and A.B. All authors have read and agreed to the published version of the manuscript.

**Institutional Review Board Statement:** Not applicable.

**Informed Consent Statement:** Not applicable.

**Conflicts of Interest:** The authors declare no conflicts of interest.

**List of Contributions:**

1. Zhang, S.; Zhang, Z.; Wang, K.; He, D.; Huang, Y. Research on the Impact of Blasting Vibration in Mining Areas on Surrounding Railway Structures. *Appl. Sci.* **2025**, *15*, 4624. https://doi.org/10.3390/app15094624.
2. Dudek, M.; Dworzak, M.; Biessikirski, A. Influence of Blasting Approaches in In-Pit Haul Road Construction on Emission Levels and Resource Management: A Case Study from the Holcim "Dubie" Open-Pit Mine. *Appl. Sci.* **2025**, *15*, 12310. https://doi.org/10.3390/app152212310.
3. Xue, X.; Qiu, J.; Zhang, H.; Yang, W.; Wan, H.; Chen, F. Analysis of Internal Explosion Vibration Characteristics of Explosion-Proof Equipment in Coal Mines Using Laser Doppler. *Appl. Sci.* **2025**, *15*, 9255. https://doi.org/10.3390/app15179255.
4. Iacob, N.; Kuncser, A.; Stanciu, A.; Palade, P.; Schinteie, G.; Leca, A.; Ghicioi, E.; Laszlo, R.; Radermacher, L.; Nicola, A.; et al. Explosion Characteristics and Lethality Degree Evaluation from Improvised Explosive Device (IED) Detonation in Urban Area: Case of the Cylindrical Geometry. *Appl. Sci.* **2025**, *15*, 11851. https://doi.org/10.3390/app152211851.
5. Pyra, J.; Żołądek, T. Application of UAVs to Support Blast Design for Flyrock Mitigation: A Case Study from a Basalt Quarry. *Appl. Sci.* **2025**, *15*, 8614. https://doi.org/10.3390/app15158614.

 https://doi.org/10.3390/app16031232

6. Maranda, A.; Markowska, D.; Kukfisz, B.; Jakubczak, W. A Comprehensive Review of the Influence of Sensitizers on the Detonation Properties of Emulsion Explosives. *Appl. Sci.* **2025**, *15*, 2417. https://doi.org/10.3390/app15052417.
7. Nachlik, S.; Pytlik, M. The Investigation of Porcelain Plates Roughness Influence on Determination of Explosives Friction Sensitivity. *Appl. Sci.* **2025**, *15*, 4478. https://doi.org/10.3390/app15084478.
8. Casale, M.; Dino, G.A.; Oggeri, C. Blasting of Unstable Rock Elements on Steep Slopes. *Appl. Sci.* **2025**, *15*, 712. https://doi.org/10.3390/app15020712.
9. Zheng, Q.; Ding, P.; Yan, Z.; Zhu, Y.; Zhang, J. Dynamic Response of Methane Explosion and Roadway Surrounding Rock in Restricted Space: A Simulation Analysis of Fluid-Solid Coupling. *Appl. Sci.* **2025**, *15*, 9454. https://doi.org/10.3390/app15179454.
10. Pei, L.; Li, H.; Wang, Z.; Zhang, G.; Gao, F.; Sun, S. Propagation Characteristics of Shock Waves and Distribution Features of Loads in T-Shaped Tunnels with Protected Door. *Appl. Sci.* **2025**, *15*, 11210. https://doi.org/10.3390/app152011210.
11. Liu, X.; Yan, P.; Zhu, J.; Yang, X.; Zhang, X.; Zhou, C.; Lu, W.; Chen, M.; Wang, G.; Wang, Y. Effect of Neighboring Hole Impacts on Inter-Hole Dynamic Presplitting Process with Consideration of Crack Width Variations. *Appl. Sci.* **2025**, *15*, 10036. https://doi.org/10.3390/app151810036.
12. Gautam, A.; Kumar, A.; Ram, S.; Skrzypkowski, K.; Zagórski, K.; Zagórska, A.; Madziarz, M.; Migda, K. Strata Control by Roof Blasting for Bord and Pillar Mining Method in Mechanized Depillaring Panels. *Appl. Sci.* **2025**, *15*, 1403. https://doi.org/10.3390/app15031403.

## References

1. Siskind, D.E.; Stagg, M.S.; Kopp, J.W.; Dowding, C.H. Structure Response and Damage Produced by Ground Vibration from Surface Mine Blasting. In *U.S. Bureau of Mines Report of Investigations RI 8507*; U.S. Department of the Interior: Washington, DC, USA, 1980.
2. Singh, T.N.; Singh, V. An intelligent approach to prediction and control of ground vibration in mines. *Geotech. Geol. Eng.* **2005**, *23*, 249–262. [CrossRef]
3. Kahriman, A. Analysis of parameters of ground vibration produced from bench blasting at a limestone quarry. *Soil Dyn. Earthq. Eng.* **2004**, *24*, 887–897. [CrossRef]
4. Nateghi, R. Prediction of ground vibration level induced by blasting. *Int. J. Rock Mech. Min. Sci.* **2011**, *48*, 899–908. [CrossRef]
5. Jimeno, C.L.; Jimeno, E.L.; Carcedo, F.J.A. *Drilling and Blasting of Rocks*; CRC Press: Boca Raton, FL, USA, 1995.



 https://doi.org/10.3390/app16031232

applied sciences

MDPI

Article

# Strata Control by Roof Blasting for Bord and Pillar Mining Method in Mechanized Depillaring Panels

**Abhishek Gautam [1,2], Ashok Kumar [1,*], Sahendra Ram [3], Krzysztof Skrzypkowski [4,*], Krzysztof Zagórski [5], Anna Zagórska [6], Maciej Madziarz [7] and Krzysztof Migda [8]**

1 Department of Mining Engineering, Indian Institute of Technology (Indian School of Mines), Dhanbad 826004, Jharkhand, India; abhishek.17dp000296@me.iitism.ac.in or abhishekbansal65@gmail.com
2 Pure Searsole Colliery, Satgram-Sripur Area, Eastern Coalfields Limited (a Subsidiary of Coal India Limited), Paschim Bardhaman 722143, West Bengal, India
3 Department of Mining Engineering, National Institute of Technology Rourkela, Rourkela 769008, Odisha, India; rams@nitrkl.ac.in
4 Faculty of Civil Engineering and Resource Management, AGH University of Krakow, Mickiewicza 30 Av., 30-059 Kraków, Poland
5 Faculty of Mechanical Engineering and Robotics, AGH University of Krakow, Mickiewicza 30 Av., 30-059 Kraków, Poland; zagkrzys@agh.edu.pl
6 Research Centre in Kraków, Institute of Geological Sciences, Polish Academy of Science, Senacka 1, 31-002 Kraków, Poland; a.zagorska@ingpan.krakow.pl
7 Faculty of Geoengineering, Mining and Geology, Wroclaw University of Science and Technology, 27 Wybrzeże Stanisława Wyspiańskiego st., 50-370 Wrocław, Poland; maciej.madziarz@pwr.edu.pl
8 Gorotech sp. z o.o., Kraszewskiego 36, 30-110 Kraków, Poland
* Correspondence: ashokmin@iitism.ac.in or ashok.bhu.min09@gmail.com (A.K.); skrzypko@agh.edu.pl (K.S.); Tel.: +91-8603401161 (A.K.); +48-12-617-21-60 (K.S.)

**Abstract:** This article discusses the challenges and remedial measures (roof blasting) adopted to deal with extremely difficult cavable roofs while working with Continuous Miner Technology (CMT) in Bord and Pillar mining method. The main objective of the roof blasting in the goaf is to induce caving to minimize the abutment load in and around the working face to prevent incidences of coal bumps, goaf swelling, pillar spalling, and air blasts. It was found that roof blasting is subjected to the nature of local roof falls, results of strata monitoring studies, and the hanging span of roof strata near the line of extraction. Efforts have been made to design different geotechnical elements and structures involved during the mechanized depillaring panel of the Tawa-I mine, using empirical methods to minimize the issues of roof overhang in the goaf. Numerical simulation studies were also performed to validate the design predicted by empirical methods. The Cavability Index indicated that the roof at Tawa-I mine is difficult to cave. To address this, roof blasting design strategies have been discussed to promote caving and minimize roof overhang, ensuring safe and efficient strata control while operating CMT in the challenging and complex geo-mining conditions of the mine.

**Keywords:** Bord and Pillar; roof blasting; continuous miner; development; coal pillar; depillaring; goaf; roof fall; spalling; instrumentation; induced caving

## 1. Introduction

Continuous Miner Technology has restored the faith of the Indian mining industry in underground coal mining methods. This technology has proved to be successful with improved efficiency and safety in the difficult and complex geo-mining conditions of Indian underground coal mines. A number of coal mines have already adopted this technology since its introduction in 2003 in the Anjan Hill mine of the Chirimiri Area of South Eastern

Coalfields Limited (a subsidiary of Coal India Limited). In this line, the Tawa-I mine in the Pathakhera Area of Western Coalfields Limited (also a subsidiary of Coal India Limited) is planning to introduce this technology. Based on the project report and sheme prepared by CMPDIL, it has been proposed by the mine management of the Tawa-I mine to introduce Continuous Miner (CM) Technology in the Bagdona Coal Seam of the mine. The BCS panels of the Tawa-I mine are being developed by the conventional drilling and blasting method with a 3.8 m gallery width up to full seam thickness (average thickness = 1.80 m).

Four coal seams, namely, Upper Workable Seam (UWS), Lower Workable Seam (LWS), Bagdona Coal Seam (BCS), and Seam IA, have been established by drilling in the Tawa-I Underground Mine. The variation in thickness of different seams and their intervening parting, as encountered in different boreholes, is mentioned in Figures 1 and 2.

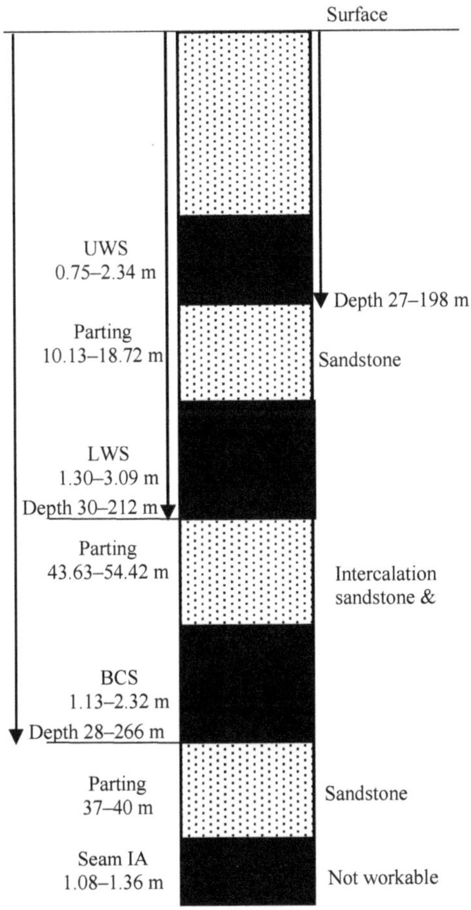


**Figure 1.** Partings between different coal seams present in Tawa-I mine.

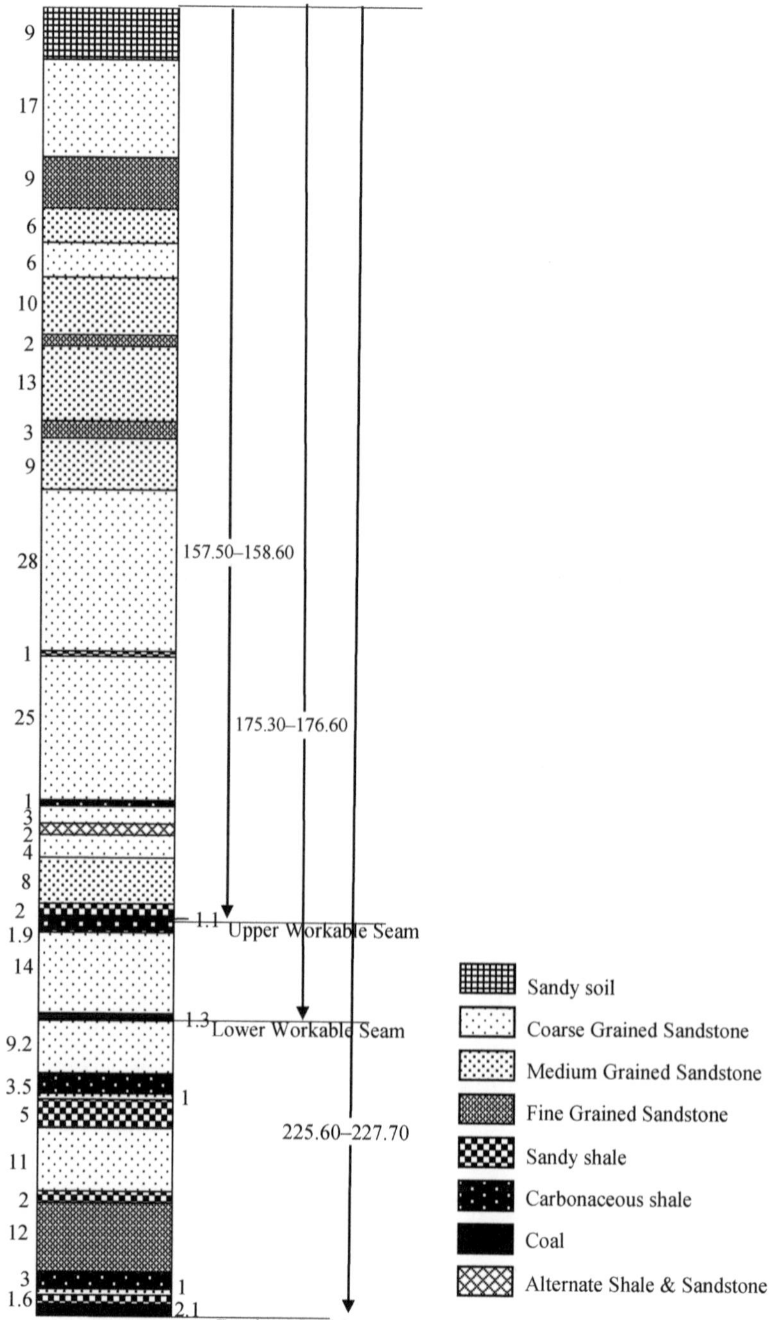


**Figure 2.** Borehole lithology above the panel of Bagdona Coal Seam based on CMPS 22.

Presently, the panel (Figure 3) is being developed on pillars (35.7 m × 31.2 m) and galleries (3.8 m × 1.8 m) by the Bord and Pillar Mining Method. It has been planned to depillar the conventionally developed pillars in the panel using mechanized Continuous

Miner (CM) technology. Furthermore, the Continuous Miner (CM) must extract the sides of the pillar to widen the existing galleries from 3.8 m to ensure adaptability.

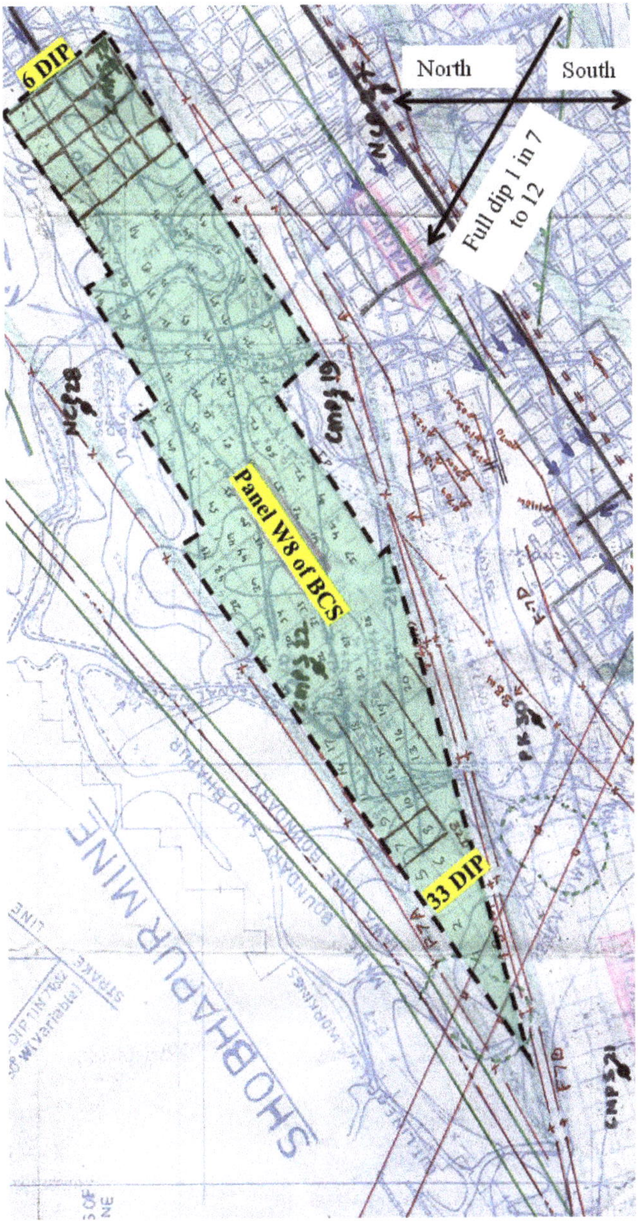


**Figure 3.** Panel in Bagdona Coal Seam of Tawa-I Underground Coal Mine.

Depillaring operation in the panel needs a preparatory arrangement for the widening of galleries by cutting the sides of the pillar using CM for its easy maneuverability and adaptability and advanced installation of geotechnical instruments. A Shale roof of varying thickness of 3–5 m is present above the coal seam in the panel. A fault with a 1.5 m throw is running from 4 Dip to 13 Dip between 63L and 64L (Figure 3) and is further surrounded by two major faults. The panel is lying below a panel that had been depillared with the caving-in LWS at an average parting of 50 m. A maximum subsidence of 1.6 m has been reported to date in the Tawa-I mine in the panel E2 section of the LWS. The maximum angle of the draw has been observed up to 11° in the E6 district of the LWS, which was depillared 11 years ago. The geo-mining conditions of the panel are mentioned in Table 1. A number of developments have been made for the design of geotechnical elements [1–4] involved in working with Continuous Miner Technology. Furthermore, researchers have also investigated methods to effectively manage the challenging cavable roof.

**Table 1.** Geo-mining conditions of the panel.

| Parameters | | Panel Details |
|---|---|---|
| Name of the coal seam | | Bagdona Coal Seam (BCS) |
| Boreholes in/around the panel | | CMPS-2, 22 |
| Grade of coal | | G-6 and G-8 |
| Degree of gassiness | | II |
| Seam thickness | | 1.77 m ≈ 1.8 m |
| Gradient of seam | | 1 in 20 along level and 1 in 16 along dip |
| Status of overlying and underlying coal seams | | Overlying LWS at a parting of 50 m has been extracted by the caving method |
| Overlying structures | | Rainy Season Nalla |
| Roof | | 1.6 m—sandy shale, 1 m—medium-gained sandstone, 3 m—carbonaceous shale, 12 m—fine-grained sandstone |
| Immediate roof | | 1.6 m—Sandy shale |
| Immediate floor | | Sandy shale with thin sandstone |
| Geological Disturbance | | Within the Panel- Local faults with a throw of 1.6 m, slips, cracks<br>Rise Side of Panel- Fault F7B with a throw of 10–20 m<br>Dip Side of Panel- Fault F7A with a throw of 28 m |
| Panel extent | | From 63L to 69L and 6D to 33D between faults F7A (upthrow 30 m) and F7B (downthrow 30 m) |
| Size of pillar | | 39.5 m × 35 m (center to center), 35.7 m × 31.2 m (corner-to-corner) during development by drilling-blasting with 3.8 m gallery width |
| Number of pillars | | 114 (may change depending upon slips and faults exposed during the development of the virgin area in the panel) |
| Panel length | | 27 pillars (27 × 35 + 4= 949 m) |
| Panel width | | Maximum 5 pillars (39.5 × 5 + 4 = 201.5 m) but vary from 1 to 5 pillars |
| Extraction height | | 1.80 m |
| Present width of the gallery by drilling-blasting | | 3.8 m during development by drilling-blasting |
| Depth cover | | 190 m to 266 m |
| RMR of seam | | 50.4 |
| Compressive strength of coal | | 23 MPa (as per NIRM report November 2012) |
| Existing Support during development by drilling and blasting | | Quick setting cement grouted roof bolts at 1.2 m × 0.8 m grid pattern |
| Incubation Period | | 24 months |
| Boundary | North | Mine boundary |
| | South | Panel No. W7 and W7 X-Cut |
| | East | Panel No. E8 |
| | West | Shobhapur abandoned coal mine |
| Existing support system | | Quick setting cement grouted roof bolts at 1.2 m × 0.8 m grid pattern |
| Waterlogging and fire | | Free from waterlogging and fire |

The literature on strata control and mining safety in coal mines presents various strategies and technologies aimed at improving safety, stability, and efficiency in underground mining operations. MSHA [5] outlines essential guidelines for managing roof and rib stability, emphasizing proactive measures such as roof bolting and rib control to prevent roof falls. Jayanthu [6] focuses on the application of geotechnical instrumentation and numerical modeling to monitor and predict strata behavior, thus, aiding in better risk assessment and safety management. Konicek et al. [7] explore the use of long-hole destress blasting to prevent rockbursts in deep coal mines, highlighting its effectiveness in reducing stress concentrations and maintaining stability in high-stress environments. Huang et al. [8] discuss hydraulic fracturing technology to manage hard roof conditions, demonstrating its potential to enhance roof stability by reducing stress and improving permeability. Xu et al. [9] examine mining-induced strata movement and roof behavior, providing insights into the importance of monitoring and analyzing strata to optimize roof support systems and enhance safety. Kang et al. [10] propose innovative intelligent mining technologies and real-time monitoring for deep coal mines, addressing challenges related to high geological stress and complex strata behavior. Guo et al. [11] investigate roof strata behavior and support resistance for ultra-thick longwall top coal caving panels, offering solutions for maintaining safe working conditions in high-stress environments. Jangara and Ozturk [12] explore the design of longwall top coal caving for thick coal seams with poor surrounding strata, providing recommendations for effective support systems and safety protocols. Kumar et al. [13] examine controlled fracturing of overlying strata to ensure the safe extraction of thick coal seams. Wang et al. [14] investigate roof presplitting techniques in non-pillar coal mining, focusing on reducing stress concentrations in hard roof conditions. Finally, Chang et al. [15] present safety and high-recovery mechanisms for the initial mining of complex coal seams with thick, hard roofs, emphasizing the need for advanced support systems and innovative technologies for safe and efficient operations. Together, these studies contribute valuable insights into the development of safer, more efficient mining practices in challenging coal mining environments.

This paper mentions the design of underground structures based on empirical and simulation studies for better strata control management with CM Technology in a trialed mechanized depillaring panel of the Tawa-I mine.

## 2. Empirical Approach to the Design of Underground Structures

Bagdona Coal Seam (BCS), after access through inseam drifts, has been extensively developed using the Bord and Pillar Mining Method (BPMM). The immediate overlying LWS was already developed and depillared many years ago. Presently, development operations are in progress in the panel using a Load Haul Dumper (LHD) and belt conveyor combination. It has been proposed to depillar these developed pillars with the combination of Continuous Miner/Shuttle Car/Quad Bolter/Feeder Breaker technology.

### 2.1. Cavebility Index

On the basis of different field experiences, the caving characteristics of overlying strata are quantified in terms of the Cavability Index ($I$) [16], which is defined as:

$$I = \frac{\sigma l^{n} T^{0.5}}{5} \tag{1}$$

where $\sigma$ is the uniaxial compressive strength in kg/cm$^2$, $l$ is the average length of core in cm, T is the thickness of the strong bed in m, and the factor $n$ has a value of 1.2 in the case of uniformly massive rocks with a weighted average of RQD of 80% and above and in all other cases, the value of n is equal to 1.

The Cavability Index of the immediate roof of the panel is estimated to be 7260, which falls under the category of roof cavable with difficulty (Table 2). Since there is no evidence of water in the overlying strata, it is anticipated that roof falls will be delayed in the goaf during work in the panel.

**Table 2.** Caving index versus caving nature of strata [16].

| Classification of Roof | Cavability Index | Remarks |
|---|---|---|
| Easily cavable roof | I ≤ 2000 | No perceptible weighting on the face |
| Moderately cavable roof | 2000 < I ≤ 5000 | Weighting but not dynamic in nature |
| Roof cavable with difficulty | 5000 < I ≤ 10,000 | Weighting may or may not be dynamic in nature |
| Cavable with substantial difficulty | 10,000 < I ≤ 14,000 | Weighting may or may not be dynamic in nature |
| Cavable with extreme difficulty | I ≥ 14,000 | Weighting will be dynamic in nature |

### 2.2. Width of Gallery

Existing galleries of 3.8 m need to be widened for adaptability of the CM (Figure 4a) by cutting the sides of the coal pillar measuring 35.7 m × 31.2 m (corner-to-corner). Considering the extent of spalling, frequency of slips/joints/discontinuities present in the panel, and available case studies of mechanized depillaring (MD) panels using CM Technology in different coalfields of India, it is proposed here to widen the 3.8 m width of the gallery to 5.5 m only. It is suggested to extract the sides of the pillars of 1.70 m in an L-shaped manner for efficient utilization of the CM in the panel. The extraction of the sides of the coal pillar will reduce its size from 35.7 m × 31.2 m (corner-to-corner) to 34.0 m × 29.5 m (corner-to-corner). Furthermore, it will also reduce the strength of the pillar. Considering the length of the machine to be 11.9 m, it will be difficult for the CM to turn around the junctions of the 5.5 m width of the gallery. A diagonal length of around 10.50 m is required around the junction for turning the entire length of the CM; therefore, it is also proposed here to reduce the corners of the pillar as shown in Figure 4b. Around 10.68 m of diagonal length will be available after the extraction of corners around the junction, which provides a sufficient turning radius for the movement of the CM. It has also been found that the sharp 90° corners become rounded ultimately due to induced stress-driven spalling. The suitability of the 5.5 m width of the gallery has also been simulated in numerical models and found to be appropriate, which is presented in the section on Numerical Modeling.

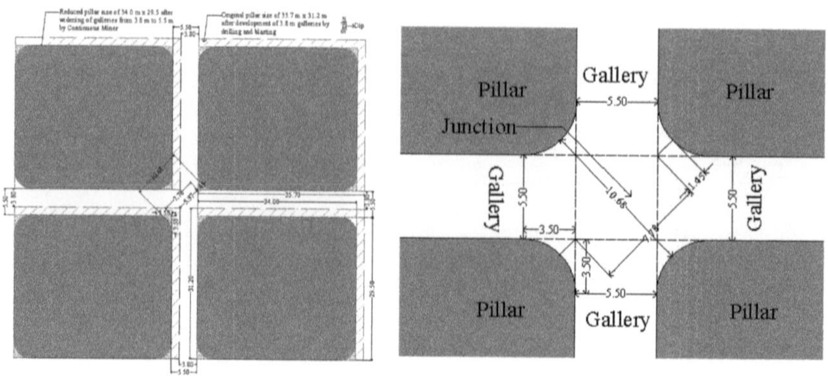


(a) (b)

**Figure 4.** (a) Widening the 3.8 m width of the gallery to 5.5 m for the adaptability of the Continuous Miner machine and (b) Extraction of 1.45 m corners of pillars around the junction for the adaptability of the 11.9 m length of the cutting machine.

2.2.1. Safety Factor of Pillar and Fender

Sheorey [17] developed an empirical formula (Equation (2)) for estimating of pillar strength, which is well established in Indian coalfields for pillar design.

$$S = 0.27\,\sigma_c h^{-0.36} + \left(\frac{H}{250} + 1\right)\left(\frac{W_e}{h} - 1\right) \text{ MPa} \quad (2)$$

where S = pillar strength (MPa), σc = uniaxial compressive strength of coal (MPa), h = working height (m), H = depth of cover (m), and we = effective width of the pillar (m) = 4A/P, where A is area and P is the perimeter of the pillar.

The Tributary Area Method [18] is used to calculate load over pillar and fender. The formation of a pillar by driving galleries all around disturbs the state of virgin stresses, keeping the total magnitude of the overlying strata weight γH constant. It is normally assumed that the entire weight overlying a tributary area is supported by solid pillars. The stress on pillar (P) is estimated using the Tributary Area Method as given in Equation (3).

$$P = \frac{\gamma H}{1 - e} \text{ MPa} \quad (3)$$

where e = recovery = $\frac{[(W_1+B)(W_2+B)-(W_1W_2)]}{(W_1+B)(W_2+B)}$, H = depth cover (m), B = width of the gallery (m), W1 = length of pillar (m), W2 = width of pillar (m), and γ = unit rock pressure (0.025 MPa/m).

The abovementioned empirical formula has been used to estimate the strength of the pillar and fender that will be formed during mechanized depillaring in the panel. Here, the fender is half of the pillar after it splits into two unequal parts (Figure 5). Unequal division has been suggested to control the cut-out distance (COD) as per the geo-mining conditions. Considering the variation in the depth of the cover of the panel, the safety factor of the pillar and fender is estimated based on the maximum 266 m depth of cover. The safety factor of the pillar is estimated using the strength and load on the pillar, as mentioned in Table 3. The estimated safety factor of the pillars (4.28) and fenders (1.64 and 2.84) for the panel seems to be competent for the pillar extraction subjected to regular caving and less disturbance due to geological discontinuities in the working area.

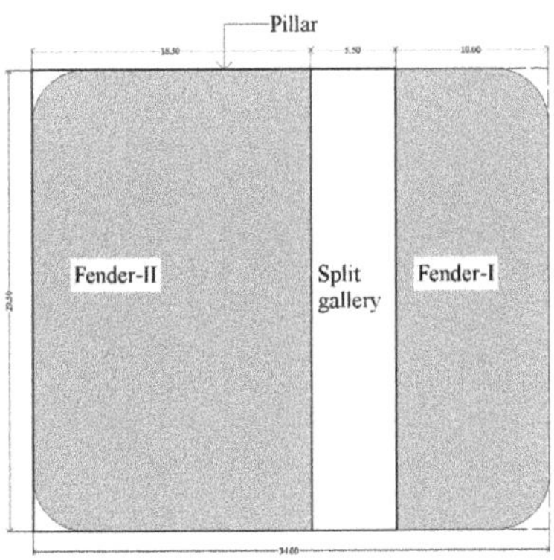


**Figure 5.** Pillar (34 m × 29.5 m) is split into two unequal fenders.

**Table 3.** Safety factors of the developed pillars by drilling and blasting and reduced pillars by widening of galleries by Continuous Miner and fenders after their splitting.

| Method | Structure | H (m) | $W_1$ (m) | $W_2$ (m) | $W_e$ (m) | B (m) | h (m) | $\sigma_c$ (MPa) | Strength (MPa) | Load (MPa) | Safety Factor |
|---|---|---|---|---|---|---|---|---|---|---|---|
| Drilling and Blasting | Pillar | 266 | 35.7 | 31.2 | 33.30 | 3.8 | 1.8 | 23 | 41.14 | 8.25 | 4.98 |
| CM | Pillar | 266 | 34 | 29.5 | 31.60 | 5.5 | 1.8 | 23 | 39.19 | 9.17 | 4.28 |
| | Fender-I | | 10 | 29.5 | 14.9 | 5.5 | | | 20.09 | 12.23 | 1.64 |
| | Fender-II | | 18.5 | 29.5 | 22.7 | 5.5 | | | 29.04 | 10.24 | 2.84 |

2.2.2. Safe Cut-Out Distance

Safe Cut-Out Distance (COD) is defined as the maximum length of the unsupported span between the advancing face and the nearest supported area, or the width between the side pillars, i.e., the width of excavation. As per the available literature review, Bieniawski [18] system (Figure 6) is found to be a suitable approach for estimating the safe COD (unsupported maximum span) during MD in the panel. The observed RMR value of the immediate roof strata of BCS is 50.40, with adjustments, as the seam was developed through solid blasting. The roof of the BCS is sandy shale, with an average value of elastic modulus of the immediate roof (E) as 2.0 GPa, and the width of the gallery has been fixed at 5.5 m.

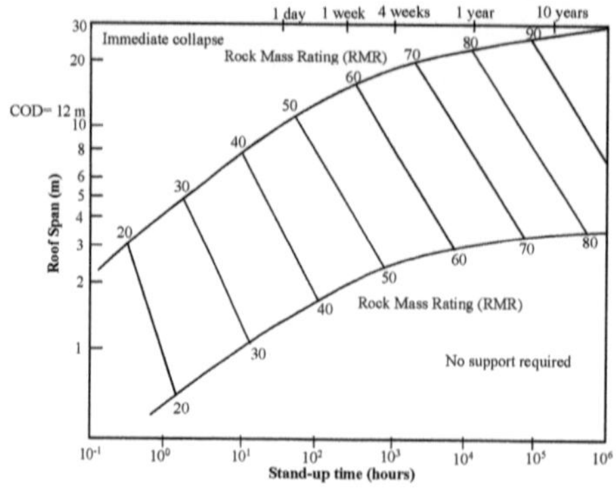


**Figure 6.** Stand-up time and unsupported span based on Rock Mass Rating (RMR) and Joint factor (Jr) [18].

Bieniawski [18] used the Rock Mass Rating (RMR) developed by Bieniawski [19–21] for stand-up time and unsupported span of roof (Figure 6). The Coal Mine Roof Rating (CMRR) developed by Molinda and Mark [22] was used to design cut-out distances in US underground mines. Bauer [23] suggested an empirical formulation (Equation (4)) to design cut-out distance based on the Coal Mine Roof Rating (CMRR), the width of the gallery, and depth.

$$D = 2.47 + 0.172R - 0.046W - 0.0008H \quad (4)$$

where D is the cut-out distance (m), R is the Coal Mine Roof Rating, W is the width of the gallery (m), and H is the depth of cover (m). Using this relationship (Equation (4)), the cut-out distance is found to be 10.67 m for an RMR of 50.40, gallery width of 5.5 m, and maximum depth of cover of 266 m.

Mandal et al. [24] proposed a relationship (Equation (5)) to estimate the safe convergence value based on COD, RMR, and width of the gallery (m).

$$C = 2.09\, W^{1.23}\, R^{-0.39}\, D^{0.11} \quad (5)$$

where C is the convergence of the immediate roof (mm), W is the width of the gallery (m), R is the Rock Mass Rating, and D is the cut-out distance (m). For a safe convergence of 5 mm, this relationship estimates a cut-out distance of 16 m for a gallery width of 5.5 m and an RMR of 50.40.

Saharan et al. [25] used the concept of Bieniawski [18] and Mark [26] to design COD for three Indian coal mines. Kumar et al. [27] attempted to define the safe and limiting value of roof sagging of 5 mm for a safe design of COD. Kumar et al. [27] proposed a relationship (Equation (6)) to estimate the COD based on elastic modulus and width of the gallery, considering a few case studies of MD in Indian coalfields.

$$D = 14.61 + 1.98E - 2.12W \quad (6)$$

where D is the cut-out distance (m), E is the elastic modulus (GPa), and W is the width of the gallery (m). Using this relationship, the cut-out distance is found to be 9 m for a 5.5 m gallery width and 2.0 GPa elastic modulus of the immediate shale roof.

Based on the (Figure 6) by Bieniawski [14], the estimated maximum unsupported span, i.e., COD for the development of galleries comes to be around 12.0 m for the existing conditions of roof rock mass above BCS. Bieniawski [18] developed the relationship to estimate the COD for a tunnel, which has a single opening. Considering the existing conditions in the panel, multiple openings, and maximum use of the CM, the COD in the panel is maintained at 11.0 m. The estimated value of the COD as 11.0 m should also be adopted during the drivage of the original/split gallery and slices in the panel. The exposed COD should be supported instantaneously after the dressing of the roof before the next cut by the CM in the original/split gallery only, not in slices. This estimated value has also been validated on numerical models through simulation. A simulation study for the geo-mining conditions of the panel is carried out to study the variation in roof deformation for different lengths of unsupported span. The cutting sequences during the development/splitting of a pillar by CM are shown in Figure 7.

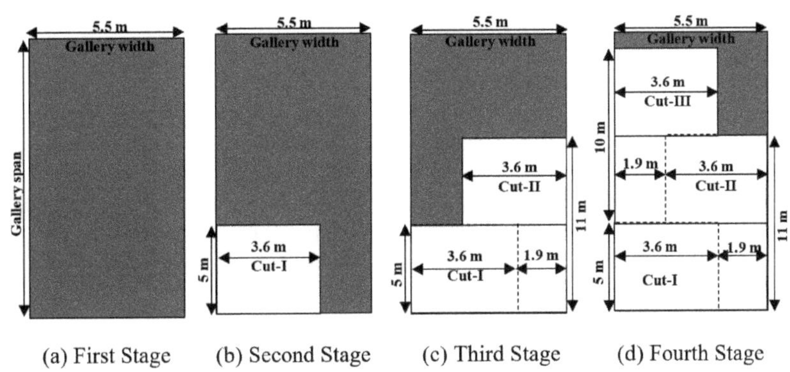


**Figure 7.** Sequence of cutting in a gallery by continuous miner for an assumed cut-out distance of 10 m.

### 2.3. Extent of Pillar Spalling

Wilson [28] developed an empirical equation to estimate the non-elastic zone based on the height of extraction and the depth of cover of the working seam. The literature survey and different field investigations in Indian coalfields have found that a higher value of vertical-induced stress is experienced by coal pillars/fenders/ribs/snooks at the goaf edge during MD. This phenomenon is more pronounced when MD is being carried out under a difficult and massive roof at a deeper cover. The deterioration of the pillar develops in the form of the non-elastic zone as side spalling on the outer part and the inner part remains under the elastic zone. Further, the non-elastic part is divided into a loose (spalled) zone and a plastic zone [29–31] (Figure 8). The spalled part of the pillar is almost dislodged, hanging, and incapable of providing any resistance against load, whereas the plastic part of the pillar remains attached to the pillar, which provides confinement against the load. Thus, it results in fracturing and spalling of corners and sides of pillars, reducing the overall size and strength of a pillar. It ultimately affects the size and shape of the rib/snook to be left and the performance of the roof bolt-based breaker line support (RBBLS) and instrumentation plan considering the location of instruments to be installed for strata monitoring around the goaf edges. An attempt has been made to estimate the extent of pillar spalling in the panel. Based on the pillar height of 1.8 m and the maximum depth of 266 m in the panel, the non-elastic zone is found to be around 2.36 m with a loose/fractured zone of 1.00 m and a plastic zone of 1.36 m. The spalled portion has been already treated during the reduction in the corners of pillars by 1.45 m for the adaptability of CM.

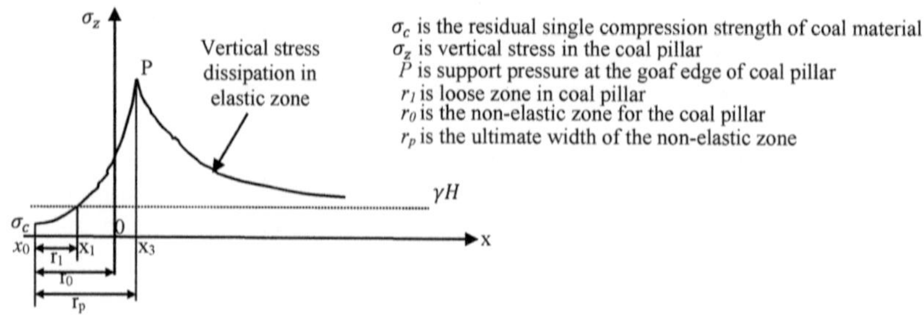


**Figure 8.** Distribution of the vertical stress for the coal pillar in the limiting equilibrium state [29–31].

### 2.4. Applied Support System in Developed Workings

An attempt has been made to design a support system for the original gallery and junction/intersection using the available empirical approaches [32], which is also studied using rock load height (RLH) on numerical models in a simulation software package, considering the site conditions. Required support resistance against the estimated rock load at the gallery and junction is calculated using the bearing/anchorage capacity of the applied supports. Supports at the original gallery and junction are designed by considering a support safety factor (SSF) (Equations (8)–(11)) of 1.5 or more for both the empirical and numerical simulation approaches. Basically, the support design depends on parameters like gallery width, rock mass rating (RMR), and unit weight of immediate roof strata.

#### 2.4.1. Rock Load

The efficacy of the support system for the original gallery and junction of developed workings of the panel is estimated using the empirical relationships given as Equations (7) and (8).

$$\textit{Rock Load in gallery } (P_g) = B\gamma\left(1.7 - 0.037\,RMR + 0.0002\mathrm{RMR}^2\right)\ \mathrm{t/m^2} \tag{7}$$

$$\text{Rock Load at Junction } (\mathrm{P_j}) = 5\mathrm{B}^{0.3}\gamma\left(1 - \frac{\mathrm{RMR}}{100}\right)^2\ \mathrm{t/m^2} \tag{8}$$

where B = gallery width (m), γ = dry unit weight of roof rock (t/m$^3$), and RMR stands for Rock Mass Rating.

2.4.2. Support Resistance

The anchorage capacity of resin grouting material required for the estimation of support resistance is tested to be 20 t during the field investigation. Support density is estimated using the number of supports (roof bolts) multiplied by the anchorage/bearing capacity of that support. Considering these parameters, support resistance is calculated using Equation (9).

$$Support\ Resistance\ (S_r) = \frac{Number\ of\ support \times Capacity\ of\ each\ support}{Area\ supported\ by\ one\ row\ of\ supports} \quad (9)$$

2.4.3. Support Safety Factor

The support safety factor is calculated using Equation (10) to establish the effectiveness of the support system.

$$Support\ safety\ factor = \frac{Support\ Resistance}{Rock\ Load} \quad (10)$$

2.4.4. Design of Applied Support

The immediate roof of the panel was examined during the field visit and found to be competent and dry. The pillars, of 39.5 m × 35 m (center-to-center), with galleries of 3.8 m, are being developed by drilling and blasting in combination with LHD for hauling and belt conveyor for the transportation of coal to the surface. The pillars were found to be intact, and there were no observations of side spalling during the field visit. Furthermore, there was no evidence of water seepage from the roof, and the shale in the roof was found to be intact within the panel during the field visit. The bearing capacity of the resin-grouted roof bolt and RMR were found to be 10 t (in 8 h) and 50.40, respectively. Rock load, support resistance, and SSF are estimated for gallery and junction using Equations (8)–(11) for suitable support design (Table 4) of developed workings of the panel by drilling and blasting. Further, it was proposed to widen the existing galleries from 3.8 m to 5.5 m; therefore, the roof bolt pattern for such widened galleries and junctions is mentioned in Table 4 and shown in Figure 9.

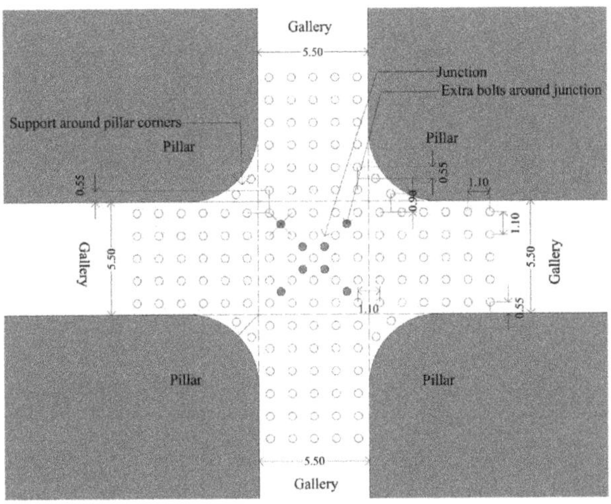


**Figure 9.** Proposed support plan in the 5.5 m widened gallery of the panel.

**Table 4.** Support design for developed panel.

| Place | Parameters | Development by Drilling and Blasting | Development by Continuous Miner |
|---|---|---|---|
| Gallery | Width (m) | 3.80 | 5.50 |
| | Dry density of rock (t/m$^3$) | 2.51 | 2.51 |
| | Rock Mass Rating | 50.40 | 50.40 |
| | Anchorage strength of resin--grouted roof bolts (t) | 10.00 | 20.00 |
| | Rock load (t/m$^2$) | 3.27 | 4.74 |
| | Spacing between two bolts in a row (m) | 1.20 | 1.10 |
| | Spacing between two consecutive rows of bolts (m) | 0.80 | 1.10 |
| | Number of roof bolts in a row | 3 | 5 |
| | Area supported by one row of bolts (m$^2$) | 4.56 | 6.05 |
| | Support resistance (t/m$^2$) | 6.58 | 16.53 |
| | Support safety factor | 2.01 | 3.49 |
| Junction | Rock load (t/m$^2$) | 4.91 | 7.11 |
| | Area supported by all bolts (m$^2$) | 14.44 | 30.25 |
| | Number of roof bolts | 11 | 33 |
| | Support resistance (t/m$^2$) | 7.62 | 21.82 |
| | Support safety factor | 1.55 | 3.07 |

### 2.5. Design of Rib/Snook

Recently, empirical formulas [1–3] have been developed to estimate the strength, area, width-to-height ratio, and equivalent width of rib/snook for Continuous Miners based on case studies of 39 mechanized depillaring panels of different Indian coalfields (Equations (11)–(14)), considering safety factor of rib/snook of 0.30–0.35.

$$S_{snook} = 15.25\frac{w^{0.58}}{h^{1.14}H^{0.17}} \tag{11}$$

$$A = 0.16\,H^{0.89}\,R^{0.14}\,h^{0.31} \tag{12}$$

$$\frac{w_e}{h} = 0.18\,H^{0.15}\,R^{0.10}\,h^{-1.03}\,A^{0.60} \tag{13}$$

$$w_e = 0.18\,H^{0.15}\,R^{0.10}\,h^{-0.03}\,A^{0.60} \tag{14}$$

where $S_{snook}$ = strength of rib/snook (MPa), w = equivalent width (m), H = depth of cover (m), and h = height of extraction (m), A = area, R = nature of roof in terms of RMR, $w_e$ = equivalent width (m).

An attempt is also made to estimate the strength of different sizes of rib/snook using formulations (Equations (12)–(14)). The strength of different sizes of rib/snook (Figure 10) formed during MD in the panel is also presented in Table 5.

**Table 5.** Strength of different sizes of rib/snook in the panel.

| Size of Rib/Snook | Area (m$^2$) | Perimeter (m) | Equivalent Width (m) | Width-to-Height Ratio | Strength (MPa) |
|---|---|---|---|---|---|
| 5.00 m × 18.50 m × 3.00 m | 101.60 | 46.84 | 8.68 | 4.82 | 10.57 |
| 4.00 m × 18.50 m × 3.00 m | 91.60 | 44.84 | 8.17 | 4.54 | 10.21 |
| 3.00 m × 18.50 m × 3.00 m | 81.60 | 42.84 | 7.62 | 4.23 | 9.81 |
| 2.00 m × 18.50 m × 3.00 m | 71.60 | 41.75 | 6.86 | 3.81 | 9.23 |
| 1.00 m × 18.50 m × 3.00 m | 61.60 | 41.75 | 5.90 | 3.28 | 8.46 |

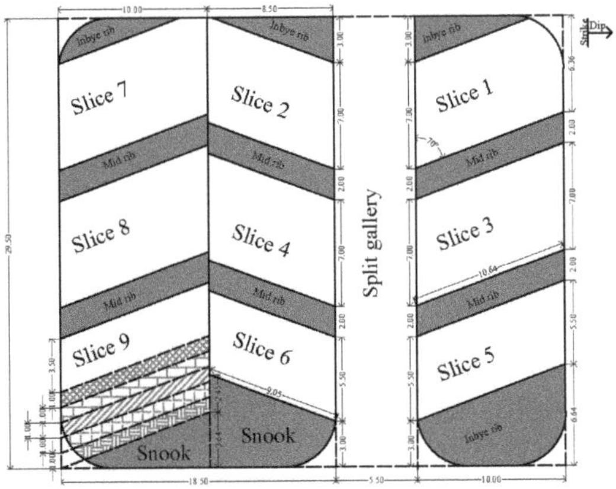


**Figure 10.** Different sizes of rib/snook considered for the panel during extraction of pillar size of 34.0 m × 29.5 m (corner-to-corner).

For the panel, H = 266 m (maximum), RMR (R) = 50.40, and h = 1.80 m, the competent area of the outby rib/snook (A) = 47.83 m$^2$, having an equivalent width of 6.16 m with a width-to-height ratio of around 3.42, is required during the extraction of pillars by CM. Therefore, a rib/snook size of 2.00 m × 18.50 m × 3.00 m, having an area of around 71.60 m$^2$ with a width-to-height ratio of around 3.81, is found to be suitable for the considered geo-mining conditions of the panel. This size of rib/snook is the minimum area to be maintained during judicious reduction, including the effects of pillar spalling (plastic and loose zone). Therefore, a rib/snook size of 3.00 m × 18.50 m × 3.00 m (area of 81.60 m$^2$, equivalent width of 7.62 m, and width-to-height ratio of 4.23) is proposed to be left during MD operation, which may be further reduced up to 2.00 m × 18.50 m × 3.00 m (area of 71.60 m$^2$, equivalent width of 6.86 m, and width-to-height ratio of 3.81) due to induced stress-driven side spalling of pillars/fenders. Further, different sizes of rib/snook in this panel are also tested during the numerical simulation study, explained in the later section.

### 2.6. Goaf Edge Support

Roof bolt-based breaker line support (RBBLS) is proposed to be installed at the goaf edges in the CM panel. Rock load at the goaf edges is estimated by multiplying the rock load height (RLH) estimated through equations mentioned below (Equations (15)–(17)) by rock density ($\gamma$). Relationships for RLH at the proposed three different positions of the RBBLS at the goaf edge have been developed [4] for different outby distances from the edge of the pillar at the goaf edge, which are given below.

For 0 m outby from goaf edge

$$RLH = 11.67\ H^{0.58}\ RMR^{-1.14} \quad (15)$$

For 1 m outby from goaf edge

$$RLH = 66.32\ H^{0.31}\ RMR^{-1.26} \quad (16)$$

For 2 m outby from goaf edge

$$RLH = 115.22\ H^{0.12}\ RMR^{-1.20} \quad (17)$$

where H is the depth of cover (266 m), and R is the RMR (50.40). Using the above equations, RLH is found to be 3.41 m at 0 m outby, 2.68 m at 1 m outby, and 2.04 m at 2.0 m outby from the goaf edge. It is not possible to use a 2 m length of roof bolts, considering the 1.8 m average thickness of the coal seam. Based on the equations and extent of spalling up to 1 m of the pillar present at the goaf edges, RLH at 1.15 m outby from the goaf edge has been used for the design of breaker line support at the goaf edge, which is shown in Figure 11. It is proposed here to install two rows of (seven bolts in each row) roof bolts of 1.70 m length at 0.75 m grid pattern located at a 1.15 m outby from the goaf edge (corner of the pillar facing the goaf line) along the middle position of the rib/snook for safe MD of the panel, as shown in Figure 11. Another two similar sets of RBBLS are to be installed in a row along the midrib in the original/split galleries. It should be mentioned that the practice of RBBLS will enhance efficiency and reduce the cost and cycle time. The normal practice of installing 1.5 m lengths of bolts should be avoided at the location of RBBLS. The design of RBBLS has also been studied on numerical models and results have been validated with the empirical formulation, presented in the numerical simulation section.

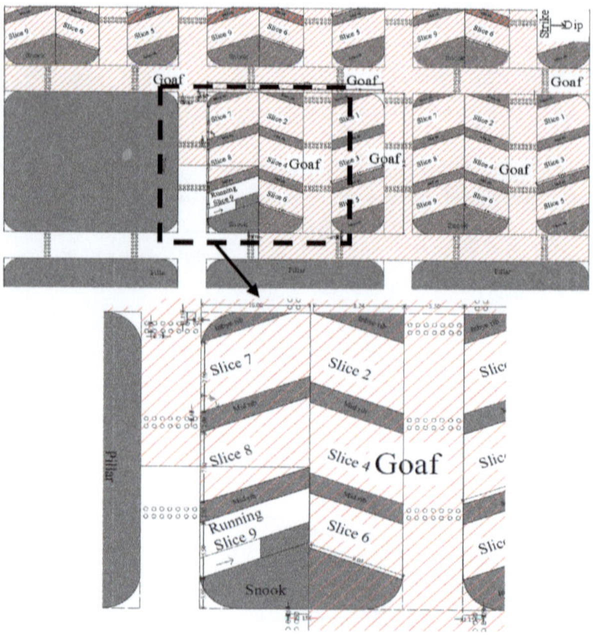


All dimensions are in meter (to-the-scale)

**Figure 11.** Proposed manner of pillar extraction and roof bolt-based breaker line support plan in panel.

## 3. Hanging Roof Span in Goaf for Caving of Overlying Strata

Van der Merwe [33] developed a relationship to estimate the roof span for roof fall (Figure 12). There are three possible fundamental modes of roof fall in the goaf. The tension failure mode for overlying strata in goaf may be triggered when the weight of the hanging roof span is more than the tensile strength of the rock, or the tensile strength across a discontinuity plane. Induced tension at the goaf edges of the roof plate may be another potential mode of failure that will occur when the mathematical sum of the induced tension and the compressive horizontal stress is greater than the tensile strength of the rock. Sliding of the rock blocks along pre-existing vertical or near-vertical joint planes may be another mode of failure that will occur when the weight of the block exceeds the cohesion and frictional resistance of the joint planes.

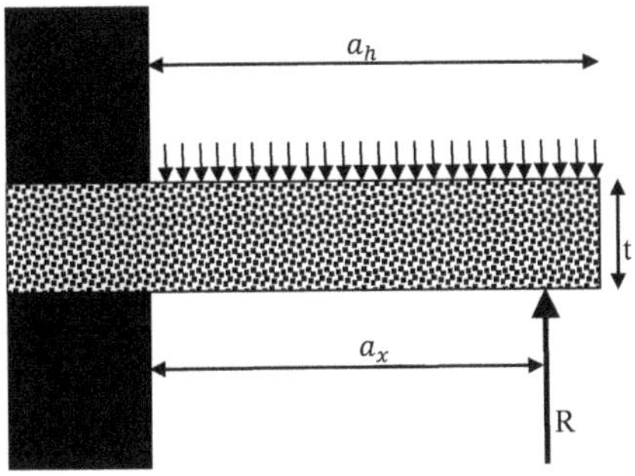


**Figure 12.** A model for an overhang of length $a_h$ supported by a snook offering resistance R at a distance $a_x$ from the closest solid pillar [33].

Most of the roof fall occurs mainly under shear failure and failure by induced tension due to roof sagging. Bending-induced tension is the most likely failure mode, which is persistent in Indian coalfields. Overlying strata in a sedimentary formation consist of a number of layers with different thicknesses and stiffnesses. The first major fall often occurs when the face advance is equal to the panel width. After the first major fall, the continuity of the roof beam is lost and the overlying strata behaves like a cantilever. After the formation of the cantilever, the continuity of strata is lost, and horizontal stresses diminish. After the occurrence of a major roof fall, the overburden plates become discontinuous, and the roof beam analogy is not valid.

The caving angle is the angle between the vertical line at the goaf edge and the inclined line to the strata over the caved zone (Figure 13). It remains almost equal to the angle of the draw in subsidence, which varies between 20 and 30° for the Indian coalfields (Figure 13). The angle of the draw of 11° has been observed in the Tawa-I mine in the E6 district of LWS, which was depillared 11 years ago. Considering the thickness, strength, and massiveness of the overlying strata of the proposed panels of the mine, the caving angle is considered to be around 11° from the vertical. As per Palchik [34], the bulking factor (BF) of different rock masses is found to range between 1.05 and 1.84, and it is found to be around 1.1 for roof rock having moderate compressive strength (20–30 MPa). Therefore, the height of caving during roof fall in the goaf is found to be around 18 m (taking 1.8 m as the height of extraction) for the panel.

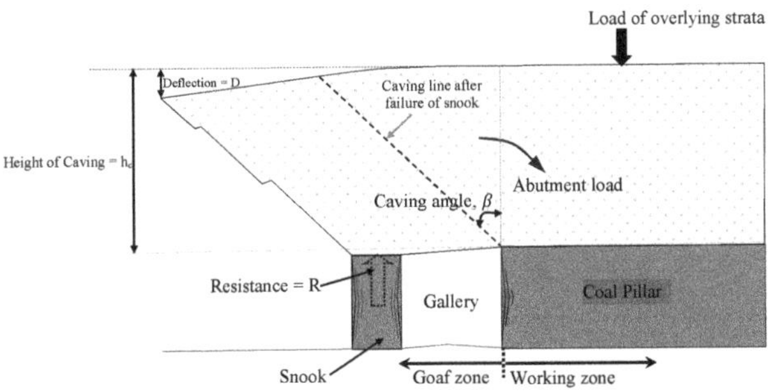


**Figure 13.** Caving mechanism at the goaf edge during MD using CM [2].

*Expected Area of Roof Falls*

For the panel, an average hanging roof span of 32 m (around 8700 $m^2$ for a panel width of 271 m) in the panel is found to be inducing the first major fall of the beam roof, forming a cantilever. Furthermore, a roof span of around 21 m (around 5687 $m^2$ for a panel width of 271 m) in the panel is found to be causing the subsequent goafing and caving of the cantilever.

Furthermore, the actual shape of the panel is asymmetrical (Figure 3) right from the start of 33DW. In the initial three rows of the panel, there is only one pillar width (33DW-31DW); thereafter, two rows of two pillars (31DW-29DW), and then two rows of three pillars (29DW-27DW). There are four pillars in the next three rows (27DW-24DW), and thereafter the actual width of five pillars (271 m) is found from 24DW. The width of the panel is an important factor for the occurrence of caving in the goaf. Roof fall will not take place in the panel during the initial five rows of pillar extraction, even after a roof span of 8700 $m^2$ or more due to insufficient panel width (33DW-29DW). The first main roof fall is expected to occur during or after the extraction of the pillar in 29DW-28DW.

The splitting of pillars is to be performed only one pillar in advance of the pillar under extraction, considering the safety of the work. A sequence of driving of split and slice galleries mentioned in Figure 11, is to be followed during the MD operation. Considering the safe COD value, the split gallery is to be driven in three cuts as first, second, and third sequences of 10 m, 10 m, and 9.5 m, respectively, in each length of cut (Figure 7).

## 4. Numerical Simulation

A systematic numerical modeling study is conducted on simulated models to visualize the performance of different empirically designed geotechnical structures under the geo-mining conditions of the panel. Taking different advantages of numerical modeling, the elastic model of the simulation package is used to simulate the rock strata at the goaf edge. Here, the rock mass failure criterion developed by Sheorey [35] is used for the analysis, which uses the Rock Mass Rating (RMR) of Bieniawski [9,20,21] for reducing the laboratory strength parameters into the corresponding rock mass values. However, this RMR is not frequently used by Indian coal mines, where CMRI-RMR [32] is commonly used. It is found that the application of CMRI-RMR in the failure criterion provides reasonable results and is used in this study.

The bulk and shear modulus are evaluated using Young's modulus and Poisson's ratio by the following equations.

$$K = \frac{E}{3(1-2v)}\ \text{GPa} \tag{18}$$

$$G = \frac{E}{2(1+v)}\ \text{GPa} \tag{19}$$

where E is the Young's modulus in GPa, K is the Bulk modulus in GPa, G is the Shear modulus in GPa, and $v$ is Poisson's ratio. Average in situ stresses [36] is used according to the following equations:

$$\sigma_v = 0.025\ \text{H MPa} \tag{20}$$

$$\sigma_H = \sigma_h = 2.4 + 0.01\ \text{H MPa} \tag{21}$$

where H is the depth cover in meters, $\sigma_v$ is vertical in situ stress, $\sigma_H$ is major horizontal in situ stress, and $\sigma_h$ is minor horizontal in situ stress. In order to assess the stability of natural support and the exposed span, safety factors are calculated using the Sheorey [35] failure criterion in the numerical models as given below:

$$\sigma_{1m} = \sigma_{cm}\left(1 + \frac{\sigma_{3m}}{\sigma_{tm}}\right)^{b_m}\ \text{MPa} \tag{22}$$

$$\sigma_{cm} = \sigma_c e^{\frac{RMR-100}{20}}\ \text{MPa} \tag{23}$$

$$\sigma_{tm} = \sigma_t e^{\frac{RMR-100}{27}}\ \text{MPa} \tag{24}$$

$$b_m = b^{\frac{RMR}{100}} bm < 0.95 \quad (25)$$

where $\sigma_{1m}$ is the triaxial strength of rock mass in MPa, $\sigma_3$ is confining stress in MPa, $\sigma_c$ is the compressive strength of the intact rock in MPa, $\sigma_t$ is the tensile strength of the intact rock in MPa, b is the exponent in failure criterion, which controls the curvature of the triaxial curve, $\sigma_{cm}$ is the compressive strength of rock mass in MPa, and $\sigma_{tm}$ is the tensile strength of rock mass in MPa.

The factor of safety is defined as

$$SF = \frac{\sigma_{1m} - \sigma_{3im}}{\sigma_{1im} - \sigma_{3im}} \text{ when } \sigma_{3im} < \sigma_{tm} \quad (26)$$

$$\text{Otherwise, } SF = \frac{-\sigma_{tm}}{\sigma_{3im}} \text{ when } \sigma_{3im} > \sigma_{tm} \quad (27)$$

where $\sigma_{1im}$ is induced major principal stress in MPa and $\sigma_{3im}$ is induced minor principal stress in MPa. From these, the rock mass shear strength, $\tau_{sm}$; the coefficient, $\mu_{0m}$ and the angle of internal friction, $\phi_{0m}$ are obtained as shown in Equations (28)–(32), which are used as Mohr–Coulomb strain softening parameters in the simulation package used during stress–strain characteristics of different sizes of rib/snook in the two panels of the Tawa-I mine.

$$\tau_{sm} = \left(\sigma_{cm}\sigma_{tm}\frac{{b_m}^{b_m}}{(1+b_m)^{1+b_m}}\right)^{1/2} \quad (28)$$

$$\mu_{0m} = \frac{{\tau_{sm}}^2(1+b_m)^2 - {\sigma_{tm}}^2}{2\tau_{sm}\sigma_{tm}(1+b_m)} \quad (29)$$

$$\phi_{0m} = tan^{-1}(\mu_{0m}) \quad (30)$$

$$\tau_{sm(residual)} = 0 \quad (31)$$

$$\phi_{0m(residual)} = \phi_{0m} - 10^\circ \quad (32)$$

It is observed that the values of shear strength, $\tau_{sm}$ and friction angle, $\phi_{0m}$ so determined as per Equations (30)–(32), needed slight adjustment to incorporate the fact that the MCSS plasticity model in the simulation package uses the linear Mohr–Coulomb criterion, while the Sheorey criterion is non-linear. To compensate the difference in nature, the value of $\tau_{sm}$ obtained from the Sheorey criterion is increased by 10% and that of $\phi_{0m}$ is reduced by 5° to use them as MCSS parameters (Figure 14).

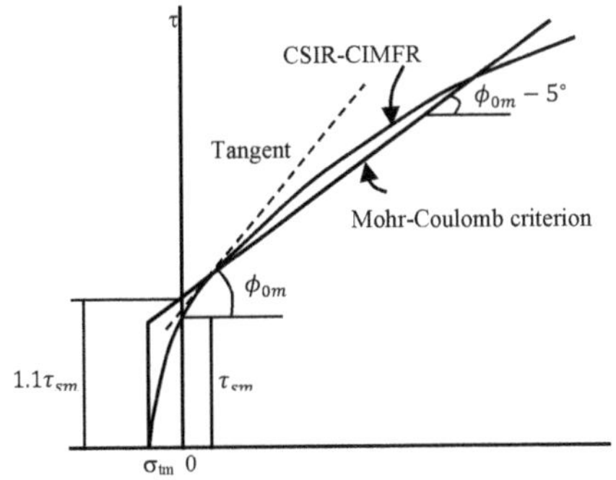


**Figure 14.** Non-linearity of Sheorey criterion against the linear Mohr–Coulomb criterion adopted in the simulation package [35].

### 4.1. Simulation of Proposed Panels

Numerical models are developed to simulate the geo-mining conditions of the panel in order to estimate the safe width of the widened gallery, cut-out distance, size of rib/snook, and pattern of roof bolts-based breaker line support (RBBLS).

#### 4.1.1. Simulation for Design of Gallery Width, Cut-Out Distance, and Rib/Snook

A simulation of a single pillar is carried out for estimation of safe split gallery width, cut-out distance, and strength of different sizes of rib/snook in the panel. Taking advantage of the existing symmetry conditions of pillar formation, a single pillar considering only three layers (floor, coal, and roof) is modeled in the simulation software for the estimation of the strength of different sizes of pillars whereas different layers as mentioned in Table 6 are considered for the simulation of safe gallery width and COD.

**Table 6.** Physico-mechanical properties of strata used in numerical modeling.

| Strata | Thickness (m) | Model | Young's Modulus (GPa) | Shear Modulus (GPa) | Bulk Modulus (GPa) | Poisson's Ratio | Density (kg/m$^3$) | Cohesion (MPa) | Friction Angle | Uniaxial Compressive Strength (MPa) | Tensile Strength (MPa) |
|---|---|---|---|---|---|---|---|---|---|---|---|
| Layer 3: Roof | 50 | Elastic | 7 | 2.8 | 4.67 | 0.25 | 2500 | 2.43 | - | - | - |
| Layer 2: Coal | 1.8 (BCS) | MCSS | 3.00 | 1.20 | 2.00 | 0.25 | 1400 | 0.181 | 39.59 | 25 | 1.67 |
| Layer 1: Floor | 50 | Elastic | 7 | 2.8 | 4.67 | 0.25 | 2310 | - | - | - | - |

Furthermore, vertical roof displacement is observed over numerical models as per Table 6 to determine a safe gallery width and a cut-out distance based on a safe limiting value of roof sagging of 5 mm. The estimated safe width of the gallery is found to be 5.5 m for the panel (Figure 15). Based on experiences of workings under similar geo-mining conditions in different Indian coal mines, the safe width of the gallery is estimated in the model, considering 5 mm [37] as the limiting value of roof sagging to induces permanent movement (plastic in nature) in the galleries to be widened/split/slice (Figure 15). It is found that the roof of the gallery is stable without a considerable amount of roof deformation after the proposed gallery width of 5.5 m (Figure 15). A pillar size of 34.0 m × 29.5 m (corner-to-corner) and a gallery width of 5.5 m are found suitable for panel, respectively, as per the numerical simulation study and found to be validating the results of empirical formulations.

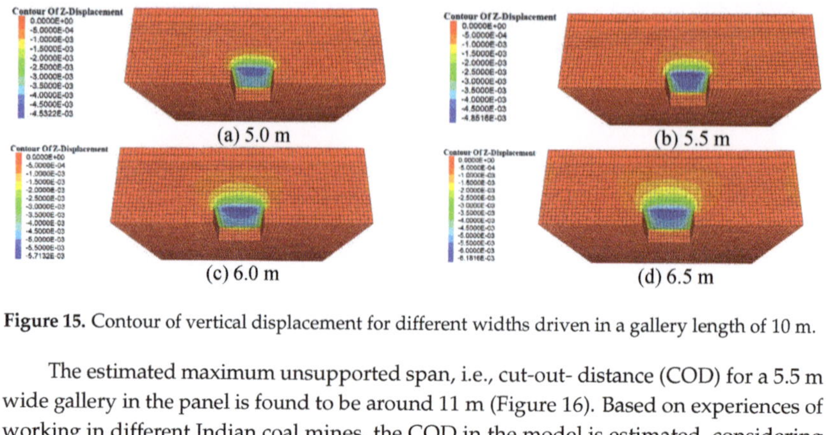

**Figure 15.** Contour of vertical displacement for different widths driven in a gallery length of 10 m.

The estimated maximum unsupported span, i.e., cut-out- distance (COD) for a 5.5 m wide gallery in the panel is found to be around 11 m (Figure 16). Based on experiences of working in different Indian coal mines, the COD in the model is estimated, considering 5 mm [37] as the limiting value of roof sagging to induce permanent movement (plastic in nature) in the split/original galleries/slice to be driven. It is found that pillars are standing without a considerable amount of roof deformation after the proposed drivage of 11.0 m, which validates the results of empirical formulations.

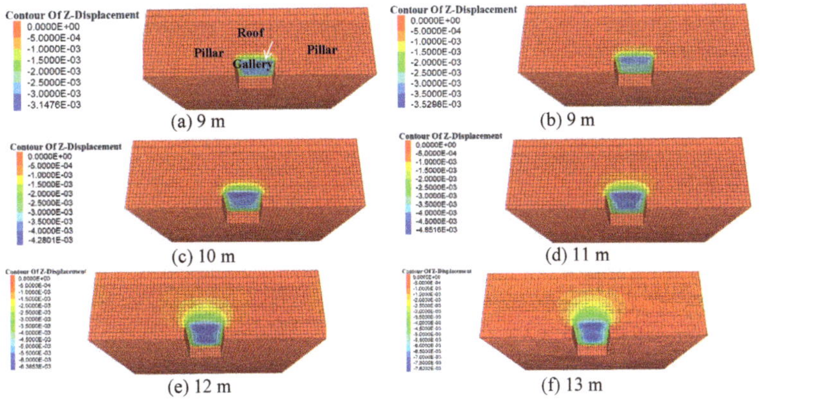


**Figure 16.** Contour of vertical displacement for different lengths driven in a gallery width of 5 m.

The upper face of the model is fixed for the estimation of the strength of ribs/snooks, with the application of a standard calibrated velocity applied in the vertical direction (Figure 17) after developing galleries. Displacements in the vertical and horizontal directions are fixed at the bottom, and four vertical symmetrical planes of the model are constrained, which are considered boundary conditions during the numerical modeling. The standard size of the model is tested during the numerical modeling, considering the limitations of the computational memory and runtime requirements; 50 m of the roof and 50 m of the floor are modeled with a 1.8 m coal seam in the panel (Figure 17). A coal seam of 1.8 m thickness is considered a strain-softening model, while the roof and floor are treated as elastic constitutive models. In situ stresses are developed in the model before studying the effects of the mining operations. It is carried out to redistribute the stresses in the model and equilibrium conditions are reached, replicating the actual site conditions. Excavation operations are carried out in the model and equilibrium conditions are reached for the estimation of strength. Vertical displacement of the top of the model is fixed and a velocity is applied in the vertical direction for estimation strength of the coal pillar. This entire process is a simple replication of the laboratory testing of rock samples in uniaxial conditions. Average vertical stress-displacement curves are plotted. The numerical model considered the material properties at 266 m depth of cover for the panel. Mohr–Coulomb Strain-Softening (MCSS) parameters have been calibrated for coal seam (Table 6) by matching the in situ stress generated in the model with the actual site and pillar strength as per Sheorey (1992).

Three layers of roof, coal, and floor are considered for the simulation as the strength of rib/snook has been found to be independent of the nature of the roof. The inputs for the rock mass properties of roof, floor, and coal are mentioned in Table 6. Different sizes of the rib/snook varied during the simulation, as the manner of pillar extraction is shown in Figure 10. The variation in the strength of different sizes of rib/snook simulated in numerical models for the panel is shown in Figure 18. The depth and progress of extraction of coal pillar induce fractured zones in the rib/snook due to standing goaf, resulting in a decrease in their strength. Therefore, the numerical simulation strength of rib/snook formed during pillar extraction is found to be different from the empirical formula strength. The strength estimated using simulation studies is found to be useful in estimating their FOS. Kumar [2,3] found that a limiting FOS of rib/snook of 0.30–0.35 is sufficient for temporary support during the slicing operation. Furthermore, such rib/snook fail in a controlled manner without inhibiting roof fall. Here, rib/snook size of 3.00 m × 18.50 m × 3.00 m (area of 81.60 $m^2$, equivalent width of 7.62 m, and width-to-height ratio of 4.23) is proposed to be left during MD operation, which may be further reduced up to 2.00 m × 18.50 m × 3.00 m (area of 71.60 $m^2$, equivalent width of 6.86 m, and width-to-height ratio of 3.81) due to side spalling of pillars based on limiting FOS of 0.30–0.35.

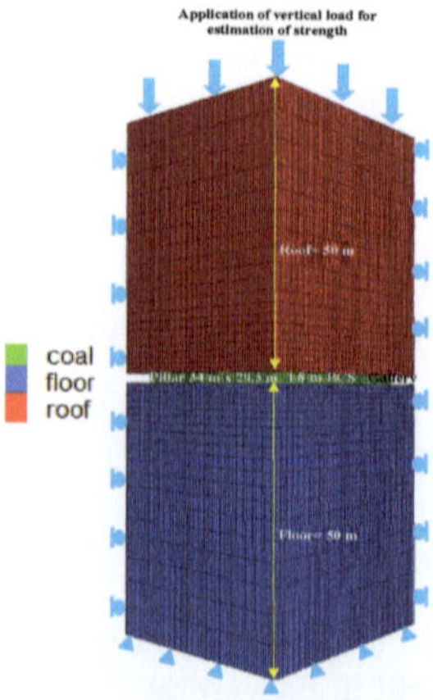


**Figure 17.** Geometry of the model and boundary conditions for simulation of rib/snook.

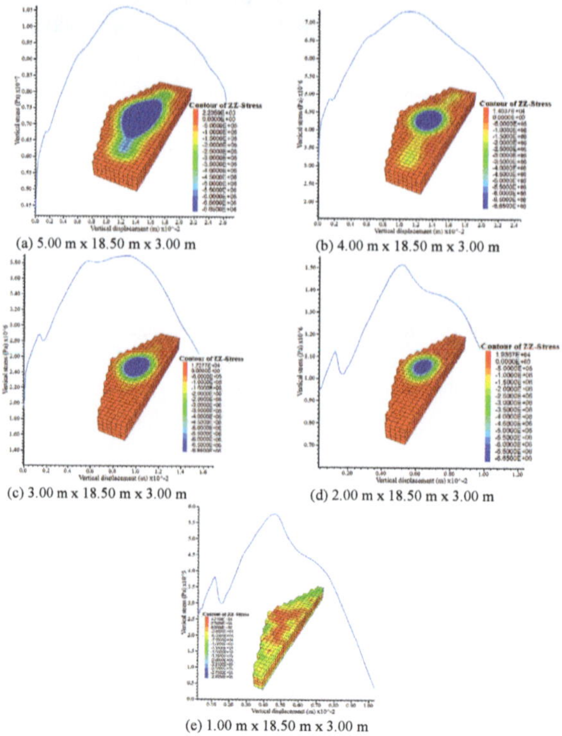


**Figure 18.** Vertical stress-displacement characteristics of different sizes of rib/snook in the panel for a pillar size of 34.0 m × 29.5 m (corner-to-corner).

4.1.2. Simulation for Estimation of RLH at Different Places in the Panel

The geometry of the generated model is 229.5 m, 271 m, and 102 m along the length (x-axis), width (y-axis), and height (z-axis), respectively, including 50 m floor (Figure 19), considering the hardware constraint (memory and runtime) of a computer. This model is incorporated with the existing geo-mining conditions and rock mass properties. The thickness of the coal seam considered in the model is 1.8 m, developed and depillared to full height. A truncated load (0.025 MPa/m × 214 m) is applied over the top of the model for the left-out unmodelled roof of 216 m. The width of original/split galleries is kept at 5.5 m, respectively. The bottom and four side walls of the model are fixed, while the top one is kept free as boundary conditions. The properties of rock mass used in the models are collected from the mines and tested in the laboratory, as mentioned in Table 7. Initially, the in situ model is developed as per the abovementioned geometry at a depth of 266 m, followed by the development of pillars and galleries, as per BPMM. This model is then depillared following a straight line of extraction (Figure 20). The height of the rock load to be supported by the roof bolt is estimated for the development and depillaring operation. Here, RLH is defined as the height of the roof strata up to the safety factor contour value of 1.0 in the model, which needs artificial support. RLH is measured at different locations, which is shown in Figures 20 and 21. Support design for the panel on the basis of numerical modeling study is given in Figure 9. RLH is measured at a distance of 1.0 m outby from the goaf edge (Figure 21) without using applied support in the model, which is found to be 1.5 m (3.75 t/m$^2$). SSF of the proposed support plan at the goaf edge is found to be 2.50 for an applied support resistance of 67.87 t/m$^2$, considering the dynamic effects at the goaf edge.

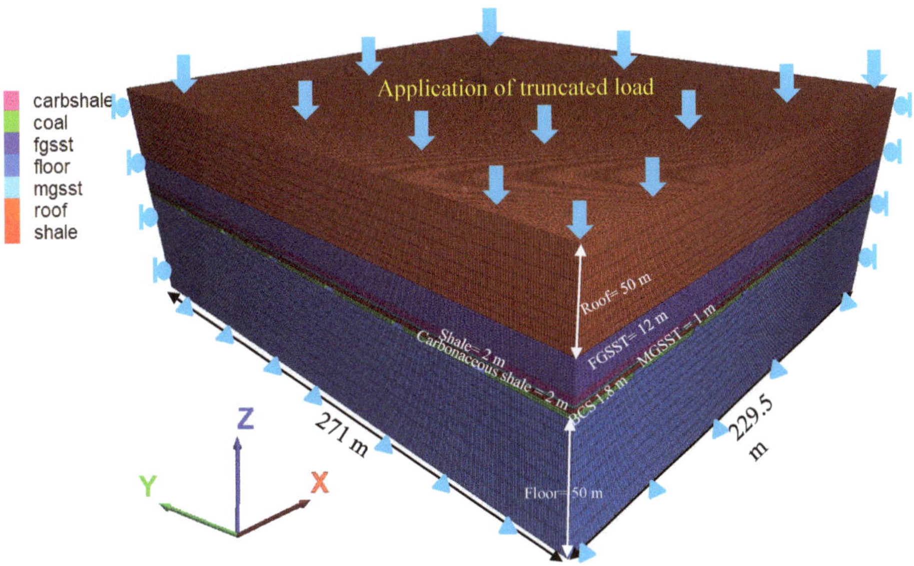


**Figure 19.** Generated block and boundary conditions in situ model of panel.

**Table 7.** Constitutive model selection and physico-mechanical properties of strata used in numerical modeling of panel.

| Strata | Constitutive Model | Thickness (m) | Young's Modulus (GPa) | Shear Modulus (GPa) | Bulk Modulus (GPa) | Poisson's Ratio | Density | RMR for FOS | Compressive Strength (MPa) | Tensile Strength (MPa) |
|---|---|---|---|---|---|---|---|---|---|---|
| Layer 1: Floor (FGSST) | Elastic | 50.0 | 7.00 | 2.80 | 4.67 | 0.25 | 2500 | 50.40 | 50.00 | 5.00 |
| Layer 2: Coal | Elastic | 1.8 | 2.00 | 0.80 | 1.34 | 0.25 | 1400 | 50.40 | 23.00 | 2.30 |
| Layer 3: Roof, Shale | Elastic | 2.0 | 2.50 | 1.00 | 1.67 | 0.25 | 1800 | 50.40 | 30.00 | 3.00 |
| Layer 4: MGSST | Elastic | 1.0 | 5.00 | 2.00 | 3.33 | 0.25 | 2200 | 50.40 | 40.00 | 4.00 |
| Layer 5: Carbona-ceous shale | Elastic | 3.0 | 2.50 | 1.00 | 1.67 | 0.25 | 1800 | 50.40 | 30.00 | 3.00 |
| Layer 6: Roof (FGSST) | Elastic | 12.0 | 7.00 | 2.80 | 4.67 | 0.25 | 2500 | 50.40 | 50.00 | 5.00 |
| Layer 6: Roof (MGSST) | Elastic | 32.0 | 5.00 | 2.00 | 3.33 | 0.25 | 2200 | 50.40 | 40.00 | 4.00 |

MGSST—Coarse Grained Sandstone, FGSST—Fine Grained Sandstone.

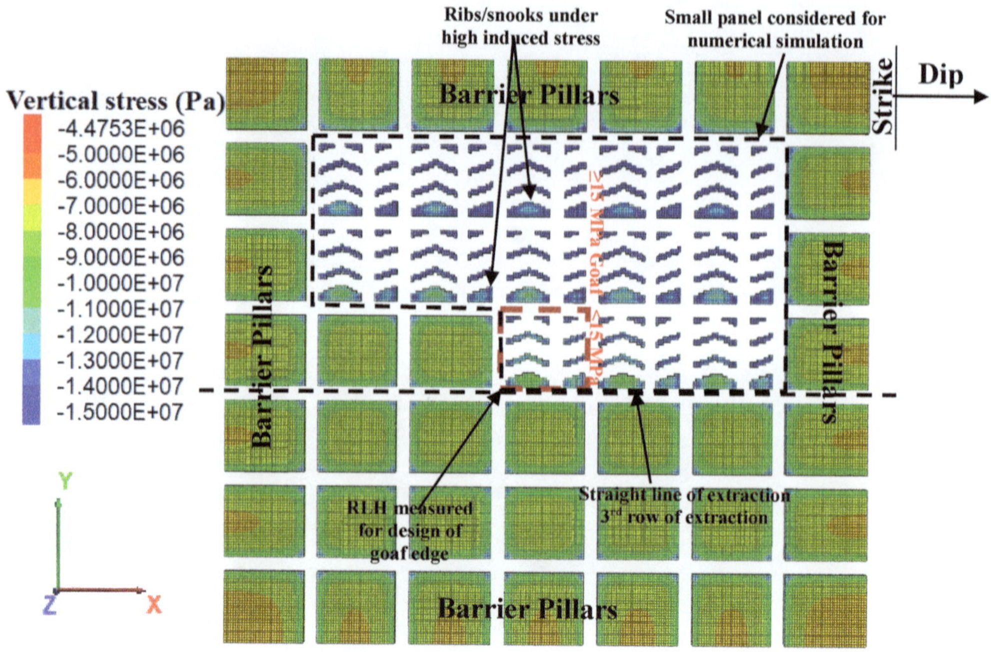


**Figure 20.** Manner of pillar extraction, their sequence and locations marked for measurement of RLH in the numerical model for the design of support system at the goaf edge in panel.

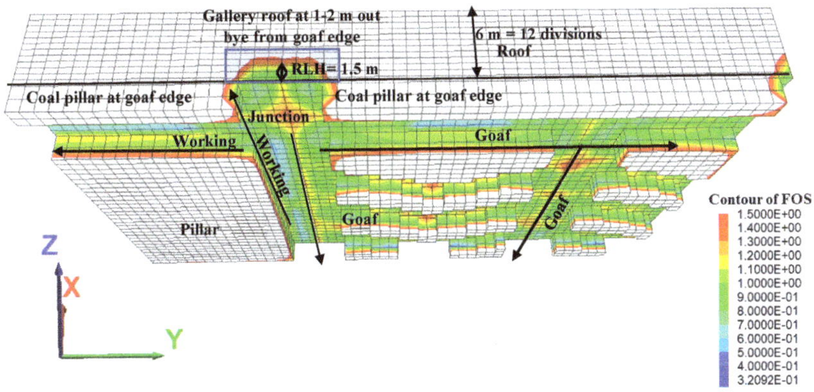


**Figure 21.** RLH at goaf edge during mechanized depillaring of panel.

## 5. Strata Control by Roof Blasting

The caving characteristics of the overlying strata of the goaf are one of the important parameters that influence the abutment loading in and around the working face and along the line of extraction. Difficulty in caving leads to the built-up of induced stresses, leading to fracturing in natural support present in and around the working face and sometimes goaf encroachment. The Cavability Index determines the difficult-to-easy level of fall of overlying strata [16] and the key strata exhibit the load transfer mechanism toward the working area. Generally, the caving nature is assessed through the Cavability Index of overlying strata of around 10–15 times of working height. In the case of easily to moderately cavable roof strata ($I \leq 5000$), generally, roof blasting for inducing caving is not required as experienced in Pinoura Colliery and Anjan Hills Mine of SECL [2]. However, overlying strata that are cavable with difficulty ($I > 5000$), need design of roof drilling and blasting to induce roof fall inside the goaf. Roof blasting induces/develops fractures in the roof rock mass, which causes the hanging roof to cave. Roof blasting for difficult-to-cave overlying strata is advantageous from a ground control point of view. However, there are some associated blasting issues of ground vibration and toxic gases that affect working conditions. The Tawa-I mine has a degree 2 gassiness (explosion chances for degree 3 gassiness); therefore, roof blasting to induce caving is to be performed with extra precautions.

### *Design of Roof Blasting for Inducing Caving*

Borehole CMPS 22 (Figure 2) is considered for the design of roof blasting to induce caving inside the goaf. Considering 10% of the bulking factor, the height of the caving (18 m) will be 10 times the height of the extraction (1.8 m). The overlying strata mostly contain 12 m of fine-grained sandstone, 1–2 m medium-grained sandstone, and 2–3 m of shale/carbonaceous shale. The average core recovery is more than 92%, and RQD is also well above 90. The Cavability Index of overlying strata of around 10 times the working height, is found to be between 7000 and 8000, which is categorized as roof cavable with difficulty in nature, and requires induced caving by the blasting of the roof in case of overhang. The fine-grained sandstone of 12 m in the roof will delay the caving of the roof in the goaf due to its massiveness of 12 m and a uniaxial compressive strength (40 MPa). It will also cause the development of higher induced stress along the line of extraction at around 8700 $m^2$ cumulative area of goaf. It is required to blast this roof horizon to induce caving for regular roof fall. Furthermore, it is reported [38] that the nature of the immediate roof strata up to 20 m plays a significant role till the caving starts. It explained the process of cantilevering after the first roof fall (break of the beam) and further collapse of the hanging cantilever of the roof strata with the progress of the extraction.

Considering the geo-mining conditions of the panel, roof blasting of up to 9 m thickness includes 2 m shale, 1 m medium-grained sandstone, 3 m carbonaceous shale, and 3 m of fine-grained sandstone roof horizon, which is required to induce fracturing in the rock mass to facilitate regular caving. Roof blasting may be conducted around the junction toward the barrier pillar right from the start of pillar extraction, as shown in Figure 22. In order to induce fractures in the roof horizon, a vertical blast hole of up to 9 m of 42 mm diameter is to be drilled toward the inby of the junction in the goaf to avoid any sort of damage to RBBLS and instruments installed at junctions, as shown in Figure 23. Details of the drilling pattern and explosive charging are shown in Figure 24 and Table 8. Six blast holes should be drilled in a row at a distance of 0.25 m from the adjacent edge of the pillar and at a distance of 1.0 m from each other. Additionally, they should be directed toward the inby of the junction in the goaf to ensure the efficacy of RBBLS and remote observations of instrument readings. Drilling and charging at the junctions should be completed before they shift into goaf. The blast hole is to be charged with a 32 mm diameter of explosive cartridge and a 6 mm detonating cord together, using plastic spacer pipes of 40 mm outer diameter, 38 mm inner diameter, and 6 m in length. The proposed method of charging blast holes is given in Figure 25.

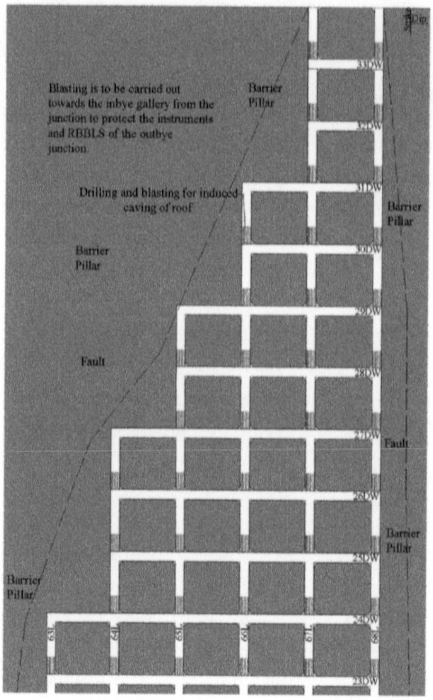


**Figure 22.** Schematic diagram of induced caving by roof blasting during extraction of second row of pillar after completion of extraction of Pillar No. 5.

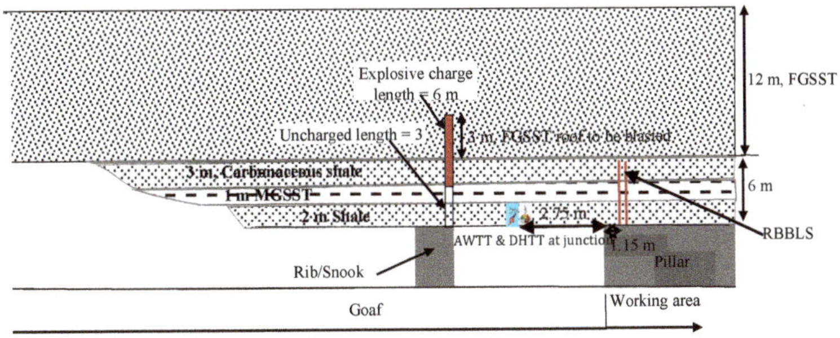


**Figure 23.** Sectional view of the drilled hole with explosive charging pattern.

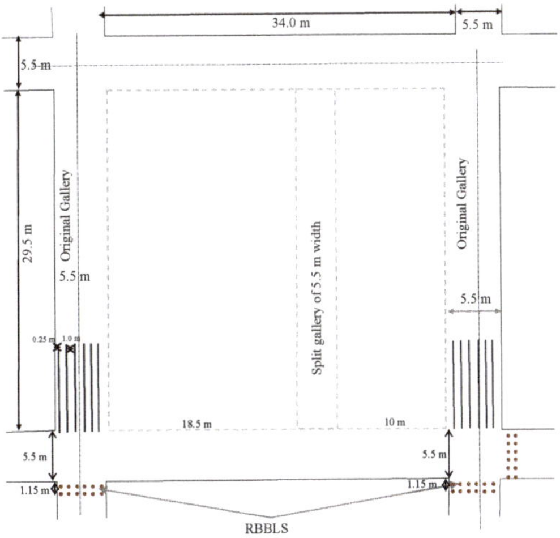


**Figure 24.** Plan view of drilling pattern for induced caving by blasting along the drilling plane.

**Table 8.** Proposed blast design parameters for induced caving.

| Blast Design Parameters | Value |
|---|---|
| Diameter of blast hole | 42 mm |
| Number of holes per ring in a single blast | 6 |
| Depth of hole | 9.0 m |
| Spacing between holes in a row | 1.0 m |
| Distance from the side of the pillar | 0.25 m |
| Explosive charge length | 6 m |
| The diameter of each emulsion permitted cartridge | 32 mm |
| Length of each permitted emulsion cartridge | 200 mm |
| Number of cartridges in one hole of 6 m length | 30 |
| Explosive charge per hole | 6 kg |
| Total explosive charge per ring | 36 kg |

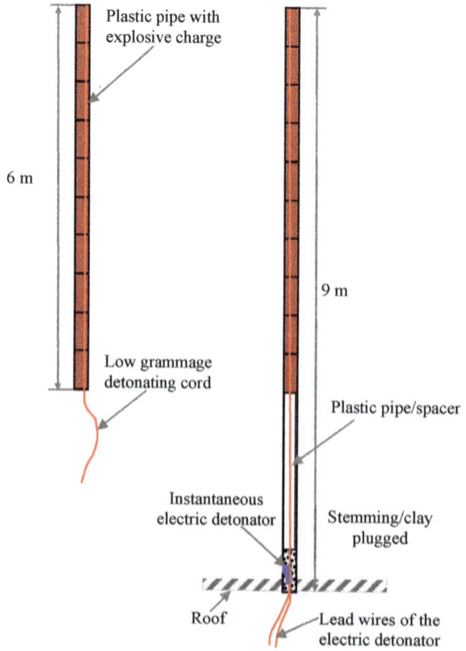


**Figure 25.** Proposed method of charging of blast holes.

## 6. Instrumentation and Monitoring

Strata behavior during the proposed mechanized depillaring of the panel needs to be studied through field instrumentation and monitoring. The purpose of the field instrumentation is to enhance the safety of the working to protect men and machinery from any risk and also to increase the production and productivity of the mine along with verification of the design of different structures involved. It may provide some quantitative values of strata movement and concentration of induced stresses in and around the workings for a better understanding of the rock mass behavior and further, in optimization of the design.

It is important to study roof sagging in galleries, with special attention to the junctions using roof extensometers. It is always good to opt for a simple approach to instrumentation and monitoring. However, attention is to be paid to the selection of durable and robust instruments, suitable for the underground coal mining environment. Attempts are made to monitor the readings of the instruments nearly continuously in time. The density of the instruments at the studied sites is optimized through available experiences of such monitoring. A preliminary instrumentation plan for monitoring the performance of underground structures in panels is shown in 26, respectively. This plan may be modified on the extent of development possible, the geological discontinuities unraveled, strata control experiences during work in the panel at different stages, and the selection of pillars extracted from the panel.

It should be noted here that extra caution is needed while correctly installing the different geotechnical instruments before commencement of MD operation. An attempt is to be made to monitor the readings of the instrument installed in the active mining zone (AMZ) frequently. Actions to be taken during depillaring operation on the basis of roof displacement are shown in Table 9. Details of the instruments proposed to be used for strata control monitoring in these panels are mentioned in Figure 26. Considering the panel for CM-based mechanized depillaring at this mine, extra caution has been taken to maintain safety during pillar extraction by the installation of a series of geotechnical instruments at suitable locations and comprehensive monitoring.

**Table 9.** Actions to be taken during depillaring operation.

| Sl. No. | Convergence Observation | Action to Be Taken |
|---|---|---|
| 1 | up to 5 mm | No action required, monitoring should be continued but must inform to Strata Control Officer and discussed. |
| 2 | 5–10 mm | Inform the Strata Control Officer, Supervisor, and Manager special attention at the observed location. |
| 3 | 10–15 mm | If sudden increase<br>Stop development work, assess conditions, and install additional support in the said zone by 1.7 m length resin grouted bolts with a 'W' strap.<br>Flexi bolts of 1.7 m length should be installed |
| 4 | >15 mm | Immediately inform the Strata Control Officer, Supervisor, and Manager.<br>Stop development work and withdraw the machine to the outby side of the work place.<br>Additional supports to be installed in the gallery based on the consultation of the mine manager with the scientific institute in the said zone.<br>All workmen and machines shall be withdrawn to a safe place within the gallery, and no work related to production shall be allowed. |

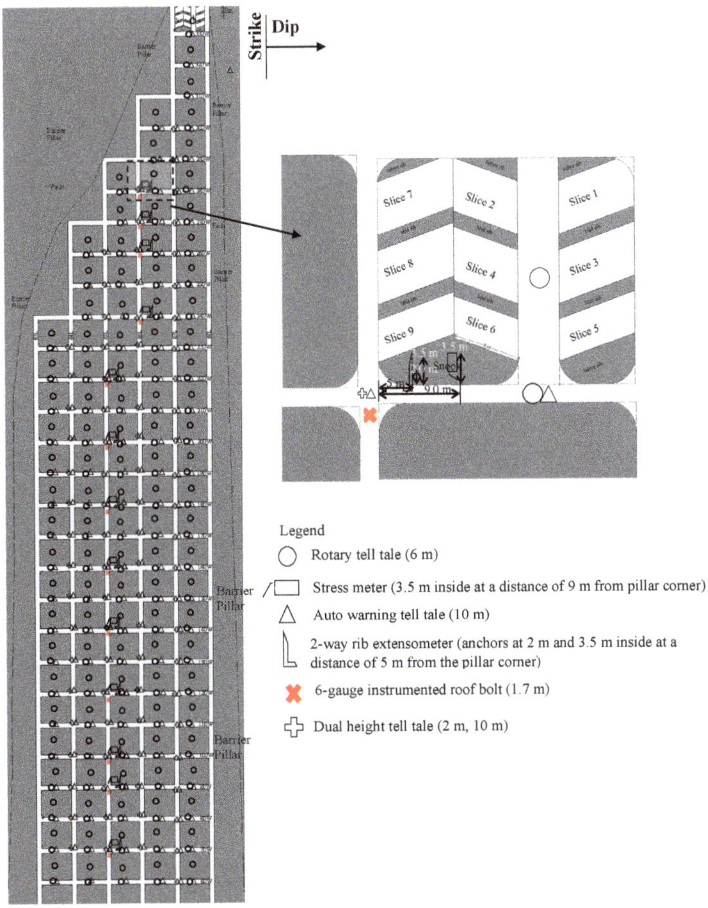


**Figure 26.** Proposed location of different instruments to be used in the panel.

The main roof fall is expected to occur at an average roof span of around 8700 $m^2$ in the goaf. Since the width of the panel is not sufficient during the first five rows (33DW-29DW), the, first main roof fall would not occur until working in 29DW. It is expected that the main roof fall will take place during or after extraction in 28DW. Considering the anticipated large roof span (>10,000 $m^2$) in the goaf before the main fall and the uncertain behavior of roof rock, instruments (stress meters, rib extensometers, and 6-gauge instrumented roof bolts) have been proposed frequently after 28DW for early indication of the first main fall in the panel. If the fall occurred as expected while working in 28DW, then the proposed instrumentation plan needs to be modified for 27DW. However, if the fall occurs while working either in 27DW or 26DW then the proposed instrumentation plan needs no modification. Further, the number of instruments may be optimized based on the observations of strata behavior at different stages of working in the panel.

## 7. Conclusions and Recommendations

Mechanized depillaring with continuous miner technology has proven its effectiveness in Indian coalfields by improving strata control management, primarily due to the faster extraction rate. An attempt has been made in this paper to predict the safe design of underground structures involved for better strata control while working with Continuous Miner Technology based on available empirical and numerical simulation studies. A numerical modeling study helped in fixing the gallery width to 5.5. m and COD of 11 m, considering 5 mm as the permissible limiting value of roof sagging to induce permanent movement (plastic/loose in nature) in the split/original galleries to be driven. A stress–strain curve was observed on the numerical models to estimate an optimum size of rib/snook such that it fails in a controlled manner without inhibiting roof falls after the completion of the slicing operation. Here, a rib/snook size of 3.00 m × 18.50 m × 3.00 m (area of 81.60 $m^2$, equivalent width of 7.62 m, and width-to-height ratio of 4.23) was proposed to be left during MD operation, which was further reduced up to 2.00 m × 18.50 m × 3.00 m (area of 71.60 $m^2$, equivalent width of 6.86 m, and width-to-height ratio of 3.81) due to side spalling of pillars based on permissible limiting FOS of 0.30–0.35. It was found to be matching with the values estimated using empirical formulations. RLH was measured at different locations for the design of breaker-line support in the numerical model at a distance of 1.0 m outby from the goaf edge without using applied supports in the model, which was found to be 1.5 m (3.75 $t/m^2$). The SSF of the proposed support plan at the goaf edge was found to be 2.50 for an applied support resistance of 67.87 $t/m^2$, and considering the dynamic effects at the goaf edge, it was found to be matching with the values estimated using empirical formulations. An empirical study estimated an average hanging roof span of 32 m (around 8700 $m^2$ for a panel width of 271 m) was found to be inducing the first major fall of the beam roof to form a cantilever. Furthermore, a roof span of around 21 m (around 5687 $m^2$ for a panel width of 271 m) was found to be causing the subsequent goafing and caving of the cantilever. However, roof falls will not take place during the initial five rows of pillar extraction, even with a roof span of 8700 $m^2$ or more, due to insufficient panel width of one/two pillars. The first main roof fall was expected to occur during or before the extraction of the pillar in the eighth row. The Cavability Index of overlying strata of around 10 times the working height was found to be between 7000 and 8000. This categorized the roof as cavable with difficulty in nature and required induced caving by the blasting of the roof to avoid overhang. Fine-grained sandstone, measuring 12 m in the roof, was expected to delay the caving of the roof in the goaf due to its massiveness of 12 m and uniaxial compressive strength (40 MPa). Roof blasting of up to 9 m in the roof horizon, which includes 2 m of shale, 1 m of medium-grained sandstone, 3 m of carbonaceous shale, and 3 m of fine-grained sandstone roof horizon, was required to be conducted to induce fracturing in the rock mass to facilitate regular caving. In order to induce fractures in the roof horizon, a vertical blast hole up to 9 m of 42 mm diameter was required to be drilled toward the inby of the junction in the goaf to avoid any sort of damage to RBBLS and instruments installed at junctions. Six blast holes were required to be drilled in a row, at a distance of 0.25 m from the adjacent edge of the pillar and at a distance of 1.0 m from each other toward the inby of the junction in the goaf to ensure the efficacy of RBBLS and remote observations of reading of instruments. Drilling and charging at the junctions were required to be completed before their shifting into goaf with the progress of pillar extraction. The blast holes were required to be charged with a 32 mm diameter of explosive cartridge and a 6 mm detonating cord together using plastic spacer pipes of 40 mm outer diameter, 38 mm inner diameter, and 6 m in length. Furthermore, the instrumentation plan proposed will be used to verify the design of roof blasting and the competence of different geotechnical structures involved. This study will prove to be helpful for mining professionals/academicians to plan safe and efficient pillar extraction using the continuous technology in Indian underground coal mines.

**Author Contributions:** Conceptualization, A.G.; methodology, A.G.; formal analysis, A.G., A.K. and S.R.; investigation, A.G.; resources, A.G., A.K. and S.R.; writing—original draft preparation, A.G., A.K. and S.R.; writing—review and editing, A.G., A.K., S.R., K.S., K.Z., A.Z., K.M. and M.M.; visualization, A.G., A.K. and S.R.; supervision, A.G., A.K., S.R., K.S., K.Z., A.Z., K.M. and M.M.; project administration, A.G., A.K., S.R., K.S., K.Z., A.Z., K.M. and M.M.; funding acquisition, A.G., A.K., S.R., K.S., K.Z., A.Z., K.M. and M.M. All authors have read and agreed to the published version of the manuscript.

**Funding:** AGH University of Krakow, scientific subsidy under number: 16.16.100.215.

**Institutional Review Board Statement:** Not applicable.

**Informed Consent Statement:** Not applicable.

**Data Availability Statement:** The data presented in this study are available on request from the corresponding author.

**Acknowledgments:** The authors express their thankfulness to the Director of IIT(ISM) Dhanbad and the Director of the Institute of co-authors for their permission to publish this paper. The cooperation provided by the management of Tawa-II Mine and M/s Aurobindo Reality and Infrastructure Private Limited at different stages of the study is thankfully acknowledged. The views expressed in the paper are those of the authors and do not necessarily reflect the opinion of their respective institutes.

**Conflicts of Interest:** Author Abhishek Gautam was employed by the company Eastern Coalfields Limited (a Subsidiary of Coal India Limited). Author Krzysztof Migda was employed by the company Gorotech sp. z o.o. The remaining authors declare that the research was conducted in the absence of any commercial or financial relationships that could be construed as a potential conflict of interest.

## References

1. Kumar, A.; Kumar, D.; Singh, A.K.; Ram, S.; Kumar, R. Development of empirical model for strength estimation of irregular-shaped-heightened-rib/snook for mechanised depillaring. *Int. J. Rock Mech. Min. Sci.* **2021**, *148*, 104969. [CrossRef]
2. Kumar, A. Development of Design Norms for Rib/Snook During Mechanised Depillaring by Continuous Miner. Ph.D. Thesis, Submitted to. Indian Institute of Technology (Indian School of Mines), Dhanbad, India, 2021.
3. Kumar, A.; Ram, S.; Kumar, D.; Singh, A.K.; Kumar, R. Development of design norms for rib/snook during mechanised depillaring by continuous miner. *Int. J. Rock Mech. Min. Sci.* **2023**, *161*, 105259. [CrossRef]
4. Ram, S.; Kumar, D.; Singh, A.K.; Kumar, A.; Singh, R. Field and numerical modelling studies for an efficient placement of roof bolts as breaker line support. *Int. J. Rock Mech. Min. Sci.* **2017**, *93*, 152–162. [CrossRef]
5. Mine Safety and Health Administration (MSHA). *Coal Miner's Handbook for Roof and Rib Control*; MSHA Publications: Washington, DC, USA, 2008; pp. 1–56.

6. Jayanthu, S. Strata control problems of underground coal mining vis-à-vis geotechnical instrumentation and numerical model studies. In Proceedings of the International Conference on Underground Space Technology, Bangalore, India, 17–19 January 2011.
7. Konicek, P.; Soucek, K.; Stas, L.; Singh, R. Long-hole destress blasting for rockburst control during deep underground coal mining. *Int. J. Rock Mech. Min. Sci.* **2013**, *61*, 141–153. [CrossRef]
8. Huang, B.; Zhao, X.; Chen, S.; Liu, J. Theory and technology of controlling hard roof with hydraulic fracturing in underground mining. *Chin. J. Rock Mech. Eng.* **2017**, *36*, 2954–2970.
9. Xu, T.; Yang, T.H.; Chen, C.F.; Liu, H.L.; Yu, Q.L. Mining-induced strata movement and roof behavior in underground coal mines. *Geomech. Geophys. Geo-Energy Geo-Resour.* **2018**, *1*, 79–89. [CrossRef]
10. Kang, H.; Wang, G.; Jiang, P.; Wang, J.; Zhang, N.; Jing, H.; Huang, B.; Yang, B.; Guan, X.; Wang, Z. Conception for Strata Control and Intelligent Mining Technology in Deep Coal Mines with Depth More Than 1000 m. *J. China Coal Soc.* **2018**, *43*, 1789–1800.
11. Guo, J.; Feng, G.; Wang, P.; Qi, T.; Zhang, X.; Yan, Y. Roof Strata Behavior and Support Resistance Determination for Ultra-Thick Longwall Top Coal Caving Panel: A Case Study of the Tashan Coal Mine. *Energies* **2018**, *11*, 1041. [CrossRef]
12. Jangara, H.; Ozturk, C.A. Longwall top coal caving design for thick coal seam in very poor strength surrounding strata. *Int. J. Coal Sci. Technol.* **2021**, *8*, 641–658. [CrossRef]
13. Kumar, R.; Mishra, A.K.; Kumar, A.; Singh, A.K.; Ram, S.; Singh, R. Importance of fracturing hard and massive overlying strata for complete extraction of thick coal seam- Case studies. *J. Geol. Soc. India* **2022**, *98*, 203–210. [CrossRef]
14. Wang, H.; Wang, J.; He, M.; Tian, X.; Ma, Z.; Liu, P. Mechanism of roof presplitting in nonpillar coal mining technology and its reasonable parameters in hard roof and longwall top-coal caving face. *Energy Explor. Exploit.* **2023**, *41*, 696–727. [CrossRef]
15. Chang, Z.; Wang, X.; Qin, D.; Yu, J.; Chen, X.; Wang, J.; Niu, Z.; Qian, C. Safety and high-recovery mechanisms and application research for initial mining of thick-coal-seam with complex structure and thick-hard roof. *Sci. Rep.* **2024**, *14*, 19638. [CrossRef] [PubMed]
16. Sarkar, S.K. *Mechanised Longwall Mining-The Indian Experiences*; Mohan Primlani for Oxford & IBH Publishing Co. Pvt. Ltd.: New Delhi, India, 1998; pp. 27–35.
17. Sheorey, P.R. *Pillar Strength Considering In Situ Stresses*; Information Circular (IC) 9315; United States Department of the Interior, Bureau of Mines: Washington, DC, USA, 1992; pp. 122–127.
18. Bieniawski, Z.T. The design process in rock engineering. *Rock Mech Rock Eng.* **1984**, *17*, 183–190. [CrossRef]
19. Bieniawski, Z.T. Classification of rock masses for engineering: The RMR system and future trends. In *Comprehensive Rock Engineering*; Hudson, J.A., Ed.; Pergamon Press: New York, NY, USA, 1993; Volume 3, pp. 553–573.
20. Bieniawski, Z.T. *Engineering Rock Mass Classifications*; Wiley: New York, NY, USA, 1989.
21. Bieniawski, Z.T. Rock mass classifications in rock engineering. In Proceedings of the Symposium on Exploration for Rock Engineering, Rotterdam, The Netherlands, 1–5 November 1976; Volume 1, pp. 97–106.
22. Molinda, G.M.; Mark, C. The Coal Mine Roof Rating (CMRR)—A practical rock mass classification for coal mines. In Proceedings of the 12th Conference on Ground Control in Mining; Peng, S.S., Ed.; West Virginia University: Morgantown, WV, USA, 1993; pp. 92–103.
23. Bauer, E.R. The Impact of Extended Depth-of-Cut Mining on Coal Mine Ground Control and Worker Safety. Ph.D. Thesis, Department of Mineral Engineering, The Penn State University, University Park, PA, USA, 1998.
24. Mandal, P.K.; Das, A.J.; Kumar, N.; Bhattacharjee, R.; Tewari, S.; Kushwaha, A. Assessment of roof convergence during driving roadways in underground coal mines by continuous miner. *Int. J. Rock Mech. Min. Sci.* **2018**, *108*, 169–178. [CrossRef]
25. Saharan, M.R.; Jha, B.K.; Sazid, M.; Kumar, R. Designing cut out distance for continuous miners operation using numerical modelling and rock mechanics instrumentation. In Proceedings of the Workshop on Application of Rock Mechanics: Tools & Techniques, Nagpur, India, 15–17 January 2010; pp. 24–31.
26. Mark, C. Application of coal mine roof rating (CMRR) to extended cuts. *Min. Eng.* **1999**, *51*, 52–56.
27. Kumar, A.; Kumar, D.; Singh, A.K.; Ram, S.; Kumar, R.; Singh, A.K.; Raja, M.; Singh, R. Rock mechanics challenges and advances in continuous miner based mechanised depillaring of coal pillars. In Proceedings of the National Conference on Advances in Mining, CSIR-CIMFR, Dhanbad, India, 14–15 February 2020; pp. 32–46.
28. Wilson, A.H. A hypothesis concerning pillar stability. *Min. Eng.* **1972**, *131*, 409–417.
29. Gao, W. Study on the width of the non-elastic zone in inclined coal pillar for strip mining. *Int. J. Rock Mech. Min. Sci.* **2014**, *72*, 304–310.
30. Suchowerska, A.M.; Merifield, R.S.; Carter, J.P. Vertical stress changes in multi-seam mining under supercritical longwall panels. *Int. J. Rock Mech. Min. Sci.* **2013**, *61*, 306–320. [CrossRef]
31. Gao, W.; Ge, M. Stability of a coal pillar for strip mining based on an elastic-plastic analysis. *Int. J. Rock Mech. Min. Sci.* **2016**, *87*, 23–28. [CrossRef]
32. Venkateswarlu, V.; Ghose, A.K.; Raju, N.M. Rock mass classification for design of roof support—A statistical evaluation of parameters. *Min. Sci. Technol.* **1989**, *8*, 97–107. [CrossRef]

33. Van der Merwe, J.N. Fundamental analysis of the interaction between overburden behaviour and snook stability in coalmines. *J. South Afr. Inst. Min. Metall.* **2005**, *105*, 63–73.
34. Palchik, V. Bulking factors and extents of caved zones in weathered overburden of shallow abandoned underground workings. *Int. J. Rock Mech. Min. Sci.* **2015**, *79*, 227–240. [CrossRef]
35. Sheorey, P.R. *Empirical Rock Failure Criteria*; Balkema: Rotterdam, The Netherlands, 1997; p. 176.
36. Sheorey, P.R. A theory of in situ stress in isotropic and transversely isotropic rocks. *Int. J. Rock Mech. Min. Sci.* **1994**, *31*, 23–34. [CrossRef]
37. Kumar, A.; Kumar, D.; Singh, A.K.; Ram, S.; Kumar, R.; Gautam, A.; Singh, R.; Singh, A.K. Roof sagging limit in an early warning system for safe coal pillar extraction. *Int. J. Rock Mech. Min. Sci.* **2019**, *123*, 104131. [CrossRef]
38. Anderson, I. *Case Studies of Buried Continuous Miners and Fatal Pillar Extraction Accidents in New South Wales*; Strata Control for Coal Mine Design, School of Mines, University of New South Wales: Sydney, NSW, Australia, 1993.

applied sciences

MDPI

Article

# Influence of Blasting Approaches in In-Pit Haul Road Construction on Emission Levels and Resource Management: A Case Study from the Holcim "Dubie" Open-Pit Mine

**Michał Dudek [1,*], Michał Dworzak [2] and Andrzej Biessikirski [2,*]**

[1] KOSD Przedsiębiorstwo Produkcyjne PP Sp. z o.o. Sp. k., Grupa Holcim, ul. Legionów Polskich 105, 32-064 Rudawa, Poland

[2] Faculty of Civil Engineering and Resource Management, AGH University of Krakow, Al. A. Mickiewicza 30, 30-059 Krakow, Poland; dworzak@agh.edu.pl

* Correspondence: michal.dudek@holcim.com (M.D.); abiess@agh.edu.pl (A.B.); Tel.: +48-12-617-20-70 (A.B.)

**Abstract:** Transportation activities can constitute up to 70% of a quarry's total operating costs, making haul roads a critical component of open-pit mine infrastructure. Generally, in-pit haul ramp construction can be accomplished through two primary blasting approaches: either peripheral blasting near the ramp location, or direct blasting at the designed ramp site. In the first method, the blasted material is transported, shaped, and compacted to form an embankment. Conversely, in direct blasting, the blast pattern is specifically designed to generate the ramp geometry, and the resulting muckpile is directed to production, eliminating the need for an embankment. Each method presents distinct operational advantages and inherent limitations. This study investigates the influence of these blasting scenarios, in particular on fume emissions (nitrogen oxides—$NO_x$—and carbon oxides—$CO_x$) and deposit management. The assessment encompasses emissions generated both from the detonation of explosives and from the operation of diesel-powered equipment. The findings indicate that the method involving peripheral blasting (bench embankment construction method) produces more than 2.5 times higher nitrogen and carbon oxides emissions compared to blasting works at the exact construction location of the ramp at the Holcim Dubie dolomite open-pit mine. In addition to emission analysis, operational factors related to ramp formation and its subsequent use were evaluated. The results demonstrate that constructing the in-pit haul ramp directly within the rock mass yields approximately $2 \cdot 10^6$ Mg less fume emissions than the embankment-based method. Furthermore, this approach facilitates the recovery of an additional 150,000 $m^3$ of dolomite for production purposes, thereby enhancing resource efficiency. Collectively, these findings suggest that the direct in-rock ramp construction technique offers superior environmental performance and operational sustainability within the context of open-pit mining practices.

**Keywords:** haul road; blasting; fumes; emission; resources; blast design

## 1. Introduction

The design of haul roads for open-pit mines is a critical engineering facet that directly impacts the operational efficiency, safety, and economic viability of mining operations. Haul roads are specifically constructed routes intended to accommodate heavy mining trucks that transport ore, waste rock, and materials from extraction areas to processing facilities or disposal sites within the mine. The complexity of haul road design stems from the need to balance several factors, including geometric alignment, structural integrity, operational safety, and cost-effectiveness, in often challenging topographical and geotechnical conditions [1–5]. A number of studies posit that haulage can be responsible for 40–70% of total operational costs [6–9], with the fuel component being considered the primary factor [8–10]. Moreover, variations in haul road design parameters can significantly influence the final geometry of the open pit. This may further result in the excavation of additional volumes of overburden and waste material, or in the restriction of access to certain ore zones. Nevertheless, alternative methodologies for haul road and pit design are primarily directed toward maximizing the net economic value of the mining operation, commonly employing diverse numerical optimization techniques [7,11–16].

Fundamentally, haul road design involves the specification of geometric parameters such as road width, gradient, curvature, and switchback arrangements, which must be tailored to the performance capabilities of the truck fleet. Poorly designed roads can increase rolling resistance, fuel consumption, and equipment wear, while also posing significant safety risks. Conversely, well-designed haul roads can reduce cycle times, improve truck utilization, and extend equipment lifespan [14]. Traditional design methods rely on empirical guidelines and engineering standards derived from field experience. For instance, studies such as [17] highlight the importance of maintaining stable gradients and ensuring safe turning radii to optimize truck maneuverability. While these approaches remain widely applied in practice, they often neglect the economic trade-offs associated with pit geometry and long-term haulage costs.

A number of studies have emphasized optimization-based approaches that integrate haul road alignment into overall pit design. Akay et al. applied linear programming techniques to optimize haul road layouts in forest and mining operations, demonstrating cost reductions by minimizing haul distances and improving accessibility [11]. Similarly, Baek and Choi developed a network-based algorithm for haul road design that accounts for truck operating constraints and pit expansion stages [13]. Advanced applications of numerical optimization have also been investigated to balance ore accessibility with the minimization of waste removal. Nancel-Penard et al. presented a mixed-integer programming model to jointly optimize pit design and haul road networks, highlighting how road placement can alter the final pit limits and, consequently, the economic value of the project [7]. Yarmuch et al. expanded this perspective by integrating stochastic optimization techniques, allowing for haul road planning under uncertainty in orebody geometry and market conditions [15]. Studies have found that, in both the cases of in-pit (roads between operation levels) and ex-pit haul roads (road within a mine that connects the pit with the processing plant or waste dump), the construction of the slope angle and bench highly depends on the physical and mechanical properties of rocks. On the other hand, the road width depends on the haul truck models applied in the open-pit mine [17–19].

Earthworks associated with haul road construction typically encompass drilling and blasting, loading, hauling, unloading, and subsequent bulldozing activities. In the context of ex-pit road design and construction, these operations are most appropriately conceptualized as an earthwork allocation problem, in which the principal optimization objective is to determine a partitioning of cut-to-fill assignments across distinct earthwork sections that minimizes material movement between them. Ramp design problems share a similar underlying objective, insofar as they require the allocation of rock mass to be executed in the most cost-effective manner. Achieving this, however, is not limited to the localized optimization of cut-to-fill balances; rather, it is situated within the broader framework of deriving an optimal ultimate pit design. As noted by Hustrulid et al., the fundamental purpose of ramp design is to secure access to the complex three-dimensional geometry of the ore body [4]. Furthermore, the spatial positioning of ramps exerts a significant influence on the achievable overall slope angle of the pit, which in turn affects the stripping ratio and, consequently, the economic performance of the mining operation, through its impact on the total volume of rock to be extracted. Moreover, it should be considered that most of the in-pit haul roads are unpaved granular roads, constructed from crushed rock or compacted natural gravel, continue to be widely employed for in-pit ramps and temporary haulage corridors because they combine rapid constructability with the capacity to utilize on-site aggregate sources. From an operational perspective, granular surfaces offer flexibility because they can be readily regraded, patched, or relocated as benches are advanced; however, their vulnerability to moisture-induced loss of strength, progressive rutting under repeated high tire pressures, and high routine maintenance demand (grading, watering or dust suppression) impose significant life-of-mine operating costs and can reduce truck availability and tire life if subgrade or base thicknesses are underspecified [3,20]. In open-pit mines exploiting deposits composed of rock with high mechanical strength, two principal blasting-based approaches can be applied to the construction of in-pit haul ramps. The first involves the direct excavation of the hauling road within the intact rock mass, while the second utilizes blasted and fragmented material from the ore body as the primary construction material for the road surface. These two methods influence both deposit management strategies and the associated environmental impacts in distinct ways.

The objective of the present study is to evaluate the application of blasting techniques in the construction of in-pit haul roads and to determine which approach is most advantageous from both operational and environmental perspectives. The analysis has been conducted on the basis of two blasting projects carried out in a dolomite open-pit mine in Poland.

## 2. Materials and Methods

### 2.1. Open-Pit Mine

The comparative assessment of in-pit haul road construction techniques was conducted at the Holcim Dubie dolomite quarry, a Middle Devonian deposit situated in the Lesser Poland Voivodeship on the north-eastern slope of Łysa Góra and the southern slope of Czerwona Góra. The exploited reserve is estimated at approximately $142{\cdot}10^6$ Mg of dolomite. Topographic and hydrogeological boundary conditions for the pit are as follows: pit rim elevation is approximately 410 m above sea level (a.s.l), pit floor elevation is at 305 m a.s.l., and groundwater occurs below the active mining horizon. The quarry comprises seven mining benches with a slope-type profile, an andesitic dike intrudes the deposit in its southern sector. Extraction is carried out using multidirectional progress of the short mining fronts. The average annual production at Dubie exceeds $1{\cdot}10^6$ Mg and yields product fractions of 0–16 mm, 0–31.5 mm, 0–63 mm, off-grade material, and semi-products for downstream processing.

Geospatial planning and blast design at Dubie employ photogrammetric surveying executed with DJI Mavic 3E and Matrice 30T (both DJI, Shenzhen, China) unmanned aerial systems (U.A.V.) in conjunction with Strayos 3D software (v. 1.0.1). Drilling and blasting are performed by in-house crews. Primary charges comprise ANFO, which was loaded by UMS (Universal Mixing System) and initiated by a non-electric detonator. Dynamite-based booster cartridges are used to prime the main charges. To mitigate adverse vibrational effects on nearby structures, regulatory limits restrict explosive mass in designated blasting sectors; total instantaneous charge masses are constrained to 1000 to 2000 kg for the total charge, and to between 24 and 125 kg for the charge per single time dela, with the quarry divided into six blasting zones accordingly.

### 2.2. Methods

Two blast-based strategies were developed for the construction of the in-pit haul road: first, bench embankment blasting, in which the transport ramp is shaped directly by sequential blasting within bench fills, followed by construction of the haul road from the blasted rock; second, full rock-block blasting, comprising production blasting of the designed ramp zone. Both approaches are established practices within Polish open-pit operations. The proposed ramp is located on the second operational level in the north-eastern sector of the excavation at an elevation of 382 m a.s.l., and will provide a connection between the first and second working levels, as shown in Figure 1.

**Figure 1.** The location of the planned in-pit haul road in Dubie open-pit dolomite mine (source: https://www.google.com/maps, access date: 1 October 2025).

Both design variants were developed using geometric data of the open-pit mine obtained through photogrammetric surveys conducted using UAVs operated by the mining facility. The acquired spatial data were processed and analyzed using Strayos 3D software (version 1.0.1), which enabled precise modeling of the excavation geometry and the planned in-pit haul road configuration. The total emissions of carbon and nitrogen oxides for each design variant were determined through a comprehensive assessment integrating multiple data sources. Emission factors for gaseous products ($CO_x$ and $NO_x$) were derived from laboratory analyses of fume emissions generated during the detonation of representative explosive samples. These per unit emissions were combined with quantitative data on explosive consumption obtained from the Strayos 3D blasting design. Additionally, emissions from diesel-powered mining machinery were estimated based on the evaluation of operational time required for each phase of haul road construction at Holcim Dubie open-pit mine, applying emission parameters consistent with greenhouse gas assessment standards for off-road diesel engines.

2.2.1. Drilling and Blasting Works

The blasting design was prepared by a certified blasting engineer from the Holcim Dubie mine, taking into account all safety constraints related to permissible explosive charges within defined safety zones and the types of explosives permitted for use at the quarry. In the initial phase, photogrammetric documentation of the quarry levels intended for haul road construction was conducted using a DJI Mavic 3E drone (DJI, Shenzhen, China). The photographic data were processed to generate a detailed three-dimensional model of the relevant section of the open pit, which served as the foundation for the blasting design, including drilling layout, explosive loading, and muckpile extraction parameters.

The design of the two blasting variants was carried out using Strayos 3D software developed by Strayos Inc. (Buffalo, NY, USA). This cloud-based, AI-assisted platform allows for the generation of high-precision three-dimensional models of mining sites from diverse geospatial data sources. The software supports the design and optimization of drilling and blasting operations by integrating on-site data and applying advanced artificial intelligence algorithms to enhance the safety and efficiency of explosive use in rock extraction. Core functionalities include integration with smart drill–rig data, measurement and correction of borehole deviation (using Boretrak data), burden geometry analysis for optimized drilling, prediction of muckpile morphology and rock fragmentation, as well as haul road optimization and related operational planning.

To ensure that the comparison between the two ramp-construction methods reflects differences arising solely from blasting strategy and equipment usage, all geotechnical and environmental boundary conditions were held constant for both variants. The analyses assume identical lithology (dolomite), bench geometry, confinement conditions, groundwater level, and burden/spacing parameters, because both variants are located within the same operational sector of the Dubie quarry. Factors such as confinement variability, explosive form, moisture content, and burden heterogeneity can influence the absolute fume emission levels, but they would affect both variants in the same direction and therefore do not alter the relative comparison. These variables were therefore not parametrically varied, as their detailed assessment lies outside the scope of this study and is more relevant to fragmentation or blast damage modeling. The present work focuses specifically on the method-driven differences in emissions and resource utilization under constant field conditions.

Two alternative blasting variants were developed for in-pit haul road formation. The first was peripheral blasting, where fragmented rock is transported, placed, and shaped to create a stable embankment that meets the required gradient (ramp with embankment), and the second related to profiling the road using blasting at the location of the in-pit ramp. The fundamental parameters of the drilling and blasting works for both variants are summarized in Table 1, while the blasting patterns are illustrated in Figures 2 and 3, and the three-dimensional views of the drilling patterns are presented in Figures 4 and 5.

**Table 1.** Parameters of drilling and blasting works for blasting with bench embankment variant and blasting directly in the rock mass variant.

| Parameter | Blasting at the Location of the Ramp (Ramp Without Embankment) | Peripheral Blasting (Ramp with Embankment) |
|---|---|---|
| Number of rows | 4 | 4 |
| Number of boreholes | 172 | 172 |
| Boreholes depth range, m | 1.9–22.1 | 21.0–23.0 |
| Subdrill, m | 0.5 | 0.5 |
| Hole diameter, mm | 89 | 89 |
| Total drilling length, m | 2012.1 | 3769.8 |
| Average face burden, m | 3.3 | 3.4 |
| Average pattern burden, m | 3.4 | 3.3 |
| Average spacing, m | 3.8 | 3.8 |
| Average bench height, m | 11.3 | 21.2 |
| Rock volume, m$^3$ | 24,964.9 | 47,476.9 |
| Total ANFO explosives weight, kg | 7883.7 | 17,478.9 |
| Number of boosters | 172 | 172 |
| Total weight of booster explosives, kg | 77.4 | 77.4 |

**Figure 2.** Shot layout of peripheral blasting (ramp with embankment).

**Figure 3.** Shot layout for blasting at the location of the ramp (ramp without embankment).

The Furukawa HCR L90-E5 (Furukawa Rock Drill Co., Ltd., Tokyo, Japan) top hammer drilling rig was used for drilling boreholes. The ANFO explosive was prepared by blending ammonium nitrate (V) with fuel oil at a weight ratio of 94.0:6.0 (wt.%), and loaded into the boreholes mechanically using a UMS. The main explosive charge was primed using 450 g dynamite cardridge boosters, which were initiated with non-electric detonators.

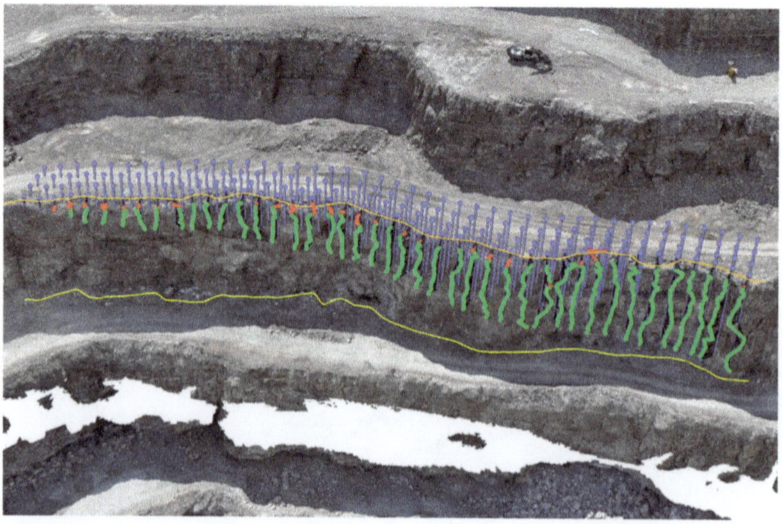

**Figure 4.** Three-dimensional drilling pattern of peripheral blasting (ramp with embankment).

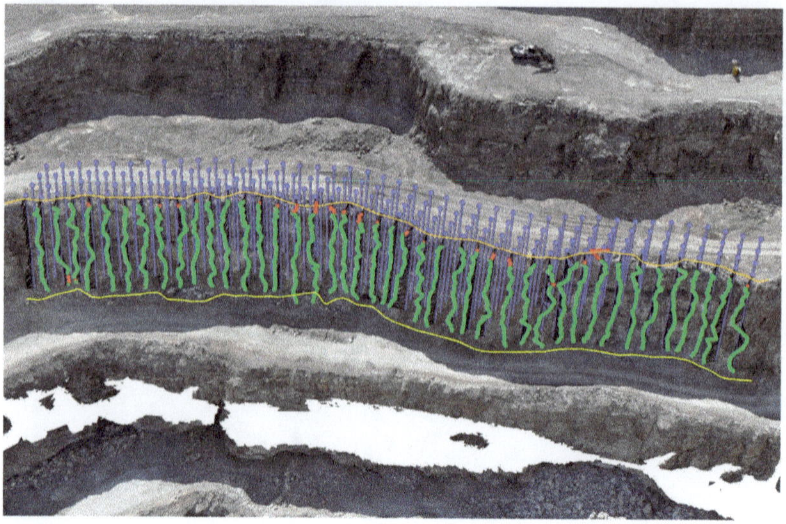

**Figure 5.** Three-dimensional drilling pattern of blasting at the location of the ramp (ramp without embankment).

2.2.2. Carbon and Nitrogen Oxides Emissions

Theoretical carbon and nitrogen oxide emissions for the explosives (ANFO and dynamite boosters) were determined based on explosive samples tested in a blasting chamber. In each experiment, a 500 g charge was detonated. ANFO charges were loaded into glass tubes with an inner diameter of 46 mm and subsequently positioned within a mortar in the blasting chamber. Dynamite charges were placed directly in the mortar without a glass tube. Initiation was achieved using a primer composed of 14 g of RDX (Royal Demolition Explosive) coupled with an electric instantaneous detonator. After each detonation, the resulting fumes were homogenized for three minutes and then extracted through a ventilation system over a 20 min sampling interval. $CO_x$ concentrations were quantified by infrared spectroscopy (MIR 25e, ENVEA, Paris, France), whereas $NO_x$ concentrations were measured by chemiluminescence analysis (TOPAZE 32M, ENVEA, Paris, France) in accordance with standard [21]. Table 2 presents average fume volume formed during the detonation process.

**Table 2.** Average fume volume formed during detonation test (experimental) [22].

| Type of Explosives | Fumes Volume, $m^3 \cdot kg^{-1}$ | | |
|---|---|---|---|
| | $CO_2$ | CO | $NO_x$ |
| ANFO | 0.1445 | 0.0164 | 0.0131 |
| Dynamite | 0.1513 | 0.0057 | 0.00139 |

In the process of forming the in-pit haul road, in addition to the energy of explosive detonation, machines were used in the phases of drilling blast holes (Furukawa HCR L90-E5 (Furukawa Rock Drill Co., Ltd., Tokyo, Japan) drilling rig), ANFO production, loading blast holes with explosives (Universal Mixing System), and muckpile extraction (Volvo EC480EL (Volvo Construction Equipment, Gothenburg, Sweden) excavator), in two variants of work.

The Furukawa HCR L90-E5 top hammer drilling rig (194 kW) has an average drilling rate of 26.4 $m \cdot h^{-1}$ and average fuel consumption during drilling process of 20.4 $L \cdot h^{-1}$. The Universal Mixing System (UMS) (353 kW) is constructed on a Scania chassis, and the operational efficiency during ANFO production and loading processes ranges between 70 and 80 $kg \cdot h^{-1}$. The average fuel consumption during the blending and loading operations was found to be between 30 and 35 $L \cdot h^{-1}$ under standard operating conditions. The Volvo EC480EL (281 kW) excavator has an average fuel consumption of 23 $L \cdot h^{-1}$ under standard operating conditions.

The machines are equipped with Tier 5 final stage emission standard compliant diesel engines (drilling rig and UMS Unit) and Tier 4 compliant engine (excavator).

Nitro and carbon oxides emissions were estimated based on Equation (1):

$$M_E = N \cdot HRS \cdot P \cdot (1 + DFA) \cdot LFA \cdot EFBase \tag{1}$$

where

$M_E$ is the mass of emissions of pollutant during inventory period in $g \cdot h^{-1}$;

$N$ is the number of engine units;

$HRS$ is the annual hours of use in h;

$P$ is the engine size in kW;

$DFA$ is the deterioration factor adjustment;

$LFA$ is the load factor adjustment;

$EFBase$ is the vase emission factor in $g \cdot kWh^{-1}$.

The equipment used in the study (drilling rig, UMS charging unit, and excavator) was nearly new at the time of assessment. Their technical condition and maintenance status were verified by the mining authority's engineering team, confirming that all machines operated within the manufacturer's nominal performance specifications. For this reason, deterioration of engine performance was assumed to be negligible. Consequently, the DFA was set at the lower bound (DFA = 0) of the Stage IV/V NRMM-recommended range, representing new or early-life engines with minimal emission degradation. This approach is consistent with EU Non-Road Mobile Machinery (EU NRMM) [23] and ICCT (International Council on Clean Transportation) modeling guidelines [24], which allow reduced DFA values when engine age and maintenance records indicate limited operational deterioration.

The quantification of carbon dioxide emissions was based on the assumption of average diesel fuel consumption, with a fuel density of 832.5 kg·m$^3$ being employed for the purpose. The estimation of emissions from machinery and vehicles was conducted using a multi-parameter approach, incorporating equipment population (N), annual operating hours (HRS), power output (P), base emission factors (EFBase), deterioration factor adjustments (DFA), and load factor adjustments (LFA). The LFA was assumed as full load (LFA = 1) for drilling operations, ANFO production and loading, and for excavation process, based on the side observation from mine authorities and the assumption that all machines are working continuously to perform designated tasks (without maneuvering).

The equipment fleet (N) was divided by technological tier and power class. Derived from this data, the study went on to create usage profiles based on age, with the annual operating hours of the equipment used as the dependent variable and the equipment's age as the independent variable. Within each subcategory, power output (P) was allocated according to typical distribution ranges. The base emission factors (EFBase) for individual pollutants were determined as functions of both technological level and power output. The EFBase were sourced from the EU Non-Road Mobile Machinery (EU NRMM) stage-specific emission factor tables [25,26] and the Environmental Protection Agency (EPA) [27], which differentiate emission levels by power class and engine tier. Therefore, drilling rigs, mixing systems, and excavators were assigned distinct EFBase values corresponding to their certified tier and rated power. Furthermore, EPA regulations comply with EU NRMM. Subsequently, deterioration factor adjustments (DFA) were applied, also as functions of technology level and power class, to account for performance degradation over time.

The fume emission assessment considered the operating time of individual machines in two operating variants. In the case of the peripheral blasting (ramp with embankment), due to charge limitations, excavation should be divided into seven blasting series. In this scenario, the excavator is assumed to operate for 120 h (collection of spoil after each blast within 12 h and formation of the road by the excavator for 48 working hours). In the case of blasting at the location of the ramp (ramp without embankment), four blasting series are required (one series of blast holes less than 6 m, and four series of blast holes longer than 6 m). The resulting rock can then be used for production.

## 3. Results and Discussion

### 3.1. Impact of Construction Method on Fumes

As noted in Section 2.2.1, the primary distinction between constructing an in-pit haul ramp with versus without embankment lies in the additional muckpile collection. For ramp construction with an embankment, the blasted material is subsequently used for ramp construction and compaction. Nevertheless, in both approaches the fume emissions can be categorized according to three stages: drilling, explosive charging, and energetic material decomposition.

Emissions from drilling were estimated based on the design drilling documentation and the TIER 5 emission standard for the drilling rig. According to the documentation, construction with a bench embankment requires a total borehole length of approximately 3770 m, whereas the direct rock mass method requires approximately 2013 m. Given an average fuel consumption of 0.0204 m$^3$·h$^{-1}$ and an average drilling efficiency of 0.0264 m·h$^{-1}$, the fume emissions were calculated, as shown in Figure 6.

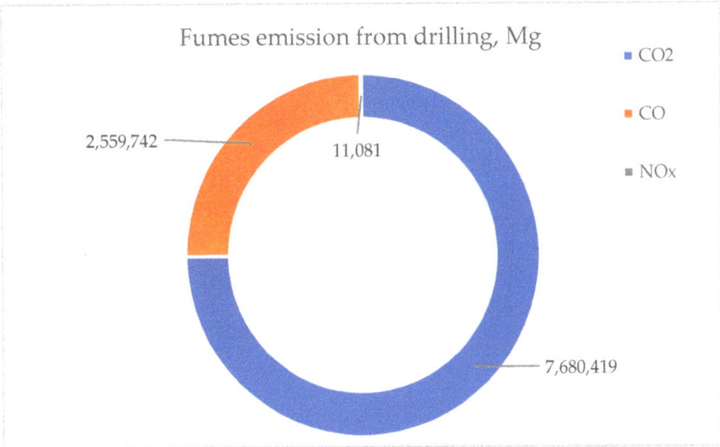


(**a**)

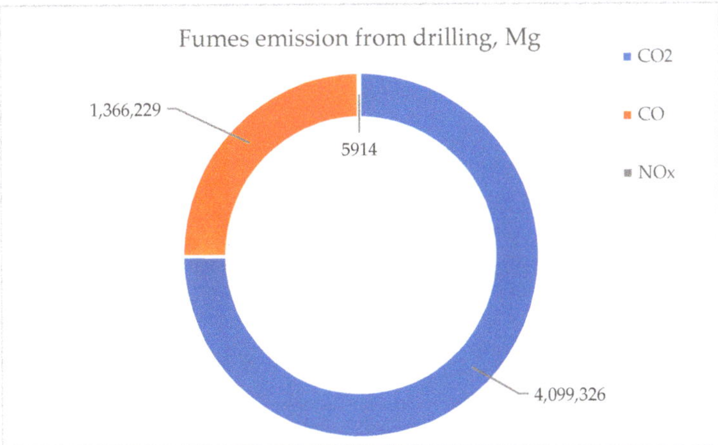


(**b**)

**Figure 6.** Fume emissions in Mg from drilling works in the case of blasting works for construction: (**a**) in-pit ramp with embankment, and (**b**) in-pit ramp without embankment (calculation).

Figure 6a,b demonstrate that, in both scenarios, carbon dioxide accounts for the majority of the emissions, approximately 75%. This finding is consistent with conventional fuel-to-$CO_2$ conversion factors and established emission inventories in which $CO_2$ dominates the exhaust mass budget [24,28]. The lowest contribution, approximately 0.11%, was attributed to $NO_x$, reflecting the effect of TIER standards and broader regulatory efforts to minimize $NO_x$ generation [29,30]. Although the proportional composition of the fumes remains constant due to the use of the same drilling equipment, the total mass of emissions varies. For peripheral blasting (ramp with embankment), the drilling process generates approximately $10.25 \cdot 10^6$ Mg of fumes, as shown in Figure 6a, whereas blasting at the location of the ramp (ramp without embankment) generates approximately $5.41 \cdot 10^6$ Mg, shown in Figure 6b. This difference is directly attributable to the duration of the drilling operations; the peripheral blasting method requires approximately 142.8 h of drilling compared with 76.3 h for the blasting at the location of the in-pit ramp. It is important to note that these evaluations were limited to the drilling process itself, and additional maneuvering emissions were not considered.

The selected blasting method also determines borehole length and, therefore, the required explosive mass. According to the design data, seven blasting series for the construction or ramp with an embankment require approximately $17.49{\cdot}10^6$ Mg of ANFO and 77.4 kg of emulsion cartridges as primers. In contrast, blasting that will result in the ramp without embankment requires approximately $7.89{\cdot}10^6$ Mg of ANFO and the same 77.4 kg of priming material. The Uniform Mixing System for ANFO production has an efficiency of approximately 85 $kg{\cdot}min^{-1}$. Consequently, the required masses of $17.49{\cdot}10^6$ Mg and $7.89{\cdot}10^6$ Mg correspond to explosive charging times of approximately 4 h and 2 h, respectively. Based on TIER 5 emission factors ($NO_x$ = 72 $g{\cdot}h^{-1}$; CO = 240 $g{\cdot}h^{-1}$; and $CO_2$ = 3.167 kg per 1 kg diesel combusted), the total emissions from the loading process were estimated at approximately $0.2–0.4{\cdot}10^6$ Mg (Table 3).

**Table 3.** Emissions from explosive charging depend on the in-pit ramp construction method (calculation and experimental).

| Parameter | Fume Emission, kg | | |
|---|---|---|---|
| | $CO_2$ | CO | $NO_x$ |
| Index value, in the case of $CO_2$ $kg{\cdot}dm^{-3}$, other: $g{\cdot}h^{-1}$ | 3167 | 240 | 72 |
| In-pit ramp with embankment | 41,171 | 1.08 | 0.29 |
| In-pit ramp without embankment | 20,586 | 0.54 | 0.15 |

However, it should be noted that in many blasting operations in Dubie, open-pit mine explosive charging is performed manually. In such cases, the emissions from the charging equipment would be 0 kg. The impact of different scenarios was broadly discussed in [31].

The overall emissions from ANFO decomposition were estimated using ballistic mortar tests and the methodology described in the methods section (Section 2.2), applying emission factors per kilogram of explosive consumed. The results, based on the total explosive mass of ANFO and emulsion primers, are presented in Table 4.

**Table 4.** Fume emission from explosives' detonation depending on in-pit ramp construction method (calculation based on experimental).

| Type of Construction | Explosive | Fume Emission, kg | | |
|---|---|---|---|---|
| | | $CO_2$ | CO | $NO_x$ |
| Ramp with embankment | ANFO | 4697.8 | 329.7 | 396.1 |
| | Priming | 21.8 | 0.5 | 0.2 |
| | Overall | 4719.6 | 330.2 | 396.3 |
| Ramp without embankment | ANFO | 2118.9 | 148.7 | 178.7 |
| | Priming | 21.8 | 0.5 | 0.2 |
| | Overall | 2140.7 | 149.2 | 178.9 |

The overall fume emission from explosive detonation depends on the in-pit ramp construction method. Table 4 indicates that in-pit road construction without embankment is more environmentally friendly compared to the method involving embankment. Blasting that results in ramp construction with embankment emits approximately 2.2 times the mass of $CO_2$, CO, and $NO_x$ compared with direct rock mass blasts (4719.6 vs. 2140.7 kg $CO_2$; 330.2 vs. 149.2 kg CO; 396.3 vs. 178.9 kg $NO_x$). Taking into consideration the total mass of explosives that would be used in seven blasting series, the scale effect is clearly visible. However, it should also be considered that, in real conditions, the confinement, explosive form (e.g., bulk, cartridge), and looser burden/spacing will produce larger absolute emissions [22]. Furthermore, Table 4 indicates that priming (dynamite cartridge charge) contribution is negligible in mass terms for both configurations; the dynamite priming contributes about 0.46–1.02% of the $CO_2$ mass and well under 1% of CO and $NO_x$ compared to ANFO charge. Primers are small in mass but critical for reliable detonation; their chemical signature can matter for local high-temperature chemistry, yet the data confirm that the bulk emissions scale with the ANFO charge. This aligns with experimental comparisons showing that small booster primer charges produce little of the total post-blast fume mass when used with much larger main charges [32–36]. Biessikirski et al. have reported that the use of different priming explosives can reduce overall fume emissions by up to 9% [22]. Furthermore, peripheral blasting (ramp with embankment) is generally carried out in the same area where the in-pit ramp will be constructed. Consequently, it is necessary to transport the blasted muckpile to form the ramp and establish the embankment. This process involves excavation (approximately 12 h) and ramp forming (approximately 48 h) for a total of approximately 60 h of additional machinery operation. Based on the TIER 5 emission indices (Table 5) and an excavator average fuel consumption of 0.0239 $m^3 \cdot h^{-1}$, the resulting fume emissions were calculated and are presented in Table 5 and Figure 7.

**Table 5.** Overall fume emission from each construction stage of in-pit haul road.

| Mining Method | Overall Fume Emission from Each Construction Stage, tons | | | |
|---|---|---|---|---|
| | Drilling | Loading | Detonation | Muckpile Extraction |
| Blasting at the location of the ramp (ramp without embankment) | 5.471 | 0.207 | 2.469 | 0.000 |
| Peripheral blasting (ramp with embankment) | 10.251 | 0.413 | 5.446 | 6.427 |

Figure 7a indicates that in the case of the peripheral blasting (ramp with embankment) method, fume emissions are largely generated by drilling ($10.25 \cdot 10^6$ Mg of fumes), muckpile extraction ($6.43 \cdot 10^6$ Mg of fumes), and detonation ($5.45 \cdot 10^6$ Mg of fumes), with smaller contributions from loading ($0.41 \cdot 10^6$ Mg of fumes). The absence of muckpile extraction emission, in the case of in-pit ramp construction without embankment, as shown in Figure 7b, reflects that muckpile is not used for construction process but only as the regular mining product. Furthermore, the construction of in-pit ramp without embankment, as shown in Figure 7b, reveals that drilling ($5.47 \cdot 10^6$ Mg of fumes) and detonation ($2.47 \cdot 10^6$ Mg of fumes) dominate, while explosive charging ($0.41 \cdot 10^6$ Mg of fumes) has minor significance. However, the open-pit mine blasting service can perform explosive charging by hand that will also slightly decrease the overall gross fumes. What is more, evaluated data indicates that in-ramp construction with embankment demands substantial drilling effort, which significantly increases fume emissions from diesel-powered rigs. The detailed analysis revealing the impact of either the detonation process or machine work is presented in Figures 8 and 9.

Figures 8 and 9 highlight a consistent pattern, showing that machinery contributes the majority of emissions. For the blasting method, machines account for $17.09 \cdot 10^6$ Mg of fumes compared with $5.45 \cdot 10^6$ Mg from explosives. For the in-pit ramp with embankment, the balance is $5.68 \cdot 10^6$ Mg from machines and $2.47 \cdot 10^6$ Mg from explosives. Although explosives are the source of the most toxic fume species (CO, $NO_X$), machinery dominates the carbon footprint ($CO_2$) through sustained fuel combustion. This shows that attempts to mitigate potential fumes should be directed to potential machine selection. The overall fume composition characteristic is presented in Figure 10.

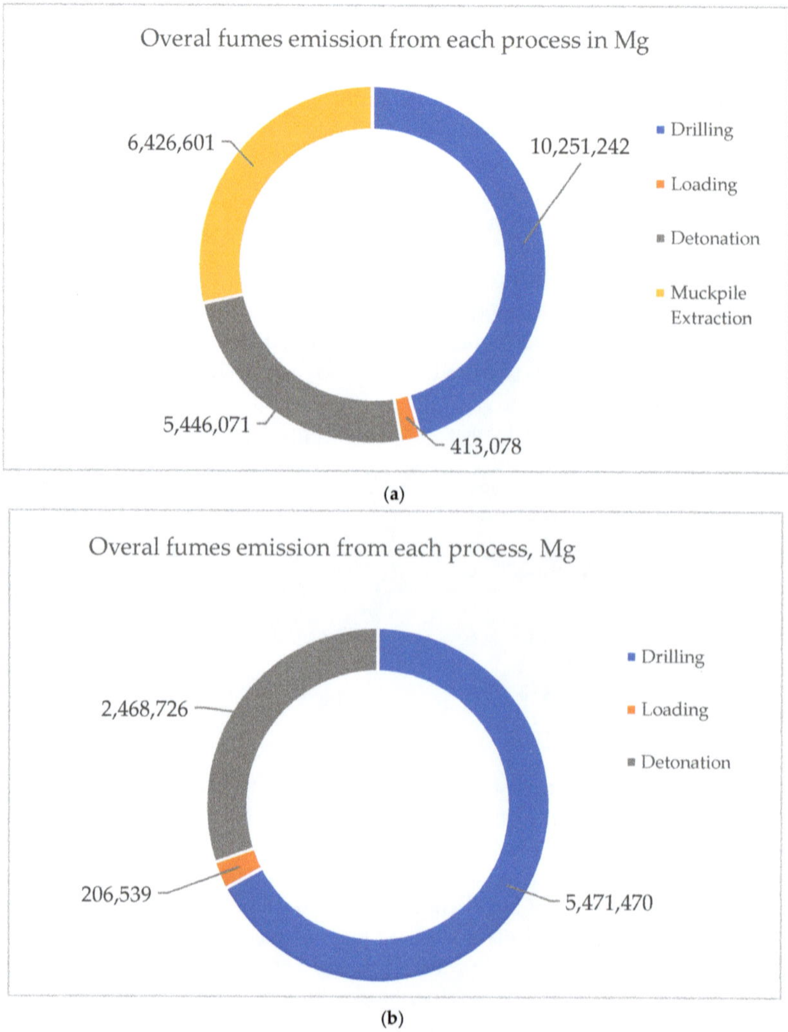


**Figure 7.** Overall fume emissions in Mg depending on the process for (**a**) in-pit ramp construction with embankment, and (**b**) in-pit ramp construction without embankment (calculation and experimental).

According to Figure 10, $CO_2$ is the dominant emission in both strategies, with approximately 19 tons for peripheral blasting (ramp with embankment) and approximately $7{\cdot}10^6$ Mg for the in-pit ramp without embankment. This reflects the major contribution of diesel combustion and detonation chemistry. The approximately 2.5 tons and 1.5 tons of CO emission is lower in tonnage than $CO_2$; however, due to the high toxicity of CO this can be a major concern in the case of the vast introduction of mining force to the blast site. Additionally, $NO_x$ also appears in much smaller quantities (smaller than $0.5{\cdot}10^6$ Mg, Figure 10), and it is also critical due to its acute toxicity. This species distribution aligns with published field measurements showing that $CO_2$ dominates mass inventories, while CO and $NO_x$ govern occupational health risk [22].

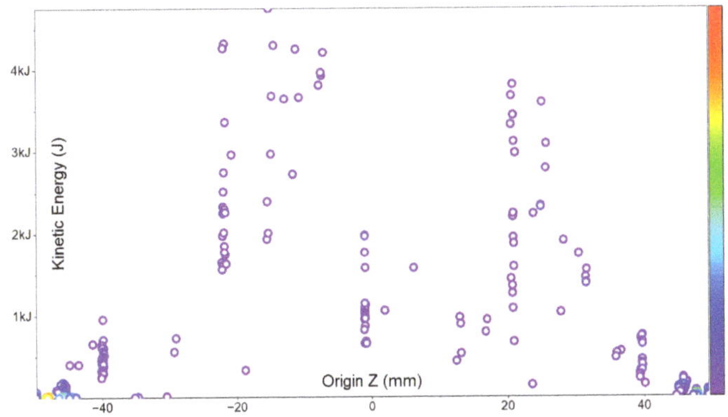


**Figure 8.** Two-dimensional representation of the kinetic energy of the fragments with respect to axial dimension of the pipe.

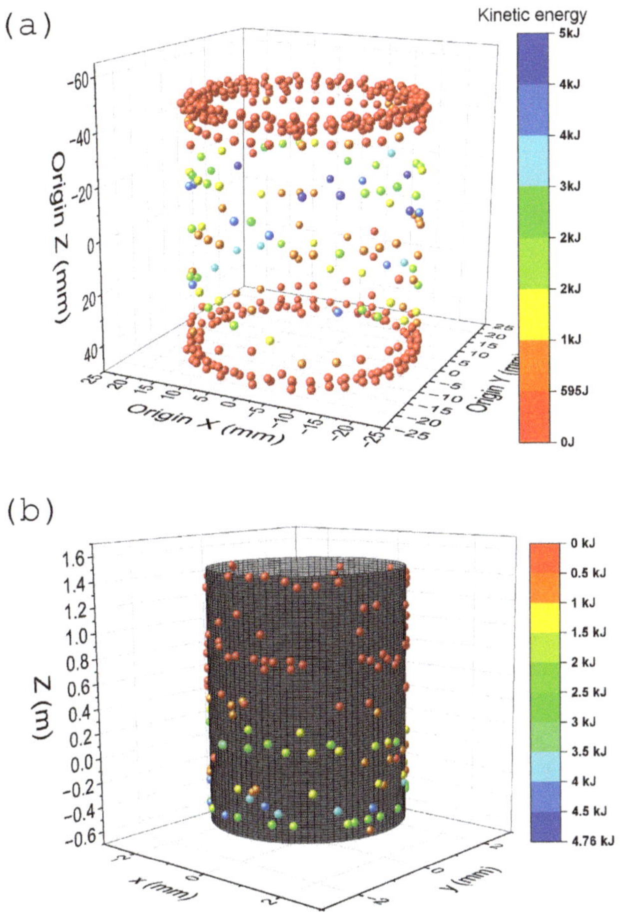


**Figure 9.** (**a**) Initial positions and kinetic energy distribution of the fragments plotted on the cylindrical surface of the steel pipe; (**b**) the spatial distribution of positions and kinetic energies of the resulting fragments on cylindrical surface of 2 m high with respect to the ground built up at 2 m distance around the detonation place.

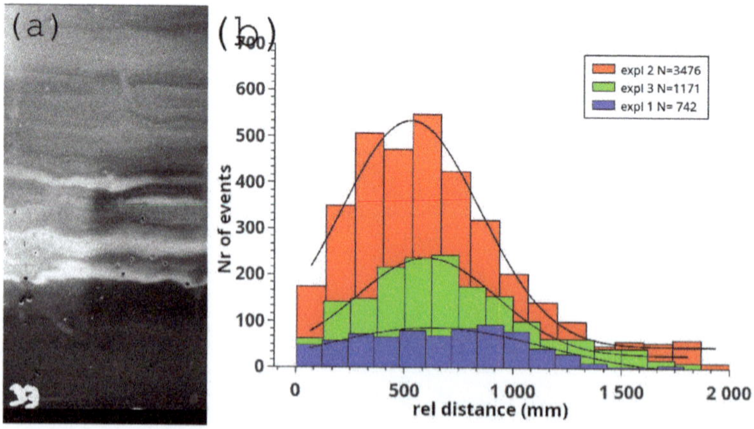


**Figure 10.** (**a**) The image of one of panel hit by fragments; (**b**) the statistical distribution calculated by the algorithm for fragment scattering.

Quantitative information related to the angular distribution of the fragments was provided by analysing the mutual distances between all pairs of perforations left by fragments on the three panels in Figure 2a. That was performed by numerical processing of each panel image recorded by a camera using a methodology based on a special algorithm implemented in a Python 3 home-designed software.

The methodology involves the fully automatic identification of fragment holes on strategically placed aluminium panels, based on image thresholding followed by flooding algorithms and centre of mass computations. The resulting data are structured in a list containing each hole size and position relative to a fixed origin.

Figure 10b shows the statistical distributions, as obtained by the algorithm, of the computed distances between all pairs of perforations in the case of the three fragment-scattering experiments: (a) 3 m, represented by the blue histogram (exp. 1); 2 m, represented by the red histogram (exp. 2); and (c) at 2 m with the difference that the metal case was mechanical modelled according to a grenade's pattern, represented by the green histogram (exp. 3).

The probability function used for fitting the experimental spatial distributions has the formula

$$y = y_0 + \frac{A * \sqrt{2/\pi}}{w} exp(-2\left(\frac{x - x_c}{w}\right))^2$$

where $A$ is the area of the distribution; $w$ is the double of the standard deviation ($\sigma = w/2$); $x_c$ is the mean value of the relative distances with corresponding values in Table 4.

**Table 4.** The probability function parameters.

| Parameter | Experiment 1 | Experiment 2 | Experiment 3 |
|---|---|---|---|
| $y_0$ | −14,805 | 45,864 | 26,366 |
| $A$ | 131,746 | 389,541 | 168,995 |
| $w$ | 1,047,203 | 637,187 | 642,156 |
| $x_c$ | 631,497 | 529,465 | 602,846 |

It is defined by specific parameters such as mean value and standard deviation and considered the fingerprint of the explosion. A significant difference in the pair distance across the three experiments is noted. The distribution obtained for the cased (uniform pipe wall) detonation with the panels located at a 2 m distance has a sharper profile, indicating a higher impact probability than in the case of panels located at a 3 m distance. The most probable relative distance is also higher in the first case (expl. 1) than in the second one (expl. 2), i.e., 360 px vs. 250 px.

Comparing the numerical distributions for the two detonations with panels at 2 m, expl. 2 and expl. 3 (uniform pipe wall and pipe wall with grenade-like patterns), one can observe lower intensity profile of the probability distribution in case of expl. 3, with the maximum kept at almost the same relative distance of about 250 px. The first observation indicates a lower number of recorded perforations on the panel whereas the second indicates almost a similar angular dispersion of the fragments.

Therefore, the only explanation is the formation of a lower number of fragments (due to the pre-engineered patterns of the case) with higher mass and a higher kinetic energy in case of expl. 3.

Models have been created in military research to assess the incapacitating effects of fragment penetration in the human body and estimate fatal and severe injuries, and are based in the probit function [35]

$$Y = 0.24 + 1.96 \cdot ln(v \cdot m^{0.4}) \tag{1}$$

where $v$ is the velocity of fragment in m/s and m is the mass of fragment in kg. The relationship between the probit function and fatality percentages was identified [36].

A first lethality effect is given by those fragments that have a high probability of hit in the proximity of the explosion, up to a few tens of meters, but there is a second lethality effect given by those fragments that fly on ballistic trajectories and can hit over long distances.

In the second case, only heavy fragments can cause serious injuries or even death. In Figure 11 the probit function (1) is plotted in the colour scale associated with fragment positions projected on the cylindrical surface in Figure 9b.

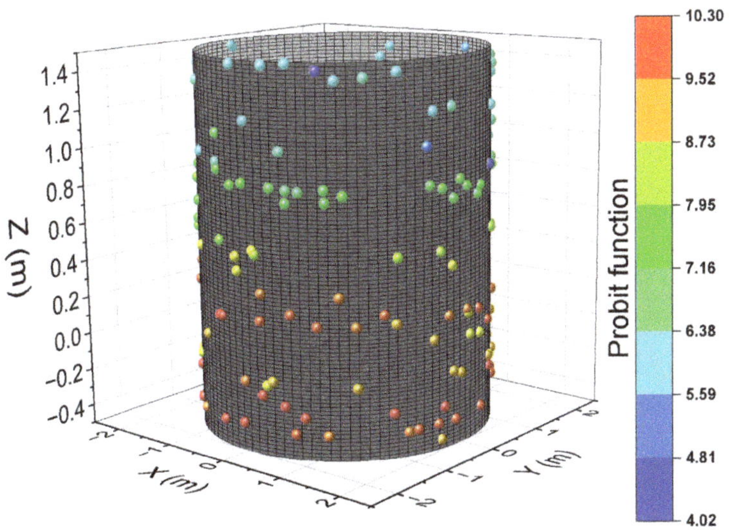


**Figure 11.** Probit function values corresponding to fragments with spatial distribution of positions on a cylindrical surface of 2 m high with respect to the ground, built up at 2 m distance around the detonation place.

As the Figure 11 shows, in the vicinity of the ground (between −0.4 and 0.4 on Z axis, Z = 0 being the axial centre of the explosive device) most of the fragments have a maximum potential of fatality (<99.9%); they have a mass over 1 g. The velocity of all the fragments with a mass over 1 g has a constant velocity distribution ranging between 1000 m/s and 1500 m/s.

## 4. Discussion

A crucial parameter used in the evaluation of a projectile's lethality is kinetic energy. The 9 × 19 mm calibre bullet has, for example, a kinetic energy around 500–600 J and a 5.56 × 45 mm calibre bullet has around 1.7 kJ. This energy is lost with distance, being strongly dependent on the projectile shape. Along with kinetic energy, projectile shape and size also are determinant factors influencing the lethality degree. It is the case of metal fragments resulting from the detonation of cased charge (grenades, projectiles and IEDs) projectiles, which may cause higher level of tissue damage because of their high kinetic energy and complex motion and, implicitly, trajectories.

The kinetic energy distribution of the metallic fragments generated in the detonation process of a cased IED with cylindrical shape has a two-lobe profile, with maximum kinetic energy reaching up to 4 kJ. This energy depends on a series of factors such as type of explosive and case material and the ratio between explosive mass and case mass.

Similar numerical simulations of steel AISI 1006-enclosed Composition B charge of length 77.8 mm, metal thickness of 3.04 mm and inner radius of 11.8 mm have shown a large velocity distribution from approx. 700 m/s from fragments belonging the cylinder's ends to more than 1400 m/s for those fragments expelled by the middle part of cylinder [12]. As in our work, the fragmentation, presented here, is powerful with a wide size distribution of larger elements in the middle part of the case, but the work does not study the mass distribution. This observation can be seen from the 3D simulation pictures. The explosive Composition B is powerful than C4 (8400 m/s with respect to 8000 m/s velocity detonation), but less stable. Another work reported numerical results based on LS-Dyna simulations in the frame of the Johnson–Cook model and Gruneisen equation for an AISI 1045 steel cylindrical case and Jones–Wilkins-Lee (JWL) equation of state for different explosives that filled the steel case [30]. The work proved the same pattern of fragmentation with larger fragments with higher velocities in the middle part of case than at its ends. Another result of [37] indicated that the fragmentation process is strongly dependent on the explosive's power. Accordingly, for the same steel and dimensions of case, the Composition B charge produced 664 fragments, the TNT charge produced 481 and Al/AP charge produced only 345 fragments. In this paper, the steel case had a length of 77.3 mm, external diameter of 29.68 mm and thickness of 3 mm. The fragments' velocity reached values over 1400 m/s. Comparison results involving a second steel case of double wall thickness showed smaller velocities that do not exceed 1000 m/s. Along with our results, we can mention other experiments involving detonation of mild steel cylinders filled with aluminised ammonium nitrate explosive, which showed a less powerful fragmentation (fragments of hundreds of grams) [37]. This is mainly caused by a different steel thickness (from 6–10 mm) in comparison with 3.2 mm in our case, but also another type of explosive was used. The fragments resulting here have higher kinetic energy values than in our case, due to their larger mass, but velocity values were under 1000 m/s. The article also proved that, at the cylinder ends, fragments' velocities were lower than in the middle part. However, this kind of fragmentation is capable to produce serious damage for civil or critical infrastructure, not only fatal injuries.

Regarding to our work, excepting the fragments originating from the case ends, all the other fragments have enough energy to cause serious damage to a human body. On the other hand, even small fragments of a mass of under 0.5 can cause injuries, especially in the proximal area around the detonation, having at least hundreds of m/s. Sensitive part of body like the eyes or neck are most predictable to be impacted from such fragments. The heavy fragments of masses over 1 gr and, implicitly, high kinetic energies are able to induce severe injuries even at large distances. The injury levels for the same parts of the body are different with respect to the fragment shape and size. Therefore, evaluation of the potential effects of detonation of such metal-enclosed IEDs is a complex work. It involves the estimation of the impact probability of the fragment in the vicinity of detonation and at different larger distances, mainly by experiments (most of the experimental fragment tracking is made in the proximity of detonation) and also of the injury level with respect to the size/shape and kinetic energies of them. In other words, extensive impact experiments and simulations based on human body models are required.

Designing ballistic protective systems for building and industrial/strategical facilities should involve structure models and materials capable of mitigating the kinetic effects of such high-energy metal fragments. Regarding to the damage level that such an event may produce for building or critical infrastructure, and corresponding protection measures, experimental and numerical simulation of impact events of fragments of different geometries and kinetic energies with various materials are also required.

Specific numerical algorithms developed here, which provide impact probability parameters, along with fragment energy distributions resulting from different experiments and simulation scenarios, can be used to evaluate both lethality degree and damage levels that such metal-enclosed IEDs can potentially induce.

## 5. Conclusions

Explosions in industrial or civil spaces are rare but highly destructive events, that can cause severe damage, economic loss and human casualties, especially in densely populated areas. This work brings new insights in the evaluation strategy of human causalities and possible infrastructural damage as a result of the intentional detonation of metallic-enclosed charges (IEDs). In this regard, this paper is focused on analysing the evaluation on the impact probability of metallic fragment distributions generated by detonation of experimental small IEDs and by numerical simulations. Therefore, experimental detonations were performed, with fragment collection and fragment scattering on special panels, and corresponding numerical simulations using Explicit Dynamics and Autodyn packages of the commercial Ansys program were conducted. The fragments resulting from the detonation experiment were visually collected in a small fraction (21.7%) from the total mass of the steel case, thus agreeing with the simulation that showed a high fraction of (over 60%) fragments with a mass under 0.3 g, difficult to be visually identified. Statistical analyses of both experimental and simulated IED fragmentation and the association between initial localisation of fragments with respect to their kinetic energy and their projection on a cylindrical surface built up at 2 m around the detonation place were performed. Experimental fragmentation was reasonably validated by simulation on the interval of 0.3–1 gr. Simulations showed a powerful fragmentation of the steel case, of about 474 fragments, most of them being generated by the ends of the steel case, which have small kinetic energies (under 1 kJ). In contrast, the fragments generated by the middle part of the case have larger masses and higher kinetic energies (up to 5 kJ), having a large potential lethality. Specific numerical algorithms were developed for analysing the correlations between all pairs of spatial coordinates of the perforations made by scattered fragments on special aluminium–wood sandwich panels. The algorithms use 2D images of the impact area and provide a histogram related to the spatial coordinates of the metallic fragments, characterised by a spatial distribution probability function (defined by specific parameters such as mean value

well as at the center of the ground 20 m from the entrance (A5). The pressure sensors in the main tunnel are installed as follows: at the cross-section 1 m away from the entrance of the main tunnel (B1–B3), on the sidewall of the back-blast surface 3 m from the entrance (B4), and on the sidewall of the front-blast surface 5 m from the entrance. In addition, five measuring points (C1–C5) are symmetrically arranged at specific locations on the protective door. Each measuring point operates with a data acquisition frequency of 1 MHz. The sensor used in this experiment are PCB pressure sensor 10113B28 (A1), 10113B27 (A2,A3) and 10113B21 (A4–A5,B1–B5,C1–C5). The sensitivity coefficients of the sensors are approximately15.489 mv/kPa, 7.692 mv/kPa and 3.662 mv/kPa, respectively.

### 2.2. Explosive Charge Design

The design of the explosion charge conditions is detailed in Table 1. The charge shape is a cylinder with an aspect ratio of 1:1, and the charge equivalents of the aluminized explosive are 0.5 kg and 1 kg, respectively. The explosion starting points are located at the entrance of the tunnel's central axis (0 m) and 1 m inside the entrance (marked as −1 m), and both are 1.2 m away from the ground. Figure 2 shows the on-site layout of the charge under the Con1 condition.

**Table 1.** Working condition.

| Condition | Type of Explosive | Equivalent (kg) | Initiation Position |
|---|---|---|---|
| Con1 | Aluminized explosive | 0.5 | 0 m |
| Con2 | | 1.0 | 0 m |
| Con3 | | 1.0 | −1 m |

Ps: −1 m indicates 1 m inside the tunnel entrance, 0 m indicates the tunnel entrance.

**Figure 2.** Loading explosives.

## 3. Results and Discussion

### 3.1. Characteristics of Shock-Wave Load Curve in the Straight Tunnel

Under different working conditions, the contour of the overpressure time-history curve in a straight tunnel remains consistent. The arrival time of the shock wave at each measuring point is directly proportional to its distance from the tunnel entrance. Furthermore, due to the influence of aluminized explosives, the overpressure waveform in the straight tunnel exhibits markedly distinct characteristics near the explosion source compared to that of TNT. Relevant studies have demonstrated that the afterburning effect of aluminum powder primarily alters the shock-wave waveform within the near-explosion region [16].

Figure 3 illustrates the overpressure time history curve at measurement point A1. Due to the combustion of metal powder, the overpressure curve waveform is distinctly divided into two batches. In the first batch, the waveform exhibits a triangular shape, with pressure rising sharply to the first peak before rapidly declining. Under Condition 3 (Con3), the explosion center is located at −1 m, relatively close to the measurement point, causing the pressure to drop to negative values. In the second batch, the curve waveforms under Conditions 1 and 2 (Con1 and Con2) display multi-peak characteristics. The maximum overpressure peaks occur at the second peak of the first batch and the first peak of the second batch, with respective peak values of 235 kPa and 219 kPa. For Con3, the overpressure curve rises steeply to its peak and then declines sharply. The maximum overpressure peak occurs at the first peak of the second batch, with a peak value of 418.6 kPa.

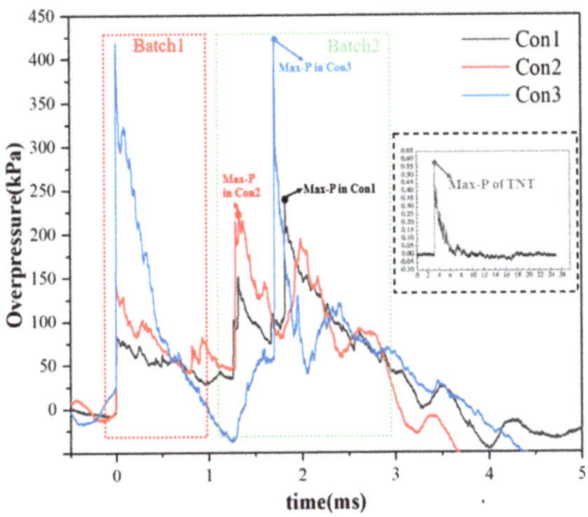


**Figure 3.** Overpressure of A1.

The small figure in Figure 3 illustrates the overpressure time-history curve of the shock wave generated by TNT explosive at a distance of 3.75 m from the tunnel entrance [21]. A comparison reveals that, unlike the triangular waveform characteristic of the TNT explosive overpressure curve, the overpressure curve of aluminized explosives is distinctly divided into two batches. Notably, under all three working conditions, the overpressure peaks are located in the second batch. Furthermore, by comparing the second-batch curves of Con2 and Con3, it can be observed that the explosion of aluminized explosives generates two prominent triangular waves. During propagation, the triangular waves in the second batch evolve into rectangular waves with two distinct peaks.

Figure 4 presents the overpressure time-history curves of measuring points A2 through A5 in the straight tunnel. From the figure, it can be observed that the overpressure waveform of the shock wave propagating in the tunnel varies systematically with time.

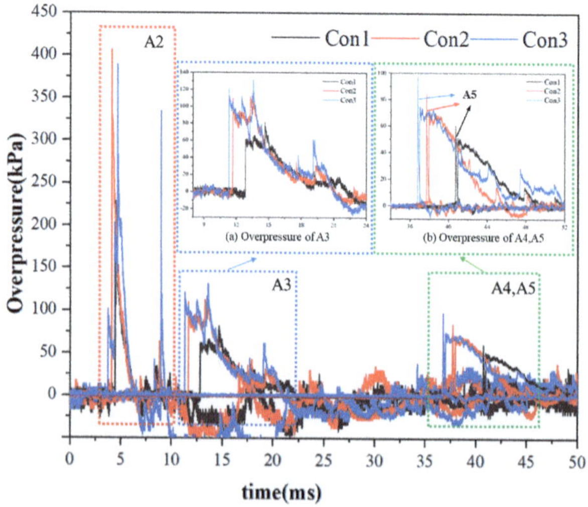


**Figure 4.** Overpressure of A2 to A5.

At the A2 measurement point, the overpressure curves under the three working conditions exhibit triangular waveforms, with peak magnitudes of 228 kPa, 406 kPa, and 389 kPa, respectively. At this measurement point, the shock wave is influenced by the enclosed space of tunnel, resulting in multiple reflections off the wall surfaces. These reflections interact, leading to mutual superposition and interference. Consequently, at the A3 measurement point, the overpressure curve waveform displays characteristics of multiple peaks, an extended positive pressure duration, a reduced magnitude of the overpressure peak, and the maximum peak occurring at the final wave peak, followed by a continuous decrease in overpressure.

With the multiple reflections of the shock wave on the tunnel wall, the initially complex shock wave gradually evolves into a plane wave. According to existing studies, the distance required for the formation of a plane wave is approximately 3 to 5 times the equivalent roadway diameter [24].

The measuring points A4 and A5 are located on the cross-section 20 m from the tunnel entrance, and the waveforms at these measurement points exhibit regular triangular patterns. The A5 measuring point is located on the floor of the tunnel, near the main tunnel entrance. Under the three working conditions, the overpressure response was first detected at the A5 measuring point, followed by the response at the A4 measuring point. The overpressure curve at the A4 measurement point exhibits a triangular wave shape, characterized by a prolonged positive pressure duration and a lower peak overpressure compared to the A5 measurement point. As the shock wave propagates through the T-shaped junction into the main tunnel, the positive pressure duration at the A5 measurement point decreases substantially relative to that at A4. Furthermore, under all working conditions, the positive pressure duration remains below 0.4 ms.

By comparing the overpressure curves at measuring points A4 and A5 under various working conditions, it is evident that the variation pattern of the overpressure time-history curve remains consistent when the explosive shock wave from an aluminized explosive reaches the T-shaped port. Changes in the detonation point and equivalent properties only influence the arrival time and peak pressure of the shock wave at the distal end of the explosion channel. The first peak overpressure and the maximum peak overpressure at each measuring point for the shock wave of an aluminized explosive explosion propagating in a straight tunnel are presented in Figure 5. As the shock wave propagates through the straight tunnel, the overall peak overpressure of the shock wave continuously attenuates, the waveform of the overall overpressure curve progressively approaches a stable triangular shape, and the first peak and the maximum peak of the shock wave gradually converge. At measuring points A4 and A5, the first response peak corresponds to the maximum peak overpressure. The comparison of shock-wave arrival times under different working conditions is presented in Figure 6. The arrival time exhibits a linear relationship with the distance from the measurement point to the tunnel entrance, indicating that the variation in shock-wave propagation speed remains relatively small.

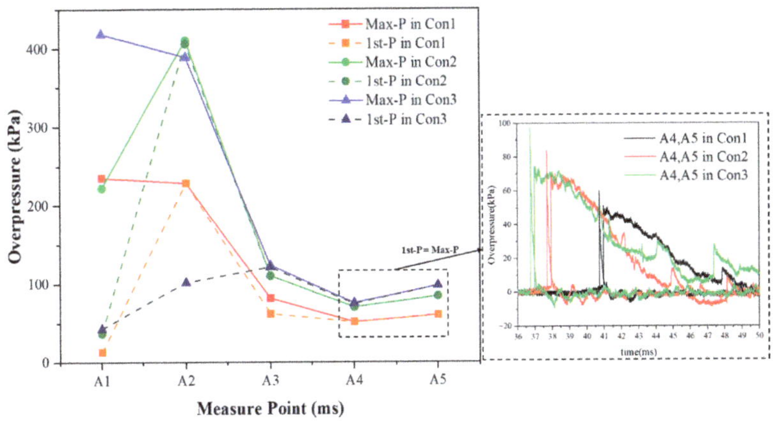


**Figure 5.** Peak overpressure of each measure point.

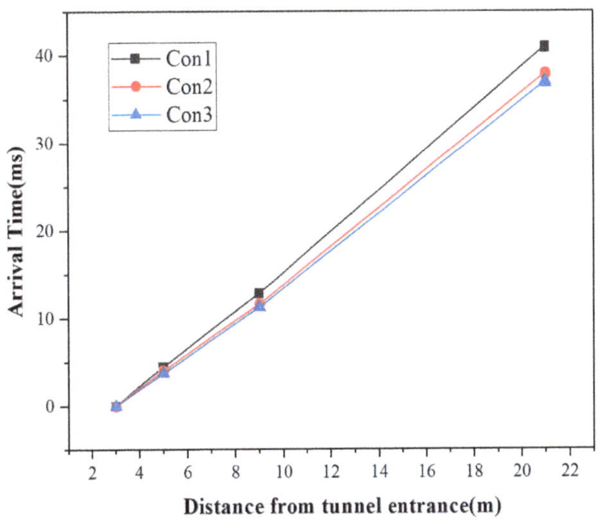


**Figure 6.** Arrival time of each measure point.

Figure 7 presents the curve graphs of momentum and positive pressure duration at each measuring point under three working conditions. By comparing the variation curves of momentum and positive pressure duration, it can be observed that the overall trends of the two curves are consistent. In conjunction with the feature shown in Figure 5, where the pressure peaks of each working condition continuously attenuate and converge as the shock wave propagates, it is evident that the positive pressure duration is the primary factor influencing the variation in impulse magnitude along the straight tunnel wall. The influence of the T-shaped junction results in a notable reduction in positive pressure duration at A5, which consequently leads to a significant decrease in the momentum.

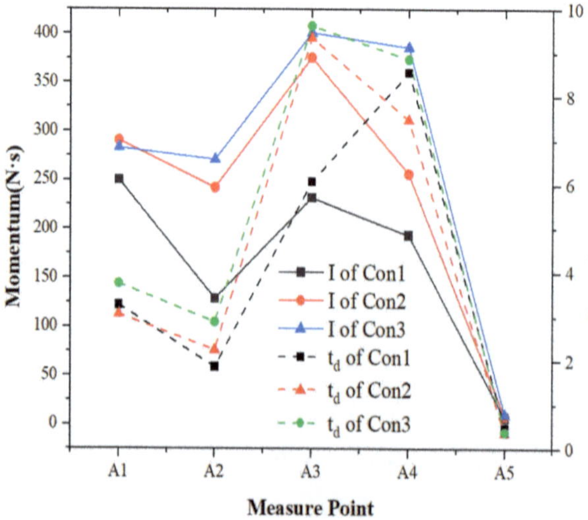


**Figure 7.** Momentum and duration time of each measure point.

### 3.2. *Load Distribution Characteristics on Main Tunnel and Protective Door*

The shock wave is influenced by the T-shaped junction, leading to complex reflection and diffraction phenomena. Although there are variations in charge equivalence and explosion center positions, the overpressure time-history curves for each condition in the T-shaped opening exhibit similar patterns.

In the main tunnel, B1, B2, and B3 are three measuring points at the same cross-section. The overpressure time-history curves of them are shown in Figure 8. According to the characteristics of the overpressure curves, the curves at the three measuring points can all be divided into four batches.

The waveform of the first batch at the B1 measurement point shows a gently varying rectangular shape, with the overpressure peak at the first wave crest. The second batch has a triangular waveform, and the maximum overpressure peak is observed at B1. The overpressure rises rapidly to its peak and then decreases continuously, approaching negative pressure. The third batch arrives with relatively low overall overpressure, showing a rectangular waveform with multiple peaks. The fourth batch also presents a rectangular waveform with multiple peaks.

The first batch of shock-wave overpressure curves at B2 exhibit waveform similarity to those at B1 but with a shorter positive pressure duration. The second batch displays a multi-peak rectangular waveform, with the maximum overpressure peak occurring at the first peak. The third batch consists of two triangular waveforms, with the overpressure peak also at the first peak.

The overpressure curve at B3 for the first shock-wave batch shows two distinct triangular peaks, with the maximum occurring at the first peak. The second batch presents an irregular multi-peak waveform, featuring three prominent peaks. The third batch displays a triangular waveform, with overpressure rising sharply to its peak before declining into negative pressure. The fourth batch exhibits multiple peaks, with the highest overpressure at the first peak.

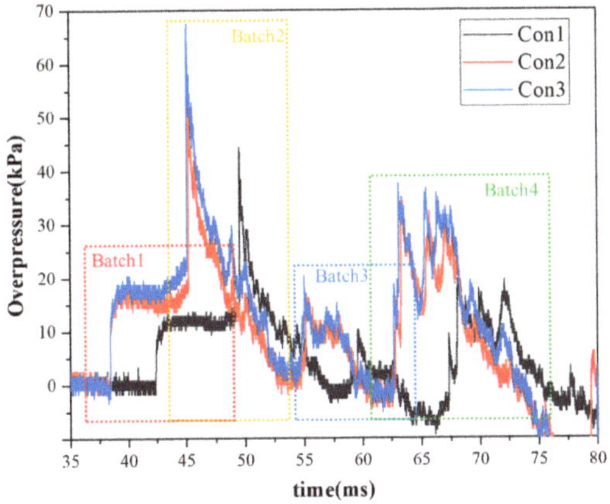


(a) B1

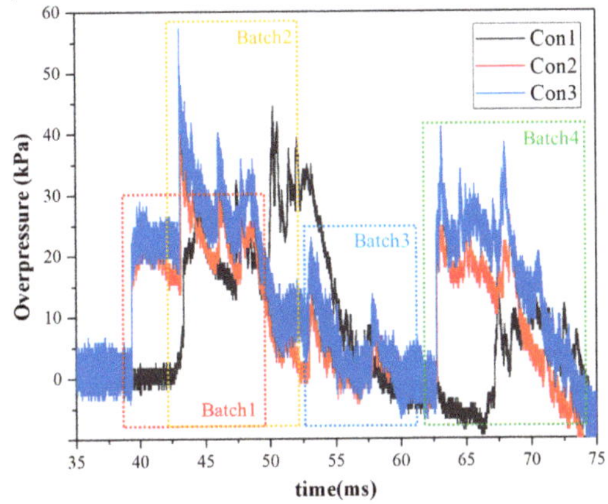


(b) B2

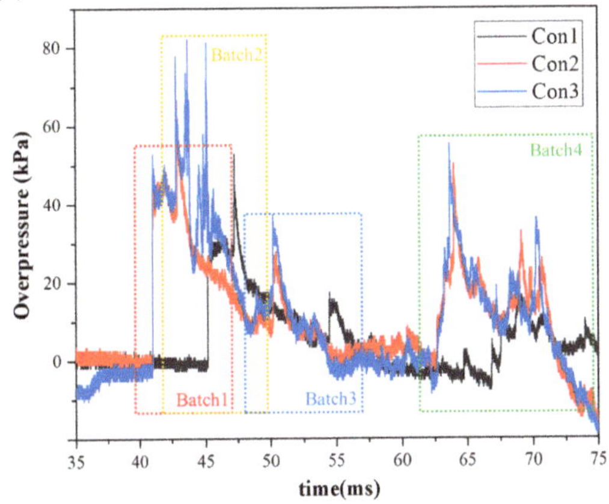


(c) B3

**Figure 8.** Overpressure of B1, B2 and B3.

By analyzing the curve characteristics of the first batch, it is evident that the overpressure waveforms at the three measuring points in the first batch exhibit multi-peak rectangular patterns, with the overpressure peak occurring at the first peak. The overall pressure variation is relatively smooth. By comparing the arrival times of the shock waves

Figure 12 illustrates the overpressure with arrival time curve at each measuring point under Condition 1. It is evident that, due to the influence of the enclosed space, the positive pressure duration at measuring point B4 on the back blast surface is significantly longer than at other points. Additionally, the maximum momentum in the main tunnel occurs at point B4. Furthermore, the momentum values within the enclosed space in front of the door are higher than those at the T-junction measuring points, and the positive pressure duration at point B5 is slightly greater than that at the T-junction measuring points.

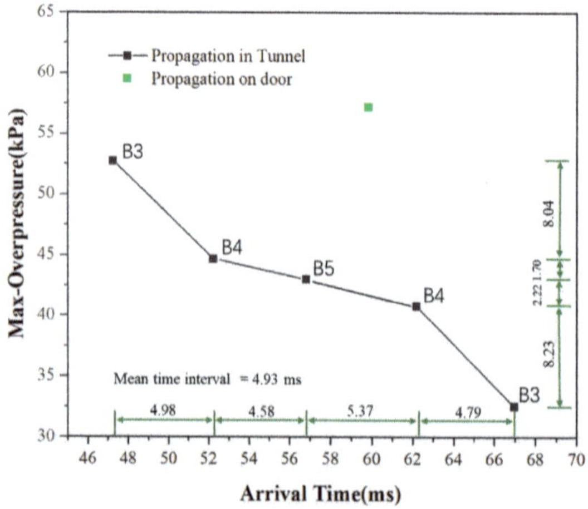


**Figure 12.** Peak overpressure with arrival time.

The overpressure peaks and corresponding arrival times on the overpressure curves for each measuring point are presented in Figure 13. It is evident that the time intervals between the arrival times of shock waves propagating along the tunnel are relatively close, with an average interval of 4.93 ms. Furthermore, the overpressure peaks of the shock waves exhibit a decreasing trend during propagation. Notably, the peak value decreases sharply between measuring points B3 and B4, whereas it decreases gradually between B4 and B5. Additionally, the magnitudes of the peak value reduction during propagation between the same measuring points are consistent.

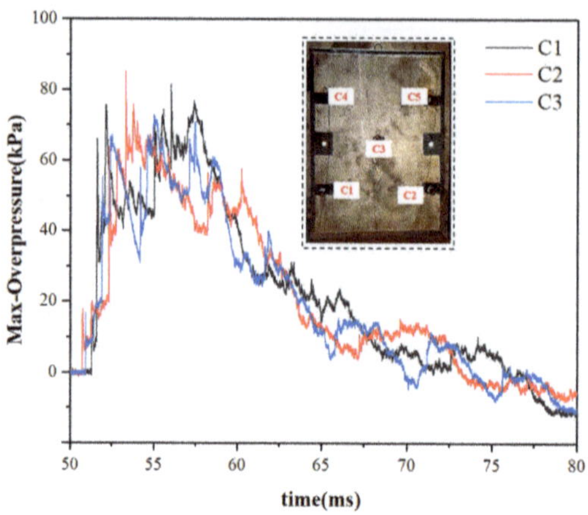


**Figure 13.** Overpressure on the protective door.

The curves of the symmetrical measuring points on the upper and lower parts of the protective door exhibit a similar pattern. Taking the C1, C2, and C3 measuring points under Condition 1 (Con1) as an example, the overpressure curve on the protective door is illustrated in Figure 13. Each measuring point on the protective door is influenced by a complex shock wave, resulting in an overpressure curve that exhibits characteristics of multiple peaks. The peak values of each wave crest are nearly identical, with the maximum overpressure peak reaching 85 kPa. The positive pressure duration of the overpressure curve on the door is consistent across all measuring points, approximately 25 ms. The originally independently propagating reflected waves superimpose at the protective door, leading to the formation of a multi-peak load characteristic. This results in the peak pressure on the protective door being higher than that on adjacent wall surfaces.

Figure 14 shows the comparison of momentum value at various measurement points on the central section of the main tunnel (point B4 is located on the reverse blasting surface). A4 and B2 have nearly the same momentum, with the momentum being around 285 N·s. Additionally, each point on the door has a relatively similar momentum. Taking the impulse in the straight tunnel as the reference of A4, the impulse increase multiples of the other four points are, respectively, 0.0264 (A5), 1.0449 (B2), 11.935 (B4) and 2.26 (C3). It can be observed that due to the reflection from the tunnel walls and the protective doors, the shock wave concentrates its effect on the rear explosion surface within the tunnel. Compared to a straight tunnel, the momentum on the doors also significantly increases.

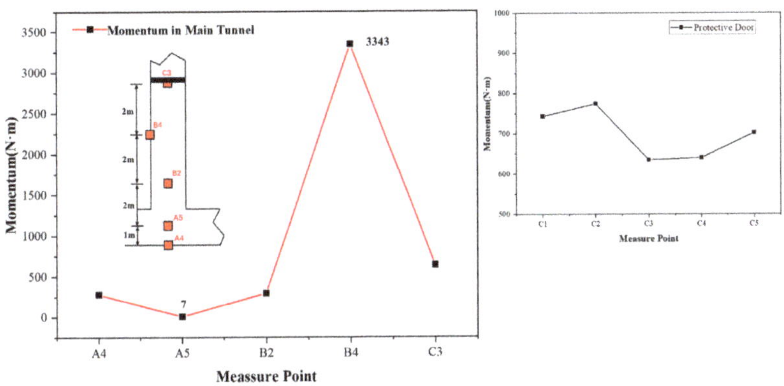


**Figure 14.** Momentum in Main Tunnel.

*3.3. The Propagation of the Shock Wave at the T-Gate*

Although the working conditions are different, the shock-wave load curves in the main tunnel exhibit similar patterns. Specifically, the shape of the overpressure curve at the same measuring point remains consistent, while the propagation law of the shock wave along the path can be clearly observed between different measuring points. To further investigate the propagation characteristics and process of the shock wave in the T-shaped main tunnel, this paper establishes a numerical model of the T-shaped tunnel using LS-DYNA numerical simulation technology.

The numerical model is designed to have the same dimensions as the experimental setup. In this model, the concrete tunnel, explosives, and air are all represented as three-dimensional solid elements. During the modeling process, the ALE algorithm is employed for both the explosives and air to enhance computational accuracy. The explosive is located at the entrance of the tunnel, 1.2 m below the ground level, and its explosive yield is 1 kg. To achieve more precise results, an obstacle-free air region is established at the tunnel entrance, with its boundaries defined as non-reflective to minimize interference. The detailed configuration of the model is presented in Figure 15 [8]. Among them, the tunnel model is a rigid body, the air grid is an ale grid, and the size of the air grid is 0.05 m [8]. The outer boundaries of all the boundaries are set as non-reflective boundaries.

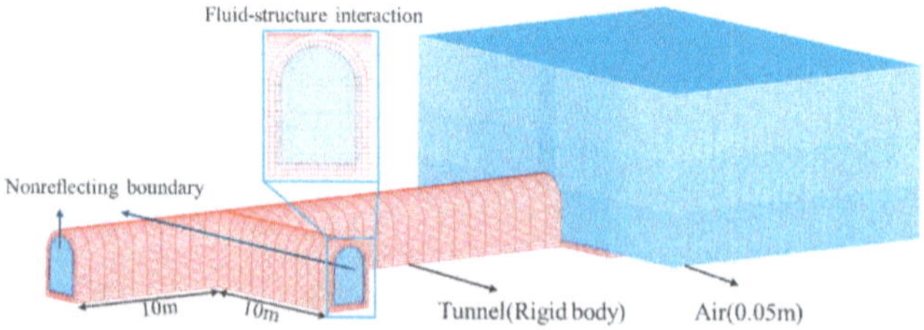


**Figure 15.** Numerical simulation model.

Air was characterized as an ideal gas without viscosity, which was defined by MAT_NULL material and EOS state equation.

$$P = \left[C_0 + C_1\mu + C_2\mu^2 + C_3\mu^3\right] + \left[C_4 + \mu + C_6\mu^2\right]E_0 \tag{1}$$

where $P$ is air pressure, $C_0$~$C_6$ are constants, $\mu$ is the rate of density change, and $E_0$ is the energy in initial state. The detailed parameters of air were given in Table 2.

**Table 2.** Parameters for air.

| $\rho$/(kg/cm$^{-3}$) | $V_0$ | $E_0$/(KJ·kg$^{-1}$) | $\gamma$ |
|---|---|---|---|
| 1.293 | 1.0 | $2.068 \times 10^5$ | 1.4 |

The explosive was defined by MAT_HIGH EXPLOSIVE BURN material and JWL state equation.

$$P_{cj} = A_1\left(1 - \frac{\omega}{R_1 V}\right)e^{-R_1 V} + B_1\left(1 - \frac{\omega}{R_2 V}\right)e^{-R_2 V} + \frac{\omega E_0}{V} \tag{2}$$

where $P_{cj}$ is the pressure generated in explosion, $E_0$ is the internal energy density, $A$, $B$, $R_1$, $R_2$ and $\omega$ are material constants [25]. The detailed parameters of explosive were given in Table 3.

**Table 3.** Parameters for explosive.

| $E_0$/(GPa) | $\rho$/(g·cm$^{-3}$) | A/GPa | B/GPa | $\omega$ | $R_1$ | $R_2$ |
|---|---|---|---|---|---|---|
| 9.4 | 1.590 | 989.085 | 11.12 | 0.396 | 5.167 | 1.046 |

The numerical model is mainly used to verify the analysis of shock-wave propagation at the T-shaped opening in this paper. As shown in Figure 16, at the wall measuring point of the curve 20 m away from the tunnel entrance, the peak values of the curves are close, and the variation laws are similar.

At present, in the research of aerodynamics, it is found that when the shock wave reaches the T-shaped junction, it will expand into the tunnel interior in a certain pattern. According to the expansion wave theory of supersonic airflow in aerodynamics [26], when supersonic airflow encounters a small external turning angle, it transitions from a high-pressure region to a low-pressure region. This transition causes the airflow to accelerate while experiencing a pressure drop, resulting in the formation of expansion waves. Each small turning angle on the various interfaces traversed by the supersonic airflow generates an associated expansion wave.

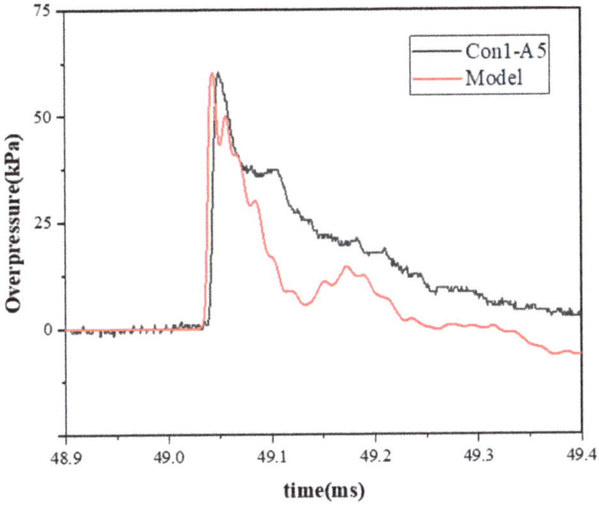


**Figure 16.** Model accuracy verification.

If the length of each external folding angle's line segment is progressively reduced, in the limit case, all the external folding angles converge into a single corner with a larger angle. When supersonic airflow passes through this configuration, it generates a fan-shaped expansion wave region centered at the turning point, as illustrated in Figure 17. Through this fan-shaped area, the airflow velocity increases while both pressure and density decrease. Similarly, when a shock wave traverses the T-shaped opening, it is influenced by the blast-facing surface of the opening, causing the cross-sectional area of the tunnel to abruptly reduce. If the right-angle corner is divided into an infinite number of infinitesimal folding angles, the compression waves generated by each folding angle will have the same direction. In the limit case, after the shock wave passes through the right-angle corner, it propagates uniformly in the same direction, forming an oblique shock wave (as shown in Figure 17).

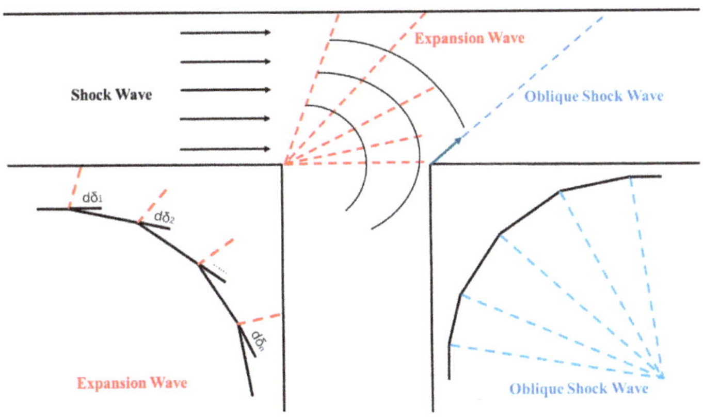


**Figure 17.** Schematic diagram of expansion wave and oblique shock wave.

Based on the analysis of the numerical simulation pressure chart (Figure 18), in conjunction with experimental data and aerodynamic theory, it can be concluded that when the shock wave enters the T-shaped port space, an expansion wave is generated at the free interface and propagates inward toward the interior of the T-shaped port. Simultaneously, the shock wave within the straight tunnel continues to expand toward the main tunnel, forming a triangular wave front that propagates along the bottom of the T-shaped port. Subsequently, this wave front undergoes its first reflection on the blast-facing surface, generating a reflected wave with relatively high pressure. Additionally, due to the sudden contraction effect of the T-shaped port section, an oblique shock wave with a specific angle forms within the straight tunnel. This oblique shock wave superimposes with the triangular wave initially transmitted from the exit of the T-shaped port, and they propagate together forward along the corridor tunnel. In the main tunnel, as the reflected wave on the blast-facing surface propagates, it undergoes a second reflection on the blast-back surface and returns to the blast-facing surface again; however, no significant third reflection phenomenon is observed. Ultimately, as the shock waves gradually catch up with and superimpose upon one another, a red high-pressure planar region propagating along the tunnel gradually forms.

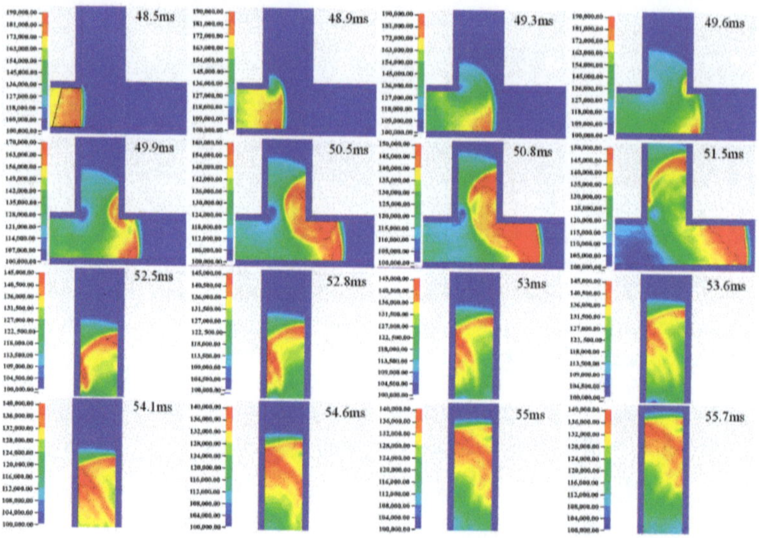


**Figure 18.** Simulation pressure chart.

According to the load characteristics of shock waves at each batch measured at points B1, B2, and B3, in conjunction with the numerical simulation propagation contour map and aerodynamic theory, the formation mechanisms and propagation features of the shock-wave curve characteristics for different batches can be systematically analyzed and determined.

Based on the analysis of Figures 9 and 10, when the shock wave initially enters the T-shaped opening, the expansion wave diffracts and reaches the back blast surface, subsequently propagating along it toward the interior of the tunnel. Given that the sensors are positioned 1 m away from the entrance section of the main tunnel, at the measurement point section, the expansion wave has already reached the back blast surface via diffraction. The peak overpressure of the first batch progressively decreases from B3 to B1, while the shock wave arrives at the B1 measurement point first, which aligns with the propagation characteristics of the expansion wave. It can thus be inferred that this shock wave corresponds to the expansion wave generated by the inward propagation of the shock wave in the straight tunnel as it interacts with the T-shaped opening. Based on the overpressure characteristics and arrival times of the second batch of shock waves, it is evident that these shock waves result from the reflection formed after the shock wave in the straight tunnel collides with the wall of the blast-facing surface. This reflected wave propagates outward spherically from the detonation point with a relatively high velocity. Consequently, the second batch of overpressure responses occur nearly simultaneously at the B2 and B3 measurement points, where they overlap with the expansion wave, leading to more intricate waveform curves.

In the experiments described in this paper, a semi-enclosed space was formed as a result of the protective door's presence. The shock wave underwent successive reflections off the blast-facing surface, the back-blast surface, and the door itself. Notably, measuring point B4 was concurrently influenced by reflections from both the back-blast surface and the door, leading to the longest positive pressure duration and the highest impulse recorded at this specific location.

### 3.4. Prediction of Momentum at Measurement Points of Main Tunnel

By comparing the impulse and peak value within the T-shaped opening, it can be found that the impulse is the main factor for evaluating the degree of the shock wave's effect on the inner wall surface of the main tunnel. A dimensional analysis was conducted on the physical model of the T-shaped tunnel with a protective door (as shown in Figure 19). Ignoring the influence of the size and shape of the explosive, the tunnel and the protective door were assumed to be rigid bodies. The width of the T-shaped opening was D, and the distance from the tunnel door to the opening of the T-shaped opening was L.

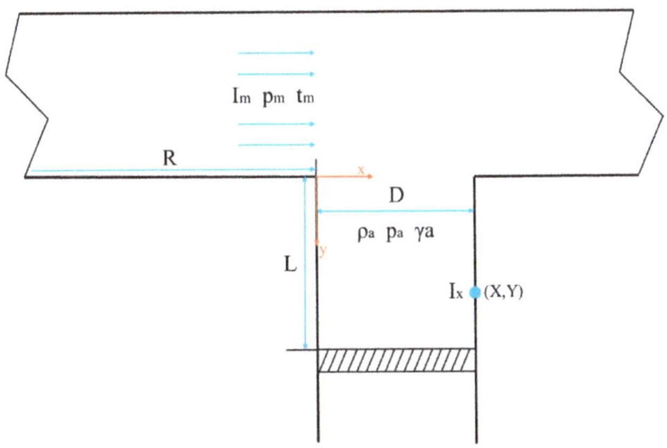


**Figure 19.** Physical model.

Considering the equivalent of the explosive, the proportional distance between the T-junction and the entrance of the tunnel is $\overline{R}$. It is assumed that the distance from the T-junction to the entrance of the tunnel is sufficient for the shock wave to form a stable triangular plate. Suppose the shock-wave pressure entering the T-junction is $p_m$, the positive pressure action time is $t_m$, the initial density of the air in the T-junction is $\rho_a$, and the initial pressure is $p_a$. The propagation of the shock wave is regarded as an adiabatic process, and the adiabatic coefficient of the air is $\gamma_a$.

Let the impulse of the shock wave entering the T-shaped port be $I_m$, the peak overpressure be $p_m$, the positive pressure duration be $t_m$, and the coordinates of a point on the inner wall surface of the T-shaped port be $X$ and $Y$. The impulse at this point is $I_x$. Ignore the influence of the tunnel cross-sectional shape.

The main determining parameter group of this problem is $p_m, t_m, I_m$, $D$, $L$, $X$, $Y$, $\rho_a, p_a, \gamma_a, \overline{R}$. The impulse at a point on the inner wall of the T-shaped port can be expressed as the following functional relationship.

$$I_m = f(p_m, t_m, D, L, X, Y, \rho_a, p_a, \gamma_a, \overline{R}) \quad (3)$$

Using the L-T-M measurement unit system, according to the Π theorem, taking the independent dimensional quantities as $p_m$, $I_m$, and $\overline{R}$, the following dimensionless combinations can be determined from the main parameter group:

$$\begin{cases} \Pi_1 = \rho_a I_m \overline{R}^3 \\ \Pi_2 = \frac{p_a}{p_m} \\ \Pi_3 = \frac{D}{p_m^{1/6} I_m^{1/2} \overline{R}^{1/2}} \\ \Pi_4 = \frac{L}{p_m^{1/6} I_m^{1/2} \overline{R}^{1/2}} \\ \Pi_5 = \frac{X}{p_m^{1/6} I_m^{1/2} \overline{R}^{1/2}} \\ \Pi_6 = \frac{Y}{p_m^{1/6} I_m^{1/2} \overline{R}^{1/2}} \\ \Pi_7 = t_m p_m^{1/3} \overline{R} \\ \Pi_I = \frac{I_x}{I_m} \end{cases} \quad (4)$$

After rearranging and simplifying Equation (4) and substituting it into Equation (3), the dimensionless functional expression for the impulse can be derived as follows.

$$\frac{I_x}{I_m} = f(\frac{X}{D}, \frac{Y}{D}, \frac{p_a}{p_m}, t_m p_m^{1/3} \overline{R}, \rho_a I_m \overline{R}^3) \quad (5)$$

The following will substitute the reference values into the above formula, conduct operations and parameter traversals using code, and provide reference suggestions for engineering applications.

Based on the analysis and prediction with the parameters of this test model, under the fixed test conditions, when the proportional distance $\overline{R}$ is constant, the parameters $p_m, t_m$ and $I_m$ at the entrance of the T-shaped port remain unchanged. Meanwhile, the shock wave propagates in the same medium (air), and $\rho_a$ and $p_a$ remain unchanged. Considering the practical application and convenience, let $p_m = 1$, $t_m = 1$, $I_m = 1$, $\rho_a = 1$, and $p_a = 1$, to obtain the simplified engineering model of the shock-wave impulse on the inner wall surface of the tunnel T-shaped port:

$$\frac{I_x}{I_m} = f(\frac{X}{D}, \frac{Y}{D}, \overline{R}) \quad (6)$$

When organizing the form of a function by experimental methods, an exponential relationship form can often be adopted in a certain range of the independent variable. Suppose the variable has the following relationship:

$$\frac{I_x}{I_m} = A(\frac{X}{D})^{\alpha} (\frac{Y}{D})^{\beta} \overline{R}^{\gamma} \quad (7)$$

Among them, $A$ is a constant.

Based on the 13 sets of impulse data in the T-shaped tunnel, the parameters of each working condition are brought in, ensuring that $A$ is a constant, and the two points (B3, B4) with the highest peak and the largest impulse in Con3 are verified.

The parameters $A$, $\alpha$, $\beta$, and $\gamma$ are obtained after screening in a traversal manner as shown in Table 4:

Table 4. Parameter prediction results.

| $A$ | $\alpha$ | $\beta$ | $\gamma$ |
|---|---|---|---|
| 0.0075 | −0.60 | 1.65 | 2.00 |

Under this parameter, the comparison of the predicted and actual impulse values of measuring points B3 and B4 is shown in Table 5. This formula is applicable for the excessive volume prediction of the internal space when a protective door is installed in a T-shaped oral cavity. However, due to the limitations of this experiment, this model is established in a circular arch-shaped tunnel with a width of 1.2 m. Considering the multiple reflection phenomena of shock waves within the tunnel, when using this formula, it is necessary to make further corrections based on the actual width of the tunnel.

$$\frac{I_x}{I_m} = 0.0075\left(\frac{X}{D}\right)^{-0.60}\left(\frac{Y}{D}\right)^{1.65}\overline{R}^{2.00} \quad (8)$$

Table 5. Comparison between the experimental and the predicted results.

| Measure Point | Momentum of Experience | Momentum of Prediction | Margin |
|---|---|---|---|
| 3-B3 | 334.89 | 337.69 | −2.80 |
| 3-B4 | 4011.56 | 3999.87 | 11.69 |

## 4. Conclusions

This paper conducted an explosion test in a T-shaped tunnel equipped with protective doors and investigated the propagation patterns and loading characteristics of the shock wave generated by the detonation of aluminized explosives. Specifically, the analysis focused on the behavior of the shock wave at the T-shaped junction and within the semi-enclosed space in front of the protective door. The primary conclusions are summarized as follows:

1. Compared to TNT explosives, the overpressure curve of aluminized explosives in the near-explosion zone has two distinct stages, with the maximum overpressure peak on the second stage. In the far explosion zone, the shock-wave overpressure waveform is a stable triangular shape.
2. The propagation of blast-induced shock waves in a T-shaped tunnel exhibits consistent patterns across four distinct batches, shaped by reflections and diffractions. The regularities of each measurement point in multiple tests show consistency, highlighting the influence laws of the geometric structure on the wave pattern and load distribution.
3. In the main tunnel of the T-shaped junction with a protective door, the maximum momentum occurs on the back-blast surface in front of the door. Due to the reflection of the shock wave from both the main tunnel wall and the door, the momentum measured on the back-blast surface in front of the door can be up to 12 times higher than that in the reflected area at the entrance of the main tunnel's blast-facing surface.
4. When the shock wave traverses the T-shaped junction, expansion waves and oblique shock waves are generated. Following two distinct reflections in the main tunnel, a stable waveform propagating along the tunnel is established. On the overpressure waveforms at each measuring point in the main tunnel, relatively smooth rectangular waves, triangular waves, and multi-peak waves are observed. Based on differences

in waveform characteristics and arrival times, the curves can be categorized into multiple groups.

5. This paper presents a dimensional analysis method for evaluating impulse at critical measurement points in the main tunnel and predicts impulse parameters under the conditions of this test. This study offers a reference for the design of relevant engineering protection measures.

**Author Contributions:** Conceptualization, Z.W. and H.L.; methodology, L.P. and H.L.; software, H.L. and L.P.; validation, L.P.; formal analysis, Z.W. and G.Z.; investigation, L.P. and H.L.; resources, Z.W. and F.G.; data curation, H.L. and Z.W.; writing—original draft preparation, L.P.; writing—review and editing, H.L. and Z.W.; visualization, L.P. and H.L.; supervision, Z.W., G.Z. and S.S.; project administration, H.L. and Z.W.; funding acquisition, Z.W. All authors have read and agreed to the published version of the manuscript.

**Funding:** National Natural Science Foundation of China: 52422808; National Natural Science Foundation of China: 52378401; National Natural Science Foundation of China: 12202494.

**Institutional Review Board Statement:** Not applicable.

**Informed Consent Statement:** Not applicable.

**Data Availability Statement:** The data that support the findings of this study are not openly available due to reasons of sensitivity and are available from the corresponding author upon reasonable request.

**Conflicts of Interest:** The authors declare no conflict of interest.

## References

1. Yan, Q.; Zhang, Y.; Sun, Q. Characteristic study on gas blast loadings in an urban utility tunnel. *J. Perform. Constr. Facil.* **2020**, *34*, 04020076. [CrossRef]
2. Chen, C.; Li, C.; Reniers, G.; Yang, F. Safety and security of oil and gas pipeline transportation: A systematic analysis of research trends and future needs using WoS. *J. Clean. Prod.* **2021**, *279*, 123583. [CrossRef]
3. Silvestrini, M.; Genova, B.; Trujillo, F.L. Energy concentration factor: A simple concept for the prediction of blast propagation in partially confined geometries. *J. Loss Prev. Process. Ind.* **2009**, *22*, 449–454. [CrossRef]
4. Benselama, A.M.; William-Louis, M.J.-P.; Monnoyer, F.; Proust, C. A numerical study of the evolution of the blast wave shape in tunnels. *J. Hazard. Mater.* **2010**, *181*, 609–616. [CrossRef] [PubMed]
5. Igra, O.; Falcovitz, J.; Houas, L.; Jourdan, G. Review of methods to attenuate shock/blast waves. *Prog. Aerosp. Sci.* **2013**, *58*, 1–35. [CrossRef]
6. Chen, D.; Wu, C.; Li, J.; Liao, K. An overpressure-time history model of methane-air explosion in tunnel-shape space. *J. Loss Prev. Process. Ind.* **2023**, *82*, 105004. [CrossRef]
7. Li, H.; Wu, H.; Wang, Z.; Zhang, G.; Li, J.; Zhou, H.; Wang, M.; He, Y. Experimental and numerical simulation of the propagation law of shock waves in corrugated steel-lined tunnels. *Process Saf. Environ. Prot.* **2022**, *168*, 1019–1030. [CrossRef]
8. Seitz, M.W.; Skews, B.W. Effect of Compressible Foam Properties on Pressure Amplification During Shock Wave Impact. *Shock Waves* **2006**, *15*, 177–197. [CrossRef]
9. Chen, T.; Tao, X.; Li, H.; Zeng, D.; Zhang, S.; Yin, S.; Zhang, Y. Engineering Calculation Method of Explosion Effect Inside Partial Confined Tunnel Structure. *Trans. Beijing Inst. Technol.* **2024**, *44*, 588–596. [CrossRef]
10. Gao, C.; Kong, X.; Fang, Q. Experimental and numerical investigation on the attenuation of blast waves in concrete induced by cylindrical charge explosion. *Int. J. Impact Eng.* **2023**, *174*, 104491. [CrossRef]
11. Ji, Y.; Zhang, G.; Li, G.; Deng, S.; Xu, T.; Yao, J.; Li, J.; Wang, M.; He, Y. Attenuation characteristics and mechanism of explosion shock wave generated by thermobaric explosive in L-shaped and gallery tunnels. *Acta Armamentarii* **2023**, *44*, 26–40.
12. Lilley, S. *Kiloton Killer*; NASA Safety Center: Cleveland, OH, USA, 2013; pp. 1–5.
13. Davis, S.; DeBold, T.; Marsegan, C. Investigation findings and lessons learned in the West Fertilizer explosion. *J. Fire Sci.* **2017**, *35*, 379–395. [CrossRef]
14. Hang, Y. *Study on Non-Ideal Detonation Behavior of Aluminum-Containing Explosives*; China Academy of Engineering Physics, Beijing Graduate School: Beijing, China, 2002.
15. Ji, Y.G.; He, Y.; Tan, Y.Z. Thermal effect and shock wave evolution of thermobaric explosive under explosion conditions near tunnel entrance. *Chin. J. Energetic Mater.* **2024**, *32*, 1287–1297.

16. Ji, Y.; Zhang, G.; Li, G.; Deng, S.; Yao, J.; Li, J.; Wang, M.; He, Y. Explosion characteristics of thermobaric explosive (TBX) detonated inside a tunnel and the related influential factors. *Explos. Shock Waves* **2024**, *44*, 032301. [CrossRef]
17. Ji, Y.; He, Y.; Tan, Y.; Li, J.; Shao, L.; Zhang, J.; Zhang, G.; Li, G. Evolution laws of the thermal effect and shock wave generated by explosion of thermobaric explosive (TBX) around the tunnel entrance. *Chin. J. Energetic Mater.* **2024**, *32*, 2024207. [CrossRef]
18. De, A. Numerical simulation of surface explosions over dry, cohesionless soil. *Comput. Geotech.* **2012**, *43*, 72–79. [CrossRef]
19. Koneshwaran, S.; Thambiratnam, D.P.; Gallage, C. Blast response of segmented bored tunnel using coupled SPH–FE method. *Structures* **2015**, *2*, 58–71. [CrossRef]
20. Tolias, I.C.; Venetsanos, A.G.; Markatos, N.; Kiranoudis, C.T. CFD modeling of hydrogen deflagration in a tunnel. *Int. J. Hydrogen Energy* **2014**, *39*, 20538–20546. [CrossRef]
21. Widanage, C.; Mohotti, D.; Lee, C.K.; Wijesooriya, K.; Meddage, D.P.P. Use of explainable machine learning models in blast load prediction. *Eng. Struct.* **2024**, *312*, 118271. [CrossRef]
22. Fu, G.; Wang, J.; Yan, M. Anatomy of Tianjin Port fire and explosion: Process and causes. *Process Saf. Prog.* **2016**, *35*, 216–220. [CrossRef]
23. Zhang, Y.; Wang, S.; Yuan, J.; Zhang, J.; Li, S. Experimental study on the propagation law of blast waves in a square tunnel. *Chin. J. Energetic Mater.* **2020**, *28*, 46–51.
24. Marty, A.; Daniel, E.; Massoni, J.; Biamino, L.; Houas, L.; Leriche, D.; Jourdan, G. Experimental and numerical investigations of shock wave propagation through a bifurcation. *Shock Waves* **2019**, *29*, 285–296. [CrossRef]
25. Wu, W.; Gao, G.; Han, Z.; Wang, Z.; Yuan, X.; He, Y. Theoretical and numerical study on blast wave propagation and afterburning effect of thermobaric explosives. *J. Nanjing Univ. Sci. Technol.* **2023**, *47*, 468–485. [CrossRef]
26. Pan, J.X. *Fundamentals of Aerodynamics (Revised Edition)*; National Defense Industry Press: Mianyang, China, 1989; Volume 1, pp. 41–50.

applied sciences

MDPI

Article

# Effect of Neighboring Hole Impacts on Inter-Hole Dynamic Presplitting Process with Consideration of Crack Width Variations

Xiao Liu [1,2], Peng Yan [1,2,*], Jin Zhu [1,2], Xiasen Yang [1,2], Xiangyu Zhang [1,2], Chao Zhou [1,2], Wenbo Lu [1,2], Ming Chen [1,2], Gaohui Wang [1,2] and Yang Wang [1,2]

1 State Key Laboratory of Water Resources Engineering and Management, Wuhan University, Wuhan 430072, China
2 Key Laboratory of Rock Mechanics in Hydraulic Structural Engineering, Ministry of Education, Wuhan University, Wuhan 430072, China
* Correspondence: pyanwhu@whu.edu.cn

**Abstract:** To analyze the effect of neighboring hole impacts on the inter-hole presplitting process, the dynamic cracking behavior under explosive load and ground stress is theoretically investigated by developing an inter-hole cracking model. Considering the variations in crack width, the influence mechanisms of stress waves from both in-hole and adjacent holes on the dynamic cracking process of presplitting blasting are examined through numerical simulations. Meanwhile, explosion tests under different detonation conditions were carried out to verify and further elucidate the dynamic effects of inter-hole presplitting blasting. The results indicate that the opening of presplitting holes can be restricted, reduced, or even closed during the cracking formation process due to the stress wave from adjacent holes when detonated simultaneously. Furthermore, there is a tendency for inter-hole cracks to be constrained in the large-angle direction, limiting crack propagation. And millisecond detonation timing can reduce the dynamic stress superposition effect between pre-cracked holes, thereby mitigating the damage to the surrounding rock. This finding is particularly applicable to presplitting blasting technology.

**Keywords:** rock excavation; dynamic cracking; presplitting blasting; crack width; high in situ stress

## 1. Introduction

With the large-scale construction of large underground power plants in China, there is an increasing need for refined control over plant safety and quality, particularly concerning the vibration and damage of surrounding rock [1–3]. In the process of blasting excavation, the propagation and evolution of blast stress waves play an important role in the rock fracture and fragmentation [4–6]. These directly influence the vibration response strength and damage area. As shown in Figure 1, the primary objective of blast fracturing is to use inter-hole blast action to promote the crack expansion in zones I, II, and III, which means a greater crushing effect.

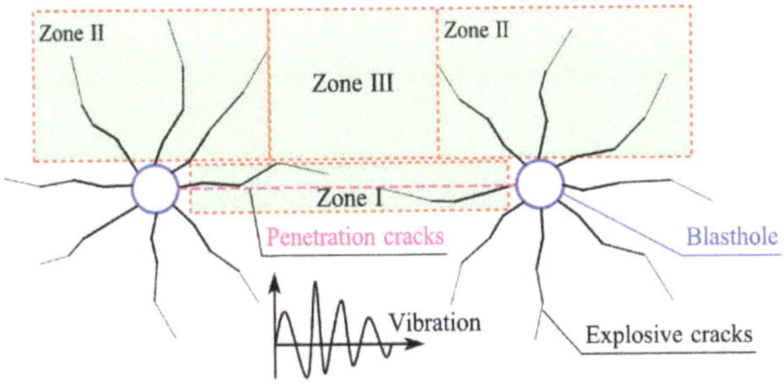


**Figure 1.** Crack extension zones for inter-hole presplitting or blast crushing.

Currently, the most widely used mechanism for inter-hole blasting crack formation is based on traditional blasting theory, which focuses on the combined effects of blast stress waves and explosive gas pressure [7–10]. For example, Hino [11] proposed a shock wave tensile failure theory based on dynamic fracture tests of rocks, suggesting that stress waves induce reflected tensile failure at free surfaces. Sazid and Singh [12] investigated the dynamic blast energy evolution for elastoplastic rock cracking through numerical simulation, emphasizing the important role of stress wave loading for blast cracking. Langefors [13,14] proposed the theory of explosive gas expansion failure, suggesting that the high-temperature, high-pressure gas products generated by detonation expand and perform work within the rock after detonation, leading to rock failure. Kutter and Fairhurst [15] described the quasi-static pressure effects of explosive gases, noting that explosive gases can form a "quasi-static pressure field" within rock or exhibit an "explosive gas wedge effect [16]". Dally et al. [17] concluded from experimental results that the work carried out by expanding blast gases in coupled charges can increase stress wave-induced fractures by 2 to 6 times. Larocque and Favreau [18] analyzed fracture distributions near boreholes through field and laboratory tests, explicitly proposing three stages of blasting fragmentation: a. radial fissures caused by stress waves; b. flaking induced by reflected waves; and c. expansion and ejection stages driven by blast gases. Daehnke et al. [19,20], based on PMMA explosion-induced fracturing experiments, concluded that blasting fractures result primarily from stress wave action (8%) and detonation gas action (92%).

Regarding the inter-hole crushing mechanism, Duvall and Petkof [21] proposed that the superposition of borehole stress waves promotes crack initiation at the midpoint between boreholes, which then propagates toward the boreholes. Yi et al. [22] discovered that the superposition of stress waves produced by adjacent blastholes contributed to the promotion of inter-hole rock fragmentation, while an increase in detonation delay decreased the quality of inter-hole fragmentation, that is, focusing on zones II and III in Figure 1. Ding et al. [23] investigated the evolution of the stress component of the double-hole blast wave with simultaneous detonation and found that the improper stress wave superposition will lessen the likelihood of crack initiation between holes. The influence of detonation delay on blasting, crushing, and vibration control was studied by blasting experiments and numerical simulations [24,25], and they both indicated the complexity of blasting processes and effects under low-delay conditions.

In contrast to inter-hole crushing, inter-hole pre-cracking blasting primarily aims to enhance the penetration of cracks along the inter-hole line (zone I) while mitigating crack propagation and minimizing damage in the retained rock mass (zones II and III), especially under complex high in situ stress conditions. He et al. [26] discovered that reflected and bypassed stress waves play an important role in crack extension by the high-speed photographic method (Digital Image Correlation, DIC), and lateral cracks are more likely to be influenced by the loading direction. Yang et al. [27] developed five crack test models based on the phenomenon that cracks between adjacent holes tend to avoid each other when they are close together. Pu et al. [28] explored the effect of crack extension behavior under two-hole blasting through numerical simulation and pointed out that the hole spacing, time delay, and pilot-hole type all have a significant effect on the inter-hole crack penetration. Xu et al. [29] pointed out that two cracks facing each other between the holes exhibit a penetration pattern of repulsion followed by attraction by digital dynamic caustic experiments and numerical simulation.

Additionally, several researchers [30–32] have highlighted the significant influence of the stress redistribution field, particularly in the directions of the major principal stresses, in inducing crack extension. Yang et al. [33] investigated the influence of in situ stress on presplitting blasting through blast experiments. Their results showed that when the direction of uniaxial compression is aligned with the blasthole layout, the crack path tends to be relatively straight. Furthermore, when confining pressure is uneven, it is recommended that the orientation of the maximum principal stress be as closely aligned as possible with the direction of the blasthole connecting line [34]. And as a result, presplitting blasting into cracks under high stress conditions was generally improved through stress adjustment of the principal stress-induced effect. For example, the lower part of an underground powerhouse is frequently excavated through the central slotting to reduce the stress concentration of the sidewall presplitting, making the local horizontal principal stress direction along the connection line of blastholes [35]. For an ultra-deep shaft, it is excavated first through the middle guide hole to achieve the stress relief of pre-cracking blasting at outer contours [36,37].

The previous research addressed the engineering requirements of crack expansion for presplitting blasting and highlighted the limited studies on the dynamic expansion process of cracks between holes. Therefore, to better understand the dynamic cracking process of inter-hole blast formation, a theoretical analysis of the cracking process in a double-hole configuration was first conducted. Subsequently, numerical simulations and explosion tests of multi-hole blasting were performed to illustrate the dynamic cracking process and demonstrate the effects of millisecond blasting between holes. The findings on the dynamic cracking process can provide valuable insights for optimizing the design of presplitting blasting.

## 2. Mechanism Analysis of Inter-Hole Dynamic Cracking Formation

### 2.1. Dynamic Cracking Tendency Under Different Loads

Figure 2a shows a double-hole blasting action model. For an explosion, taking the cylindrical charge structure as an example, the pressure–time curve (Figure 2b) will be applied to the blasthole wall. The stress–time curves of blasting load were obtained from the Laplace transform [22] and the Dubner method [38]. The detailed solution process is omitted here. Notably, the blast-induced hoop stress initially experiences compression and then rapidly transitions into a state of significant tensile stress (Figure 3b), which may be the main cause of rock tensile damage.

For the effect of initial stress, under the primary stress condition of a uniform stress field, the analysis focuses on the influence of in-plane stress on crack propagation, temporarily disregarding the effects of complex stress fields caused by lateral pressure coefficients and localized damage. In Figure 2, the initial stress field of the double hole appears as a stress concentration at the blasthole and is affected by the adjacent hole's stress field. An example is given in Figure 3 (hole diameter 90 mm, hole distance 0.6 m, and in situ stress 10 MPa). The specific calculation is omitted here. Near the hole, the hoop stress generates a stress concentration, which progressively decreases to the far-field stress as the distance from the blast increases.

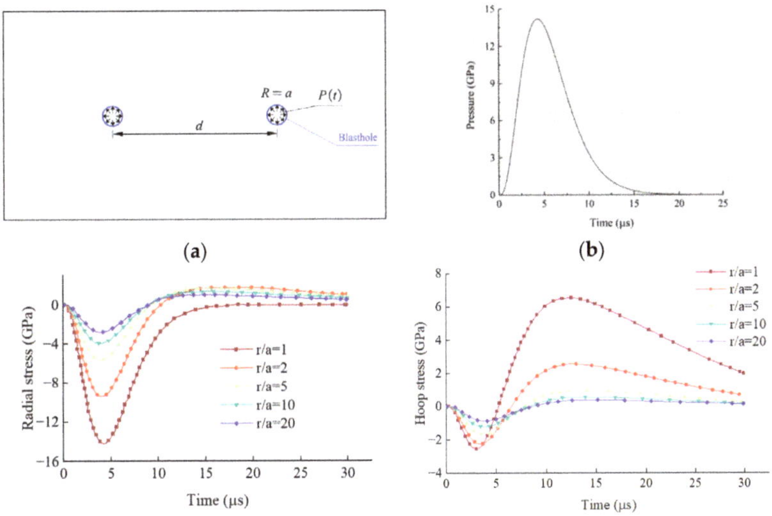


**Figure 2.** Stress–time history of typical in-hole detonation: (**a**) double-hole blasting action model; (**b**) blasting load curve of $P(t)$; (**c**) radial stress of single blast; (**d**) hoop stress of single blast.

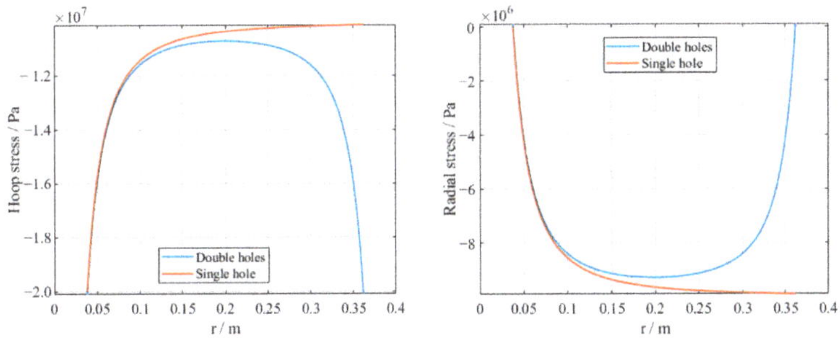


**Figure 3.** Static stress field characteristic of double-hole blasting: (**a**) hoop stress; (**b**) radial stress (single hole corresponds to a single hole at r = 0 m; double holes have two holes at r = 0 and 0.4 m).

For deep rock, the initial stress field and explosive load together influence the surrounding rock's cracking process. Taking into account the delay time of multiple blasting holes, when considering the effect of stress waves from the closest adjacent hole on one side, the stress state at any point of rock can be expressed as follows:

$$\begin{cases} \sigma_r(r,t) = \sigma_{rs}(r) + \sigma_{rb1}(r,t) + \sigma_{rb2}(r,t-\Delta t) \\ \sigma_\varphi(r,t) = \sigma_{\varphi s}(r) + \sigma_{\varphi b1}(r,t) + \sigma_{\varphi b2}(r,t-\Delta t) \end{cases} \tag{1}$$

where $\sigma_r(r,t)$ and $\sigma_\varphi(r,t)$ are dynamic radial stress and dynamic hoop stress under the coupling action of double-hole blasting, respectively; $\sigma_{rs}(r)$ and $\sigma_{\varphi s}(r)$ is the initial radial stress and hoop stress of double-hole blasting, respectively; $\sigma_{rb1}(r,t)$ and $\sigma_{\varphi b1}(r,t)$ are the dynamic radial stress and dynamic hoop stress excited by the first-hole explosion load, respectively, while $\sigma_{rb2}(r,t-\Delta t)$ and $\sigma_{\varphi b2}(r,t-\Delta t)$ are those excited by the second-hole explosion load, respectively; and $\Delta t$ is the delay time between inter-hole blasting.

### 2.2. Inter-Hole Crack Dynamic Driving Model Based on Explosion Load and Ground Stress Transient Unloading

As theoretical calculations are not applicable in the event of discrete cracking of rocks, the present work firstly offers an equation for the description of cracking in type I tensile fractures between pre-cracked holes, as illustrated in Figure 4. It depicts a dynamic seaming model for presplitting blasting that takes the effect of inter-hole blasting into consideration. This analysis is based on elastic assumptions, primarily examining the crack propagation trend of type I tensile cracks in conjunction with the previously analyzed elastic stress fields (explosion shock and in situ stress). It also investigates the influence of adjacent explosion stress fields on existing cracks, without yet addressing elastoplastic behavior or complex crack failure mechanisms.

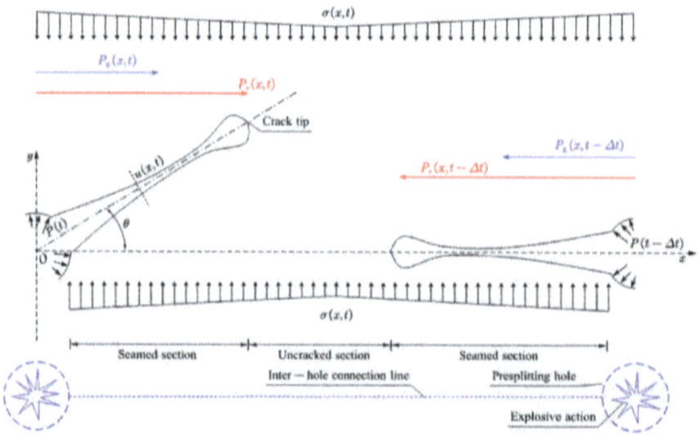


**Figure 4.** Inter-hole dynamic cracking propagation process and analysis model.

The dynamic stress intensity factor $K_I^{\mathrm{dyn}}(t)$ is employed to quantify the magnitude of the stress field near the crack tip under conditions where $r \to 0$. It can be expressed by the following relation:

$$\begin{cases} K_{\mathrm{I}}^{\mathrm{dyn}}(t) = \lim\limits_{r \to 0}\left[\sqrt{2\pi r}\sigma_{yy}(r,0,t)\right] \\ K_{\mathrm{I}}^{\mathrm{dyn}}(t) \geq K_{\mathrm{I}d} \end{cases} \tag{2}$$

Here, $K_{\mathrm{I}d}$ denotes the dynamic fracture toughness of the material. For crack initiation under explosive tensile loading in the presence of an initial stress field, the crack propagation driven by stress wave must overcome the closure effect caused by the initial stress, as described by the following criterion:

$$\begin{cases} K^{\mathrm{dyn}}(t) = K_{\mathrm{s}}^{\mathrm{dyn}}(t) - K_{\mathrm{s}}^{\mathrm{dyn}}(t - \Delta t) - K_{\sigma}^{\mathrm{dyn}}(t) \\ K^{\mathrm{dyn}}(t) \geq K_{\mathrm{I}d} \end{cases} \tag{3}$$

Here, $K_{\mathrm{s}}^{\mathrm{dyn}}(t)$ represents the dynamic stress intensity factor at the crack tip induced solely by the stress wave, while $K_{\mathrm{s}}^{\mathrm{dyn}}(t - \Delta t)$ denotes the component caused specifically by the stress wave from an adjacent blasthole. $K_{\sigma}^{\mathrm{dyn}}(t)$ is the contribution due to the in situ stress, and $K^{\mathrm{dyn}}(t)$ corresponds to the resultant dynamic stress intensity factor after superposition of all components.

Owing to the distinctly earlier initiation of the blast-induced shock wave compared to the expansion of the gaseous products, a significant temporal phase shift exists between these two loading mechanisms. Moreover, the in situ stress field remains active throughout the processes of rock fracture and crack propagation. Consequently, the relative magnitude and timing of these three factors lead to variations in the dynamic crack propagation behavior. Figure 5 presents a stress analysis performed on a rock mass element at the crack surface, in the direction normal to the crack plane following crack initiation.

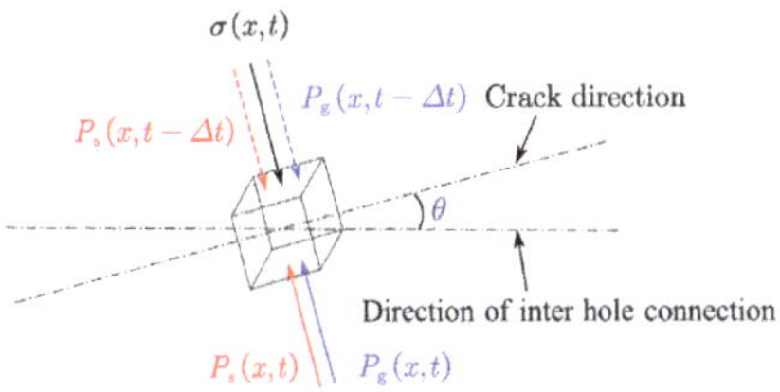


**Figure 5.** Stress analysis of the surface rock element in the vertical direction of the crack considering the stress wave of the adjacent hole.

For the fractured section, variations in crack opening displacement can significantly influence the transmission and efficiency of explosive energy. The evolution of the crack aperture is governed by the stress conditions—specifically, the dynamic in situ stress $\sigma(x,t)$ and the blast-induced loading $P(t)$—as well as the rock deformation characteristics. The blast loading applied to the surface rock mass element comprises two components: the residual hoop stress $P_s(x,t)$ resulting from the explosive stress wave and the hoop stress $P_g(x,t)$ due to the detonation gases' pressure. The mechanical response of the rock unit on the fracture surface can be characterized by an equation that incorporates the effects of stress waves originating from adjacent blastholes.

$$\sigma(x,t) + P_g(x,t) + P_g(x,t-\Delta t) + P_s(x,t) + P_s(x,t-\Delta t) = f(x,t) \tag{4}$$

With reference to the unit acceleration analysis,

$$ma_t/\mathrm{s} = f(x,t) - T_\mathrm{d} \tag{5}$$

The acceleration of the rock unit on the crack surface can be determined at any given time, from which its circumferential velocity and displacement may be derived, thereby capturing the complex evolution of crack opening.

As indicated by the foregoing analysis, during the formation of inter-hole cracks via presplitting blasting, the propagation of initial radially oriented cracks—induced by the blast—governs the inter-hole fracture process. Subsequent crack evolution occurs under the combined influence of the explosive stress wave, the pressure of detonation gases, and the dynamic in situ stress field. The mechanical model describing the dynamic crack process is represented by the following expression:

$$\begin{cases} K_\mathrm{s}^{\mathrm{dyn}}(t) - K_\mathrm{s}^{\mathrm{dyn}}(t-\Delta t) - K_\sigma^{\mathrm{dyn}}(t) \geq K_{\mathrm{I}d} & \text{Crack tip;} \\ \sigma(x,t) + P_g(x,t) + P_g(x,t-\Delta t) + P_s(x,t) + P_s(x,t-\Delta t) = f(x,t) & \text{Fractured section;} \end{cases} \tag{6}$$

Specifically, for the fracture extension of zone $\mathrm{II}_3$ and zone III in Figure 1, the neighboring hole blasting stress wave will clearly affect the dynamic crack expansion, which is more similar to the free-surface transmissive effect as the angle $\theta$ becomes closer to 90 degrees in Figure 5. Nevertheless, analytical solutions are only feasible for a limited number of idealized scenarios. In most practical cases, it is necessary to resort to numerical methods. Accordingly, this study utilizes a finite element approach to simulate the dynamic rock fracturing process under in situ stress conditions, investigate the law of dynamic crack propagation, and validate the theoretical model previously proposed.

## 3. Numerical Simulation

### 3.1. Numerical Model

The dynamic cracking process induced by inter-hole presplitting blasting was studied. To this end, a calculation model based on ANSYS/LS-DYNA for inter-hole presplitting blasting was established (see Figure 6). The model's dimensions are 2.5 m × 1.5 m × 0.005 m (length × width × thickness). The model features two presplitting holes that are symmetrical in the medial plane, accompanied by a radial uncoupled charging structure. The spacing between holes was measured to be 0.8 m, with a diameter of 76 mm for each hole. The charge diameter was determined to be 32 mm. The rock type was identified as granite, and the explosive material used was emulsion explosive. The rock is designated as a Lagrange unit, while the explosive and air are classified as ALE units. The total number of mesh grids is approximately 163,000. For the purpose of facilitating the simulation, the initial damage to the rock caused by drilling is not taken into account in this calculation.

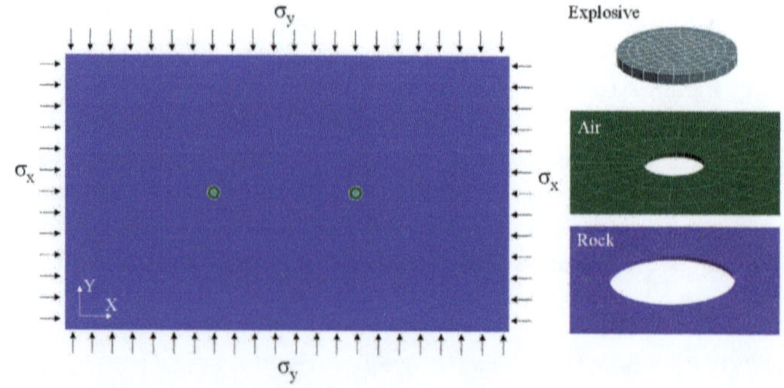


**Figure 6.** Calculation model of inter-hole blasting.

### 3.2. Parameter Selection and Verification

The precise parameters for the model are associated with granite rock, which is categorized as brittle rock. As demonstrated in Table 1, the material identity and parameters [36] are considered excellent for the purpose of characterizing the excavation disturbance dynamic behavior of hard, brittle rock.

**Table 1.** JH-2 parameters of rock input into LS_DYNA.

| Parameters | Values | Parameters | Values |
|---|---|---|---|
| Density | 3471 kg/m$^3$ | Shear modulus | 32.09 GPa |
| Strain rate parameters | 1.0 | Parameter for strain rate dependence | 0.005 |
| Pressure component, K1 | 46.6 GPa | Pressure component, K2 | −18 GPa |
| Pressure component, K3 | 3980 GPa | | |
| Intact normalized strength parameter, A | 0.70 | Fractured normalized strength parameter, B | 0.23 |
| Intact strength parameter, N | 0.61 | Fractured strength parameter, M | 0.61 |
| Maximum tensile pressure strength, T | 54 MPa | Maximum normalized fractured strength | 0.25 |
| Hugoniot elastic limit | 4.5 GPa | Pressure component of Hugoniot elastic limit | 2.578 |
| Bulk factor, β | 0.5 | | |
| Damage coefficient, D1 | 0.005 | Damage coefficient, D2 | 0.7 |

To simulate explosive events, the Fluid Structure Coupling algorithm of LS-DYNA was employed to model the dynamic impact of explosives. The fundamental principle underlying this methodology is to facilitate the movement of materials within the grid. This enables the management of substantial deformation issues, including fluid flow and explosion impact, thereby ensuring the effective resolution of complex structural challenges. The material constitutive model MAT_HIGH_EXPLOSIVE_BURN and the

Jones–Wilkins–Lee (JWL) equation are utilized to characterize the relationship between pressure and volume during explosive detonation. The equation is as follows [39]:

$$p = A_{HE}\left(1 - \frac{\omega}{R_1 V}\right)e^{-R_1 V} + B_{HE}\left(1 - \frac{\omega}{R_2 V}\right)e^{-R_2 V} + \frac{\omega E}{V} \tag{7}$$

where p is the pressure, V refers to the relative volume, and the parameters include $A_{HE}$, $B_{HE}$, $R_1$, $R_2$, and $\omega$. The empirical constant of a certain type of explosive is determined by experiments. The parameters selected in this paper are shown in Table 2 below [40,41].

**Table 2.** Explosive parameters.

| Density (kg/m$^3$) | Detonation Velocity (m/s) | P (GPa) | A (GPa) |
|---|---|---|---|
| 1130 | 4805 | 7.4 | 252 |
| B (GPa) | R1 | R2 | $\omega$ |
| 15.6 | 6.08 | 2.05 | 0.25 |

Owing to the decoupled charge configuration in presplitting blasting, the *MAT_NULL material model is employed to represent the air-filled interval. This model characterizes the pressure evolution in the air medium using a multilinear equation, given by

$$P = C_0 + C_1\mu + C_2\mu^2 + C_3\mu^3 + \left(C_4 + C_5\mu + C_6\mu^2\right)E \tag{8}$$

where P is the air pressure, $C_0 = C_1 = C_2 = C_3 = C_6 = 0$, $C_4 = C_5 = 0.4$, $\mu = \rho/\rho_0$, $\rho_0$ and $\rho$ are the initial density and current density of air, respectively, and the relevant parameters are shown in Table 3. PC is the pressure step truncation coefficient, MU is the dynamic viscosity coefficient, and $C_1$~$C_6$ are the polynomial equation coefficients.

**Table 3.** Parameters for the equation of state of air.

| ρ (kg/m$^3$) | PC | MU | $C_0$ | $C_1$ | $C_2$ | $C_3$ | $C_4$ | $C_5$ | $C_6$ |
|---|---|---|---|---|---|---|---|---|---|
| 1.18 | −1 | $1.7456 \times 10^{-5}$ | 0 | 0 | 0 | 0 | 0.4 | 0.4 | 0 |

Additionally, blast damage testing was employed to confirm the validity of granite parameters. A granite slab measuring 40 cm × 40 cm × 2 cm was used for the test. A hole of 15 mm in diameter was drilled in the center of the slab, and it was then filled with 4 g of explosive while it was kept in a stress-free environment. The results are displayed in Figure 7, and it can be noted that the numerical simulation results under the current granite settings are largely comparable with the explosion test results. Both crushing zones have a diameter of about 15 cm, and they both have obvious radial cracks. As a result of the boundary's reflection, there is also clear reflecting tensile damage near the boundary and clear circumferential cracks around 3~5 cm away from the boundary. These findings support the validity of the adopted granite parameters in characterizing cracking damage in granite rock under blast loading. Although the same rock type also exists, there is variation in parameters such as wave velocity, but it does not affect the overall hard and brittle rock law; the hard rock of dolomite and granite, as well as both the stress–strain curve and other properties or parameters exhibited by the damage characteristics and cracking law, is also similar [42–44].

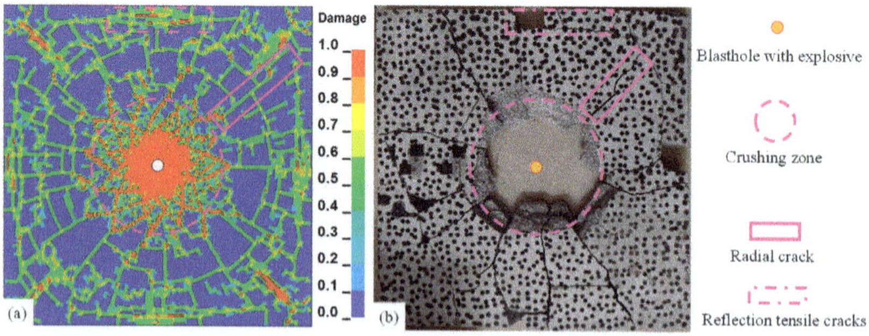


**Figure 7.** Comparison of granite explosion cracking under 4 g explosive initiation conditions: (**a**) the numerical simulation results; (**b**) the explosion test results.

### 3.3. Result Analysis

#### 3.3.1. Dynamic Cracking Process of Inter-Hole Presplitting

Figure 8 depicts the pressure propagation and damage crack extension caused by double-hole blasting under a $\sigma_x = \sigma_y = 10$ MPa condition. The blast-generated pressure gradually decays with the continuous propagation of the blast shock wave, and after the superposition of the inter-hole stress field appears at about 86 μs, the crack continues to expand with the influence of the blast stress field of the neighboring hole. This result is consistent with the effects of propagation and superposition of porous explosive stress waves described in existing research results [45,46]. In terms of crack extension morphology, the cracks induced by a single blasthole exhibit uniform expansion in all directions prior to being influenced by adjacent holes, as illustrated at 50 μs and 86 μs in Figure 8. Additionally, the cracks between holes, whose expansion direction crosses the direction of the stress wave propagation at a larger angle, appear to a certain degree of extension length limit when affected by nearby hole stress wave propagation, such as 150 μs and 1 ms in Figure 8.

To investigate the dynamic propagation of presplitting blast-induced cracks between holes and the effect of adjacent blasts, monitoring sections A, D, and E were selected as representative cases. They correspond to outer cracks, non-penetrating cracks, and inter-hole penetration cracks, respectively. Sections A, B, and C represent burst distances of 0.20 m, 0.25 m, and 0.30 m, respectively, as shown in Figure 9.

Figure 10a firstly presents the displacement–time history of the crack unit and crack surface units on both sides of the crack. Additionally, it also shows the normal stress–time curve of section A. Upon rock fracture, the rock unit A2 experiences stress dynamic unloading, with the tensile stress quickly being unloaded to zero. In contrast, the adjacent rock units A1 and A3 experience a continued increase in tensile stress before rapidly unloading, transitioning to compressive stress, and eventually converging to an equilibrium state. For a crack outside the hole, Figure 10b depicts the progression of the crack opening at 0.20, 0.25, and 0.30 m from the center of gravity. The crack opening is rapidly opened in conjunction with the rock cracking process and rapid unloading of its unit stress, followed by a certain degree of rebound under the action of ground stress. And the crack opening subsequently continues to rise under the action of blast pressure. For section D, the adjacent hole blast shock wave clearly limits the crack opening, as illustrated in Figure 10c,d. At 128–186 μs, the crack opening where section D is located is squeezed closed and stops extending.

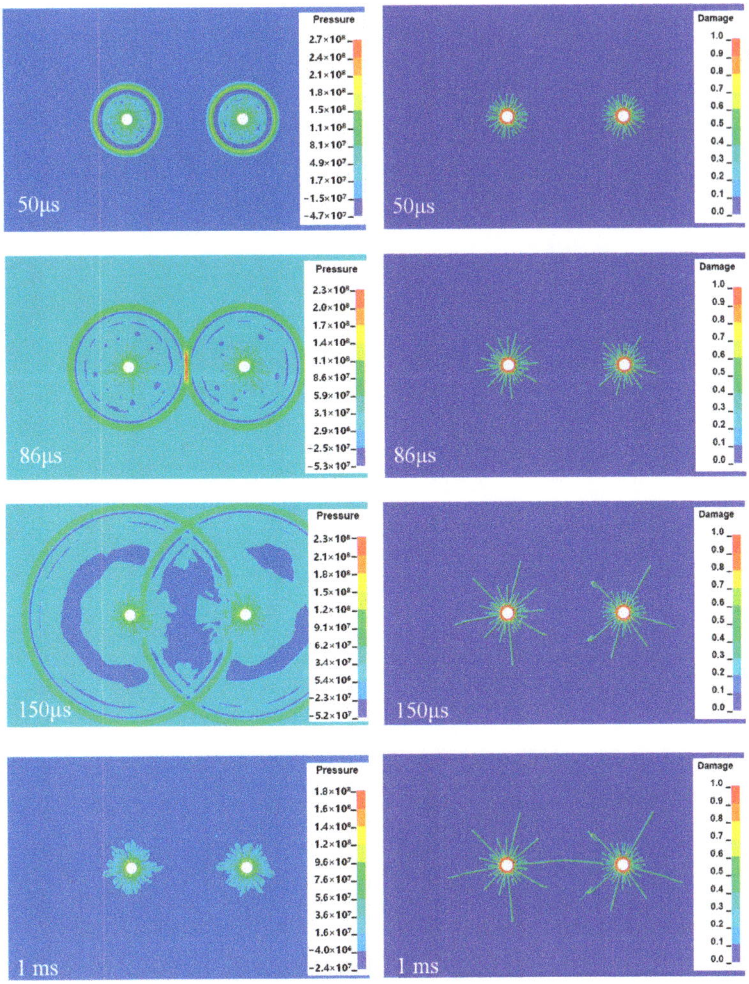


**Figure 8.** Inter-hole presplitting formation process under 10 MPa stress.

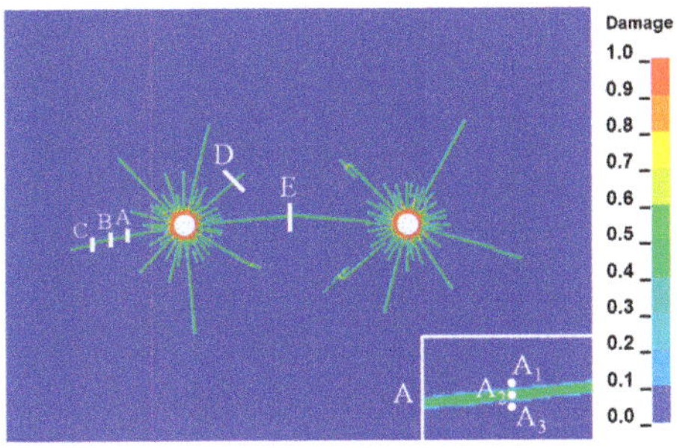


**Figure 9.** Typical cross-section and unit schematic.

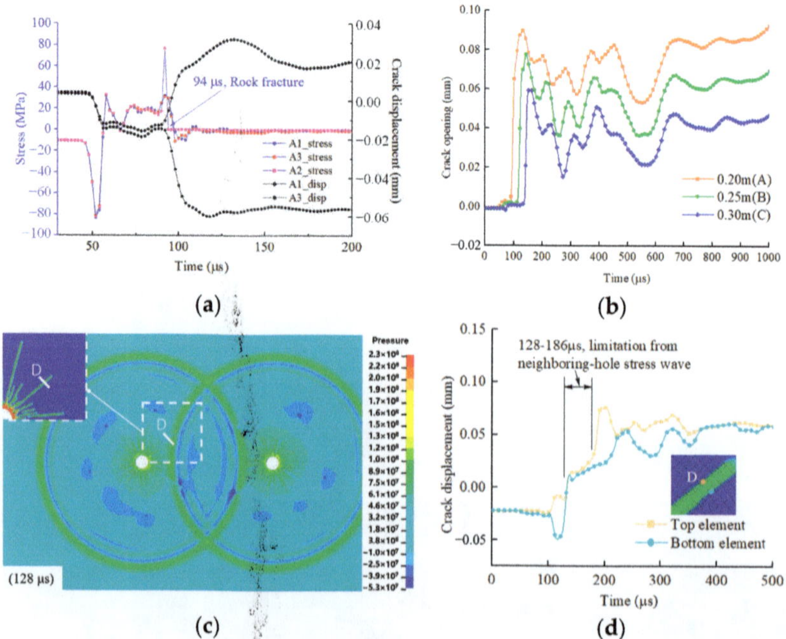


**Figure 10.** Crack displacement- and stress–time curves for typical cross-sections under double-hole blasting conditions: (**a**) typical stress–time curve of section A; (**b**) crack displacement of sections A, B, and C; (**c**) explosive stress wave contour plot at 128 μs; (**d**) crack opening of cross-section D. The letters A, B, C, D refer to the cross sections in Figure 9.

### 3.3.2. Influence of Stress Conditions on Inter-Hole Cracking Process

The complex in situ stress conditions present at the actual engineering sites significantly influence the inter-hole crack expansion of presplitting blasting. As shown in Figure 11, the crack patterns resulting from inter-hole blasting under various in situ stress environments are presented. It is discovered that when the stress level increases from 0 MPa to 20 MPa, the crack expansion reduces from about 0.75 m to 0.24 m. The inter-hole crack expansion range is lower than the crack expansion range on the exterior of the hole due to the influence of the adjacent hole stress wave, especially under lower stress levels (Figure 12). Under medium–high stress, however, the inhibitory impact of neighboring hole stress waves from adjacent holes on inter-hole fractures is reduced. Cracks around the blasthole consequently exhibit a confined pattern of uniform expansion.

For the dynamic crack expansion process (Figure 13), they are essentially consistent with the previous results under $\sigma_x = \sigma_y = 10$ MPa. For unpenetrated cracks outside the holes (Figure 13a), as the ground stress level increases, the maximum value of the initial crack opening appears significantly lower, and the crack unloading rebound increases strongly, and then tends to close. Once a crack initiates and coalesces at the midpoint between two blastholes, the crack apertures typically increase during this process, forming a penetrated crack. And the lower the ground stress, the greater the openings (Figure 13b).

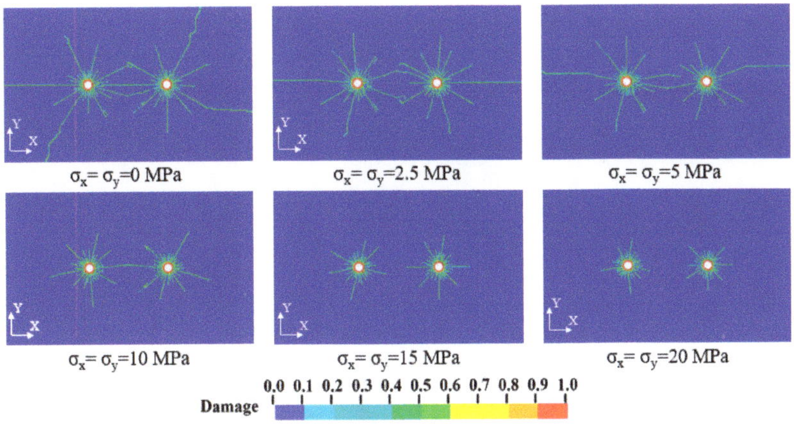


**Figure 11.** Crack expansion of double-hole cracks at different stress levels.

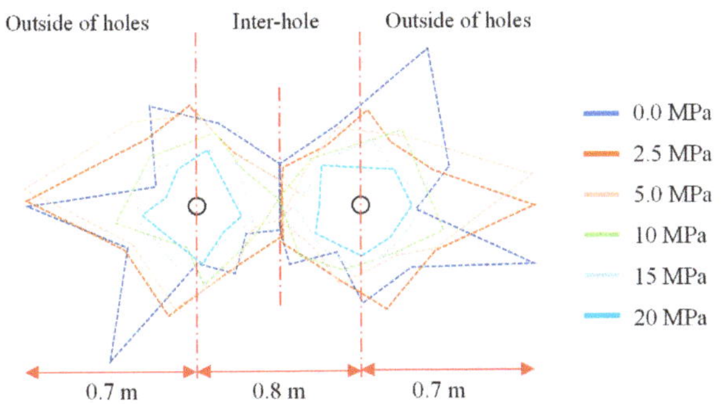


**Figure 12.** Comparison of double-hole explosion-induced crack expansion at different stress levels.

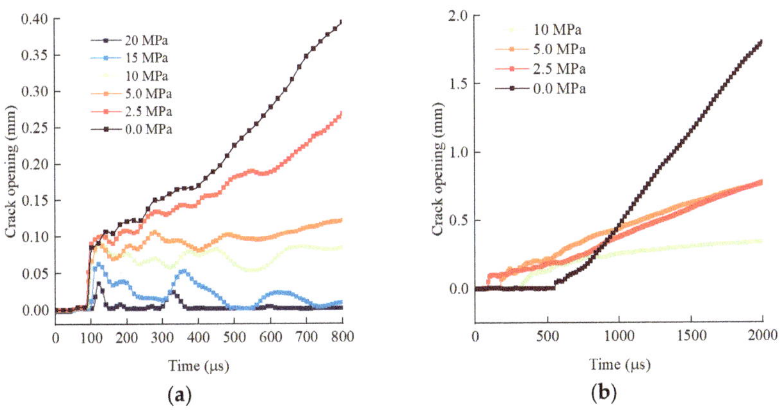


**Figure 13.** Dynamic expansion process of inter-hole and outer-hole cracks at different stress levels: (**a**) unpenetrated cracks of the outer-hole at 0.2 m of blasting center distance; (**b**) penetrated cracks of the inter-hole at the connecting-line midpoint.

Furthermore, Figure 14 illustrates fracture propagation at various lateral pressure coefficients for a blasting center distance of 0.3 m. A higher lateral pressure coefficient results in longer crack extension along the direction of maximum principal stress. These cracks also exhibit larger apertures during dynamic propagation, indicating more efficient blast energy transmission. At lateral pressure coefficients ranging from 1.3 to 4.0, the inter-hole cracks appeared as penetrated, curved interlayer, and unpenetrated, respectively. And as the lateral pressure coefficient increases, the crack opening becomes higher.

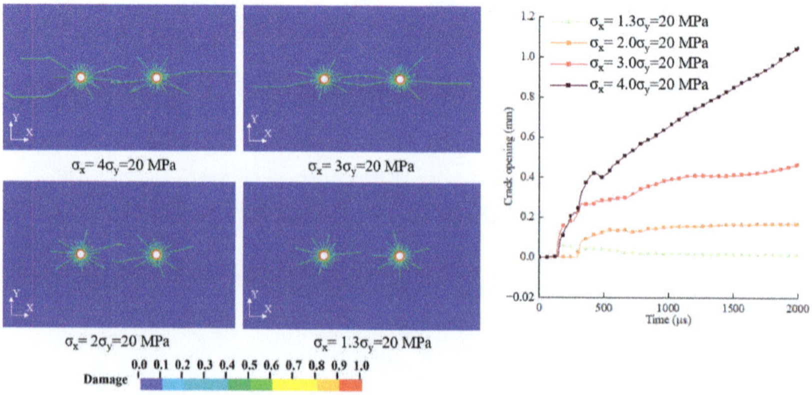


**Figure 14.** Dynamic crack expansion with different lateral pressure coefficients.

### 3.3.3. Influence of Blasting Delay Time on Inter-Hole Cracking Process

The primary goal of presplitting blasting includes vibration isolation and crack resistance, particularly in high in situ stress conditions. Given the superiority of larger lateral pressure coefficients for inter-hole crack penetration, a simulated crack formation of presplitting blasting with a two-hole model was performed under $\sigma_x = 20$ MPa, $\sigma_y = 0$ MPa to investigate the effect of blast delay time on the dynamic seam formation process, distribution pattern, and generated vibration between holes. Due to the numerical modeling, where the left hole did not crack continually and clearly after 1 ms, the right hole detonation delay is set to account for this in the numerical analysis.

Figure 15 compares the crack dynamic expansion under delayed and simultaneous presplitting blasting conditions. In the case of simultaneous detonation (Figure 15c), cracks sprout and spread around both holes in the direction of the major stresses, eventually intersecting midway. The crack patterns of the two holes are basically symmetrical along the centerline of the holes. Under delayed detonation, the left hole explosion first induces cracks gradually in all directions more uniformly and extends to the right hole (Figure 15a). Upon detonation of the right hole, the damage was intensified at the fractures previously generated from the left hole, and the width of the penetration cracks was greatly enlarged (Figure 15d). Notably, large-angle cracks from the right hole between holes essentially did not appear, in addition to the intersection penetration of the existing cracks on the left hole (Figure 15b). The fact that the outside of the hole is essentially the same as the left side helps to explain the induction of existing cracks' role in the explosion. Given the constraints of numerical modeling for finite element characterization following rock cracking, an explosive test will be conducted to further examine the molding effect of delayed presplitting blasting.

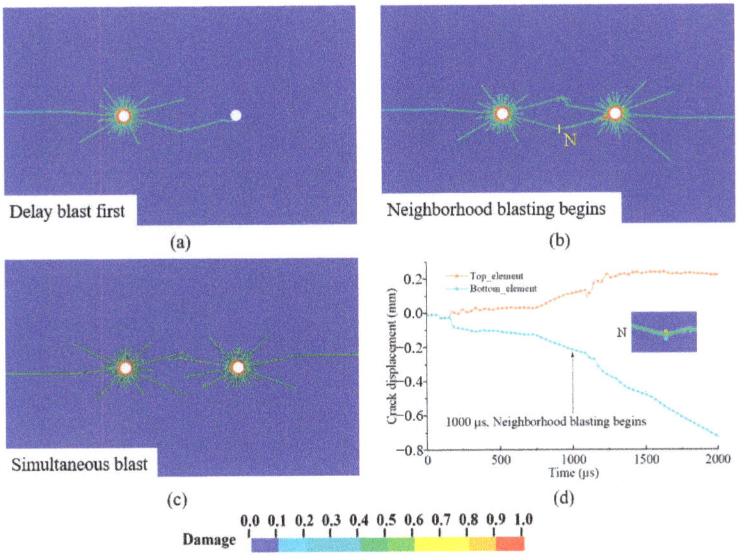


**Figure 15.** Comparison of the effect of dynamic extension of oblique cracks under microtremors: (**a**) first blast of delay blast; (**b**) neighborhood blasting begins; (**c**) simultaneous blast; (**d**) crack displacement of delay blast.

## 4. Inter-Hole Explosion Verification Test

To verify the influence of in situ stress on crack propagation and the effects of adjacent blasts, explosion tests were conducted under uniaxial loading conditions. Firstly, rock cracking is given for single-hole blasting in a uniaxial stress environment of 3 MPa. The dynamic crack extension effect and molding control discrepancies between holes are subsequently demonstrated using millisecond blasting under uniaxial stress. The size of the granite samples is 400 mm × 400 mm × 18 mm. Stress boundaries are achieved by one or two hydraulic jacks, as well as two uniform steel plates and steel structures, as shown in Figure 16. When there is no need to apply pressure in a certain direction, the jack is just in contact with the rock slab, resulting in a fixed limitation and no stress loading. Strain gauges are mainly used for dynamic monitoring of rock samples.

(1) Single-hole explosion test: A hole was drilled in the center with a hole diameter of 15 mm, and emulsion explosives were used with 2.5 g (Figure 16c,d).

(2) Multi-hole explosion test: Three blastholes with a hole diameter of 15 mm were drilled. Two of them have presplitting blastholes spaced 100 mm apart and symmetrical to the slab's center. Following presplitting holes, the third hole is made up of side blasting holes to test the crack-resistant effect of inter-hole seam formation (Figure 16e,f). Emulsion explosives were used with 2 g explosives in the presplitting hole and 2.5 g explosives in the side blasting hole. Two sets of tests were designed according to different blasting delay times.

Firstly, Figure 17 illustrates the crack pattern resulting from single-hole blasting under 3 MPa uniaxial stress. The results indicate that the unidirectional principal stress predominantly governs the direction of the crack extension. A through crack with a width of approximately 1 mm forms in the principal stress's direction, and a surface spalling range with a radius of approximately 5 cm forms in the blasthole's proximal area. Because the location of the strain gauges mentioned above is essentially close to the explosion, some of them were destroyed in the explosion. The crack extension circumferential stress curve obtained by monitoring YBθ_2 (bursting center distance of approximately 10 cm, circumferential arrangement) is provided here. After the initial compression and tensile action, a considerable tensile stress condition was exhibited. This finding is also compatible with the numerical simulation of crack opening with the continuous action of tensile stress, as described previously.

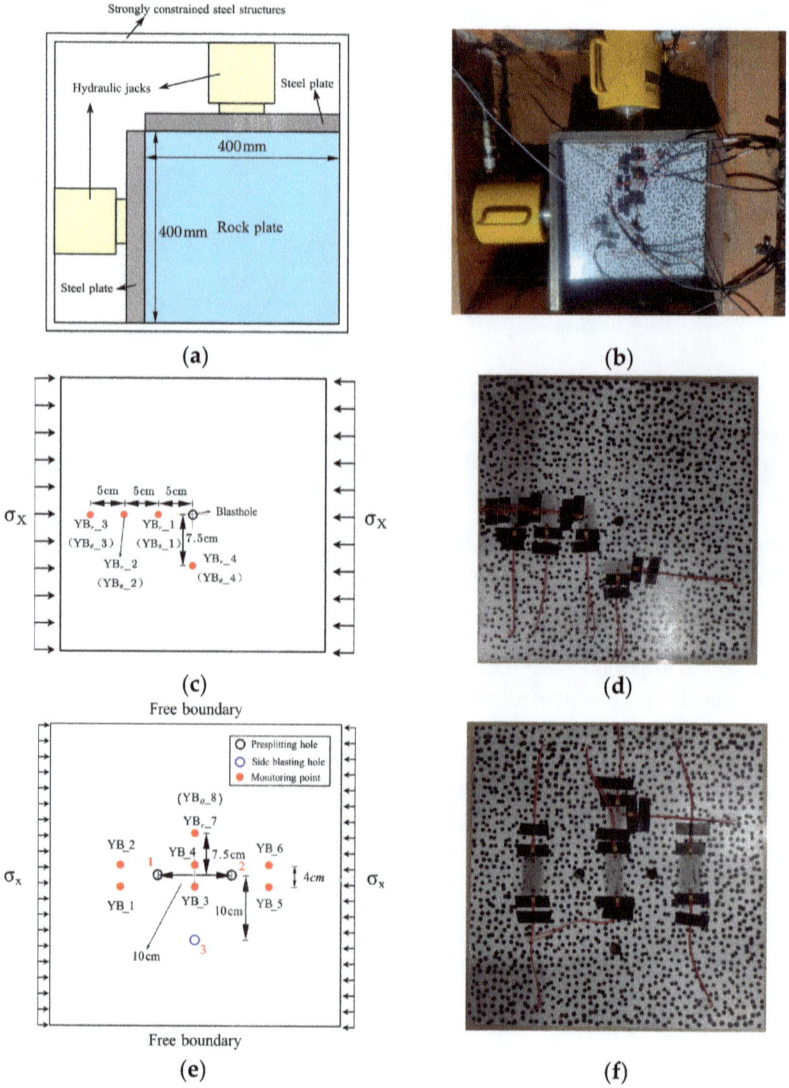


**Figure 16.** Explosive test design and monitoring arrangement: (**a**) loading device design; (**b**) on-site loading device arrangement; (**c**) single-hole design; (**d**) arrangement of single-hole rock slabs; (**e**) multi-hole design; (**f**) arrangement of multi-hole rock slabs.

In the multi-hole explosion test, comparative blasting designs are shown in Table 4. Given that the explosion at the site severely destroyed the strain-gauge monitoring lines, the following analysis is primarily focused on post-blast impacts.

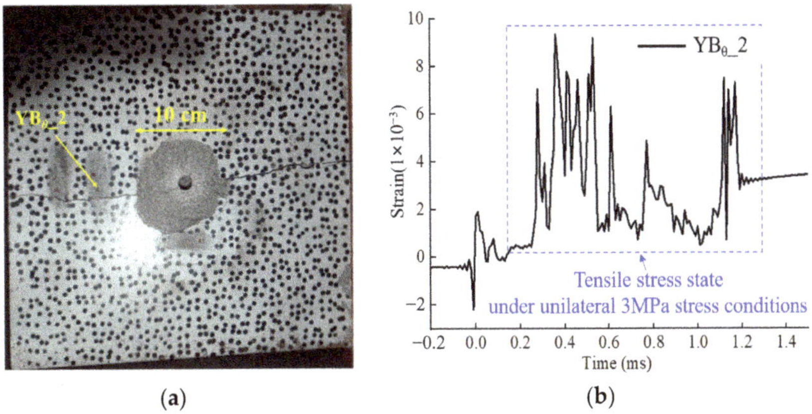


**Figure 17.** Results of normal strain monitoring of cracks under different stress conditions: (**a**) rock cracking induced by single-hole blasting under unidirectional stresses; (**b**) typical hoop stress curve of YB$_\theta$_2 on the crack extension side.

**Table 4.** Comparative designs of millisecond initiation in multi-hole explosion test.

| Test Conditions | Presplit Hole 1 | Presplit Hole 2 | Side Blasting Hole 3 |
|---|---|---|---|
| Simultaneous presplitting | MS1 | MS1 | MS3 |
| Millisecond presplitting | MS1 | MS3 | MS5 |

Figure 18 illustrates the blasting effects observed in rock plate cracking under simultaneous presplitting blasting. In this configuration, radial cracks propagate radially from the presplitting holes in multiple directions, leading to a significant enhancement in the fragmentation of the reserved rock mass on both sides. Compared to single-hole blasting scenarios discussed earlier, the control over principal stress direction appears to exert less influence on crack guidance in simultaneous presplitting blasting. The phenomenon is attributed to the superposition of stress waves from simultaneous dual-hole detonation, wherein the explosive action in the rock mass between the holes is significantly amplified. This enhanced inter-hole fragmentation overwhelms the crack guidance capability of the comparatively low in situ stress field (3 MPa), thus negating any distinct advantage for crack propagation along the major principal stress direction. Specifically, in zone III, located between the two presplitting holes, cracks originating from the right hole did not extend into this region. Instead, Crack 1 initially developed toward the inter-hole region and subsequently diverged outward, while Crack 2 extended at a considerable angle relative to the connecting line between holes, eventually propagating toward side hole 3. In contrast, the left presplitting hole experienced complete failure, with fractured material ejected toward the left free surface. This crack propagation pattern aligns well with the outcomes of the numerical simulation of simultaneous detonation presented earlier, which predicted shorter crack lengths between holes compared to outward directions, or outward deflection of inter-hole cracks. These observations effectively demonstrate the dynamic interaction between stress waves generated by adjacent blastholes and their inhibiting effect on crack development within the inter-hole region.

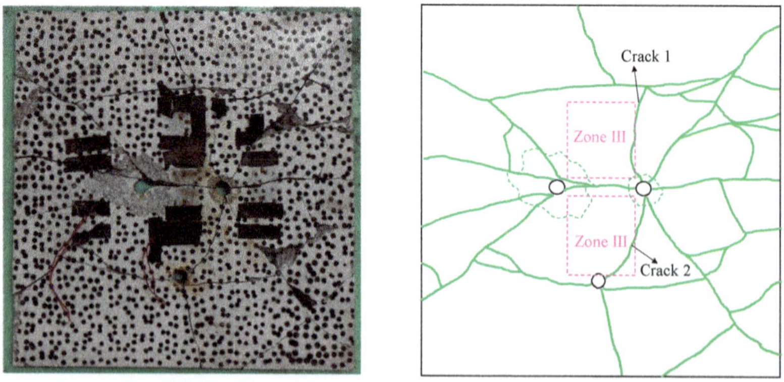


(a) (b)

**Figure 18.** The blasting effect of rock plate cracking under simultaneous blasting: (**a**) rock slab fragmentation; (**b**) crack extension pattern. The green lines represent only the location of the cracks.

Figure 19 shows the blasting effect of rock plate cracking under millisecond blasting designs. Well-defined through-going cracks are observed along the direction of the connecting line between holes (the direction of the large principal stress). In contrast, the rock mass remains intact in the direction of the minor principal stress, with no evidence of cracking induced by presplitting. It indicates an effective presplitting outcome. Meanwhile, the pre-cracks formed by millisecond blasting could effectively block the influence of adjacent blasting holes on the retained rock mass, and the retained rock mass has good integrity without obvious fractures.

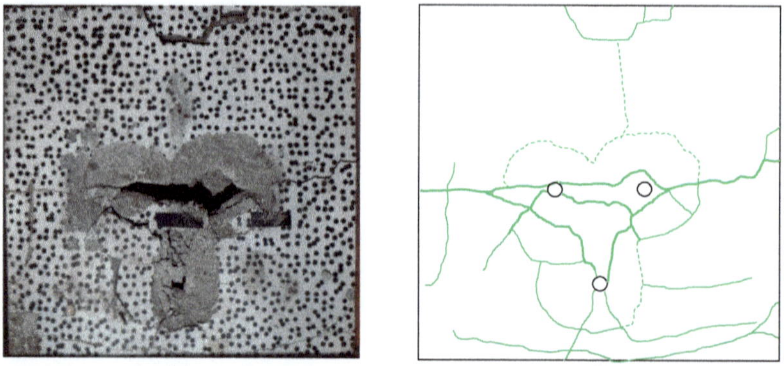

(a) (b)

**Figure 19.** The blasting effect of rock plate cracking under millisecond blasting: (**a**) rock slab fragmentation; (**b**) crack extension pattern. The green lines represent only the location of the cracks.

As a result, based on the ability to achieve rock cracking, millisecond blasting can reduce the impact of blasting stress waves, allowing for the expansion of inter-hole cracks along the direction of the main stress, whereas simultaneous blasting may cause the superposition of stress waves to become the main influence on crack expansion, and pre-cracking control in a seam does not fit.

Due to the limited number of tests conducted in this study, the results may be subject to chance variations. Additional experiments will therefore be conducted in the future to investigate the dynamic propagation of cracks. At the same time, the experimental process above used high-speed photography for monitoring. However, due to the explosive detonation process accompanied by a large amount of smoke and dust, the sight of high-speed photographic monitoring was blocked. It cannot achieve effective experimental monitoring of the dynamic expansion of cracks in the near-area blast. These challenges will be addressed in subsequent experimental studies.

To demonstrate the validity and novelty of the conclusions, the findings are compared and discussed with those of other researchers [23,28,47,48], as shown in Figure 20. These results align with the crack distribution characteristics induced by dual-hole (multi-hole) blasting in existing studies, as can be observed. Specifically, crack propagation length is significantly shorter at large angles between holes than at the outer sides of the two holes. This indicates that explosive stress waves between holes have a pronounced influence on crack propagation. While prior studies have focused on crack morphology, they have not yet conducted in-depth investigations into the differences between inter- and extra-hole cracks. This paper, however, explores the mechanism by which explosive stress waves influence the dynamic propagation of pre-existing cracks, thereby demonstrating the innovation of the work and the validity of its conclusions.

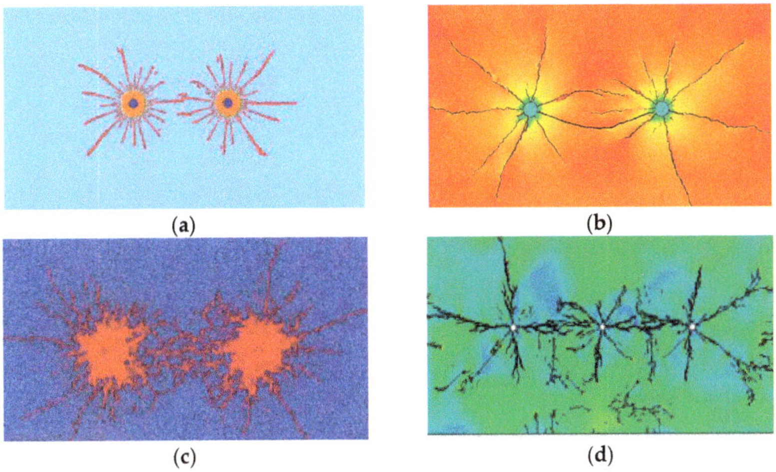


**Figure 20.** Comparison of the effect of simulation or testing of double-hole explosion cracking with existing results [23,28,47,48]: (**a**) Double-hole blast Cracks with simultaneous detonation based on finite element method; (**b**) Blast-induced crack using continuous-discontinuous element method; (**c**) cracks of simultaneous detonation given by Particle Flow Code; (**d**) Three-hole explosion crack distribution.

## 5. Conclusions

To investigate the dynamic cracking mechanism between presplitting holes in deep rock, a theoretical model was first established to analyze the inter-hole cracking process. Numerical simulations were then employed to examine the dynamic cracking behavior under various stress conditions, revealing the influence mechanism of stress waves from adjacent holes on crack propagation. Furthermore, blasting tests involving single and multiple holes were conducted, which verified the dynamic expansion patterns of inter-hole fractures and demonstrated the significant advantages of micro-deferential blasting in presplitting applications. The main conclusions are as follows:

(1) High ground stress, blast stress waves, and blast-generated gas all affect the dynamic cracking process of multi-hole blasting, and the suggested two-hole dynamic crack model may well depict the dynamic effects of stress waves from adjacent holes on crack expansion. The fracture opening shows a dynamic change law of increasing, then decreasing, and then increasing again.

(2) Neighboring hole stress waves can have a major restricting effect on dynamic fracture opening, particularly for inter-hole fractures that cross the inter-hole line at a wide angle, resulting in a restricted propulsive effect of burst-generated energy on fracture expansion.

(3) One of the most crucial ways to achieve good pre-cracking in complex stress fields is to align the direction of the local maximum principal stress with the direction of inter-hole cracking. Additionally, delay blasting can be used to improve the cracking base for the expansion and connection of cracks in neighboring holes. This further demonstrates the practicality of delay blasting.

This work was conducted using a combined approach of elasticity theory, numerical simulation, and explosion testing. The theoretical model still exhibits significant discrepancies with field conditions, and its completeness requires further refinement through future research on complex crack geometries and three-dimensional crack propagation.

**Author Contributions:** Conceptualization, X.L. and P.Y.; methodology, X.L.; software, X.Y.; validation, J.Z. and X.Z.; formal analysis, C.Z.; writing—original draft preparation, X.L.; writing—review and editing, W.L. and M.C.; visualization, Y.W.; supervision, G.W.; project administration, P.Y.; funding acquisition, X.L., P.Y. and W.L. All authors have read and agreed to the published version of the manuscript.

**Funding:** The authors gratefully appreciate the support from the National Natural Science Foundation of China (No. 52379108), 2023 Postdoctoral Innovative Research Positions in Hubei Province (Postdoctoral No. 325469), and the Key Program of the National Natural Science Foundation of China (No. 51939008). The support from the National Natural Science Foundation of China (Grant No. 52109164) is also appreciated.

**Institutional Review Board Statement:** Not applicable.

**Informed Consent Statement:** Not applicable.

**Data Availability Statement:** The original contributions presented in this study are included in the article. Further inquiries can be directed to the corresponding author.

**Conflicts of Interest:** The authors declare no conflicts of interest.

## References

1. Li, X.; Huang, J.; Luo, Y.; Dong, Q.; Li, Y.; Wan, Y.; Liu, T. Numerical simulation of blast vibration and crack forming effect of rock-anchored beam excavation in deep underground caverns. *Shock Vib.* **2017**, *2017*, 1812080. [CrossRef]
2. Zhang, T.; Huang, J.; Li, X.; Liu, T.; Bian, X.; Luo, Y. Characteristics of the elevation amplification effect of vibration velocity in rock surrounding underground cavities under different stress conditions. *Soil Dyn. Earthq. Eng.* **2023**, *165*, 107704. [CrossRef]
3. Fan, Y.; Yang, G.D.; Cui, X.Z.; Zhao, X.H.; Wu, J.G.; Wang, L.H. Low-Frequency Characteristics of Vibrations Induced by Transient Unloading of In Situ Stress and Its Influence on Safety of a Deeply Buried Tunnel. *Int. J. Geomech.* **2024**, *24*, 04023288. [CrossRef]
4. Banadaki, M.D.; Mohanty, B. Numerical simulation of stress wave induced fractures in rock. *Int. J. Impact Eng.* **2012**, *40*, 16–25. [CrossRef]
5. Saadatmand Hashemi, A.; Katsabanis, P. The effect of stress wave interaction and delay timing on blast-induced rock damage and fragmentation. *Rock Mech. Rock Eng.* **2020**, *53*, 2327–2346. [CrossRef]
6. Fan, Y.; Chen, J.; Leng, Z.; Yang, G.; Liu, X.; Tian, B. Punching Mechanism of Air-Deck Stemming for Drilling Blasting and Its Influence on Rock Fragmentation. *Rock Mech. Rock Eng.* **2024**, *57*, 3917–3935. [CrossRef]
7. Córdova, E.; Gottreux, I.; Anani, A.; Ferrada, A.; Contreras, J. Blasting and preconditioning modelling in underground cave mines under high stress conditions. *J. S. Afr. Inst. Min. Metall.* **2021**, *121*, 71–80. [CrossRef]
8. Sanchidrián, J.A.; Segarra, P.; López, L.M. Energy components in rock blasting. *Int. J. Rock Mech. Min.* **2007**, *44*, 130–147. [CrossRef]
9. Ye, Z.; Chen, M.; Yi, C.; Lu, W.; Yan, P. Quantitative Study of the Action on Rock Mass Failure under the Shock Wave and Gas Pressure in Bench Blasting. *Int. J. Geomech.* **2023**, *23*, 04023135. [CrossRef]
10. Trivino, L.F.; Mohanty, B. Assessment of crack initiation and propagation in rock from explosion-induced stress waves and gas expansion by cross-hole seismometry and FEM–DEM method. *Int. J. Rock Mech. Min. Sci.* **2015**, *77*, 287–299. [CrossRef]

11. Hino, K. Fragmentation of rock through blasting and shock wave theory of blasting. In Proceedings of the 1st US Symposium on Rock Mechanics (USRMS), Golden, CO, USA, 23–25 April 1956.
12. Sazid, M.; Singh, T.N. Two-dimensional dynamic finite element simulation of rock blasting. *Arab. J. Geo-Sci.* **2013**, *6*, 3703–3708. [CrossRef]
13. Langefors, U.; Kihlström, B. *The Modern Technique of Rock Blasting*; Wiley: New York, NY, USA, 1963; Volume 405.
14. Langefors, U. Fragmentation in rock blasting. In Proceedings of the 7th Symp. on Rock Mechanics, Pennsylvania State University, University Park, PA, USA, 14–16 June 1965; pp. 1–21.
15. Kutter, H.K.; Fairhurst, C. On the fracture process in blasting. *Int. J. Rock Mech. Min. Sci.* **1971**, *8*, 181–202. [CrossRef]
16. Hagan, T.N. Rock breakage by explosives. *Acta Astronaut.* **1979**, *6*, 329–340. [CrossRef]
17. Dally, J.W.; Fourney, W.L.; Holloway, D.C. Influence of containment of the bore hole pressures on explosive induced fracture. *Int. J. Rock Mech. Min. Sci.* **1975**, *12*, 5–12. [CrossRef]
18. Larocque, G.; Favreau, R. Blasting research at the Mines Branch. In Proceedings of the 12th US Symposium on Rock Mechanics (USRMS), Rolla, MO, USA, 16–18 November 1970.
19. Rossmanith, H.; Daehnke, A.; Nasmillner, R.; Kouzniak, N.; Ohtsu, M.; Uenishi, K. Fracture mechanics applications to drilling and blasting. *Fatigue Fract. Eng. Mater. Struct.* **1997**, *20*, 1617–1636. [CrossRef]
20. Daehnke, A.; Rossmanith, H.; Knasmillner, R. Blast-induced dynamic fracture propagation. In *Rock Fragmentation by Blasting*; CRC Press: Boca Raton, FL, USA, 2020; pp. 13–18.
21. Duvall, W.I.; Petkof, B. *Spherical Propagation of Explosion-Generated Strain Pulses in Rock*; US Department of the Interior, Bureau of Mines: Washington, DC, USA, 1959.
22. Yi, C.; Johansson, D.; Nyberg, U.; Beyglou, A. Stress Wave Interaction Between Two Adjacent Blast Holes. *Rock Mech. Rock Eng.* **2016**, *49*, 1803–1812. [CrossRef]
23. Ding, C.; Yang, R.; Feng, C. Stress wave superposition effect and crack initiation mechanism between two adjacent boreholes. *Int. J. Rock Mech. Min. Sci.* **2021**, *138*, 104622. [CrossRef]
24. Zhang, X.; Yan, P.; He, H.; Lu, W.; Liu, B.; Zhu, J.; Cheng, Y. Experimental investigations of the effect of millisecond-delay time on the blast vibration reduction with electronic detonators. *J. Vib. Control.* **2023**, *29*, 4204–4215. [CrossRef]
25. Yi, C.; Sjöberg, J.; Johansson, D.; Petropoulos, N. A numerical study of the impact of short delays on rock fragmentation. *Int. J. Rock Mech. Min.* **2017**, *100*, 250–254. [CrossRef]
26. He, C.; Chen, D.; Xiao, J.; Lu, L.; Guo, Z. Experimental study of crack propagation and failure around a horseshoe tunnel during nearby blasting. *Int. J. Rock Mech. Min. Sci.* **2021**, *139*, 104628. [CrossRef]
27. Yang, L.; Wang, Q.; Xu, L.; Yang, R.; Chao, Y.J. Fracture path of cracks emigrating from two circular holes under blasting load. *Theor. Appl. Fract. Mech.* **2020**, *108*, 102559. [CrossRef]
28. Pu, C.; Yang, X.; Zhao, H.; Chen, Z.; Xiao, D. Numerical investigation on crack propagation and coalescence induced by dual-borehole blasting. *Int. J. Impact Eng.* **2021**, *157*, 103983. [CrossRef]
29. Xu, P.; Yang, R.; Guo, Y.; Chen, C.; Yang, Y.; Zuo, J. Investigation of the interaction mechanism of two dynamic propagating cracks under blast loading. *Eng. Fract. Mech.* **2022**, *259*, 108112. [CrossRef]
30. Li, X.; Liu, K.; Qiu, T.; Sha, Y.; Yang, J. Study of presplit blasting under high in-situ stress. *Eng. Fract. Mech.* **2023**, *288*, 109360. [CrossRef]
31. Li, X.; Liu, K.; Yang, J.; Qiu, T.; Sha, Y. Effects of in-situ stress on rock cracking in presplit blasting. *Comput. Geotech.* **2024**, *168*, 106124. [CrossRef]
32. Liu, K.; Li, X.; Huang, L.; Sha, Y.; Yang, J.; Zhao, X.; Ma, S.; Hong, Z. Investigation on Rock Fracturing in Presplit Blasting Under Various Initial Stresses. *Rock Mech. Rock Eng.* **2024**, *57*, 6927–6950. [CrossRef]
33. Yang, L.; Yang, A.; Chen, S.; Fang, S.; Huang, C.; Xie, H. Model experimental study on the effects of in situ stresses on pre-splitting blasting damage and strain development. *Int. J. Rock Mech. Min.* **2021**, *138*, 104587. [CrossRef]
34. Yang, L.; Chen, S.; Yang, A.; Huang, C.; Xie, H. Numerical and experimental study of the presplit blasting failure characteristics under compressive stress. *Soil Dyn. Earthq. Eng.* **2021**, *149*, 106873. [CrossRef]
35. Liu, X.; Yan, P.; Lu, W.; Lu, A.; Zhang, X.; Chen, M.; Wang, G. Numerical investigation of an improved deep-hole presplitting method based on notched blasting for deep-buried high sidewall structures. *J. Build. Eng.* **2023**, *70*, 106310. [CrossRef]
36. Liu, K.; Li, Q.; Wu, C.; Li, X.; Li, J. Optimization of spherical cartridge blasting mode in one-step raise excavation using pre-split blasting. *Int. J. Rock Mech. Min.* **2020**, *126*, 104182. [CrossRef]
37. Liu, K.; Li, X.; Hao, H.; Li, X.; Sha, Y.; Wang, W.; Liu, X. Study on the raising technique using one blast based on the combination of long-hole presplitting and vertical crater retreat multiple-deck shots. *Int. J. Rock Mech. Min.* **2019**, *113*, 41–58. [CrossRef]
38. Dubner, H.; Abate, J. Numerical Inversion of Laplace Transforms by Relating Them to the Finite Fourier Cosine Transform. *J. ACM* **1968**, *15*, 115–123. [CrossRef]
39. Hallquist, J. *LS-DYNA Theory Manual 19. Livermore Software Technology Corporation (LSTC)*; ANSYS Inc.: Canonsburg, PA, USA, 2019.

40. Kucewicz, M.; Baranowski, P.; Małachowski, J.; Trzciński, W.; Szymańczyk, L. Numerical Modelling of Cylindrical Test for Determining Jones—Wilkins—Lee Equation Parameters. In Proceedings of the 14th International Scientific Conference: Computer Aided Engineering, CAE 2018; Rusiński, E., Pietrusiak, D., Eds.; Lecture Notes in Mechanical Engineering. Springer: Cham, Switzerland, 2019. [CrossRef]
41. Baranowski, P.; Kucewicz, M.; Gieleta, R.; Stankiewicz, M.; Konarzewski, M.; Bogusz, P.; Pytlik, M.; Małachowski, J. Fracture and fragmentation of dolomite rock using the JH-2 constitutive model: Parameter determination, experiments and simulations. *Int. J. Impact Eng.* **2020**, *140*, 103543. [CrossRef]
42. Kang, P.; Zhaopeng, L.; Quanle, Z.; Zhenyu, Z.; Jiaqi, Z. Static and dynamic mechanical properties of granite from various burial depths. Rock Mech. *Rock Eng.* **2019**, *52*, 3545–3566. [CrossRef]
43. Kucewicz, M.; Baranowski, P.; Małachowski, J. Determination and validation of Karagozian-Case Concrete constitutive model parameters for numerical modeling of dolomite rock. *Int. J. Rock Mech. Min. Sci.* **2020**, *129*, 104302. [CrossRef]
44. Liu, X.; Yan, P.; Lu, W.; Zhu, J.; Zhang, X.; Lu, A.; Chen, M.; Wang, G. Investigation of dynamic crack formation mechanism based on a new crack dynamic driving model. *Comput. Geotech.* **2023**, *159*, 105471. [CrossRef]
45. Sazid, M.; Singh, T.N. Numerical assessment of spacing–burden ratio to effective utilization of explosive energy. *Int. J. Min. Sci. Technol.* **2015**, *25*, 291–297. [CrossRef]
46. Tao, J.; Yang, X.-G.; Li, H.-T.; Zhou, J.-W.; Fan, G.; Lu, G.-D. Effects of in-situ stresses on dynamic rock responses under blast loading. *Mech. Mater.* **2020**, *145*, 103374. [CrossRef]
47. Yuan, W.; Liu, S.; Wang, W.; Su, X.; Li, Z.; Li, J.; Wen, L.; Chang, J.; Sun, X. Numerical study on the fracturing mechanism of shock wave interactions between two adjacent blast holes in deep rock blasting. *Earthq. Eng. Eng. Vib.* **2019**, *18*, 735–746. [CrossRef]
48. Fukuda, D.; Moriya, K.; Kaneko, K.; Sasaki, K.; Sakamoto, R.; Hidani, K. Numerical simulation of the fracture process in concrete resulting from deflagration phenomena. *Int. J. Fract.* **2013**, *180*, 163–175. [CrossRef]

applied sciences

MDPI

Article

# Dynamic Response of Methane Explosion and Roadway Surrounding Rock in Restricted Space: A Simulation Analysis of Fluid-Solid Coupling

Qiangyu Zheng [1], Peijiang Ding [1], Zhenguo Yan [2], Yaping Zhu [2] and Jinlong Zhang [2,*]

1 College of Arts, Xi'an University of Science and Technology, Xi'an 710054, China; y13572288566@163.com (Q.Z.); d17792583521@163.com (P.D.)

2 College of Safety Science and Engineering, Xi'an University of Science and Technology, Xi'an 710054, China; yanzg@xust.edu.cn (Z.Y.); 19809458584@163.com (Y.Z.)

* Correspondence: 22220226186@stu.xust.edu.cn

**Abstract:** A methane-air premixed gas explosion is one of the most destructive disasters in the process of coal mining, and the dynamic coupling between the shock wave triggered by the explosion and the surrounding rock of the roadway can lead to the destabilization of the surrounding rock structure, the destruction of equipment, and casualties. The aim of this study is to systematically reveal the propagation characteristics of the blast wave, the spatial and temporal evolution of the wall load, and the damage mechanism of the surrounding rock by establishing a two-way fluid-solid coupling numerical model. Based on the Ansys Fluent fluid solver and Transient Structure module, a framework for the co-simulation of the fluid and solid domains has been constructed by adopting the standard $k-\varepsilon$ turbulence model, finite-rate/eddy-dissipation (FR/ED) reaction model, and nonlinear finite-element theory, and by introducing a dynamic damage threshold criterion based on the Drucker–Prager and Mohr–Coulomb criteria. It is shown that methane concentration significantly affects the kinetic behavior of explosive shock wave propagation. Under chemical equivalence ratio conditions (9.5% methane), an ideal Chapman–Jouguet blast wave structure was formed, exhibiting the highest energy release efficiency. In contrast, lean ignition (7%) and rich ignition (12%) conditions resulted in lower efficiencies due to incomplete combustion or complex combustion patterns. In addition, the pressure time-history evolution of the tunnel enclosure wall after ignition triggering exhibits significant nonlinear dynamics, which can be divided into three phases: the initiation and turbulence development phase, the quasi-steady propagation phase, and the expansion and dissipation phase. Further analysis reveals that the closed end produces significant stress aggregation due to the interference of multiple reflected waves, while the open end increases the stress fluctuation due to turbulence effects. The spatial and temporal evolution of the strain field also follows a three-stage dynamic pattern: an initial strain-induced stage, a strain accumulation propagation stage, and a residual strain stabilization stage and the displacement is characterized by an initial phase of concentration followed by gradual expansion. This study not only deepens the understanding of methane-air premixed gas explosion and its interaction with the roadway's surrounding rock, but also provides an important scientific basis and technical support for coal mine safety production.

**Keywords:** methane-air premixed gas explosion; numerical simulation of fluid-solid coupling; shock wave propagation characteristics; damage mechanism of tunnel enclosure; dynamic damage threshold criteria

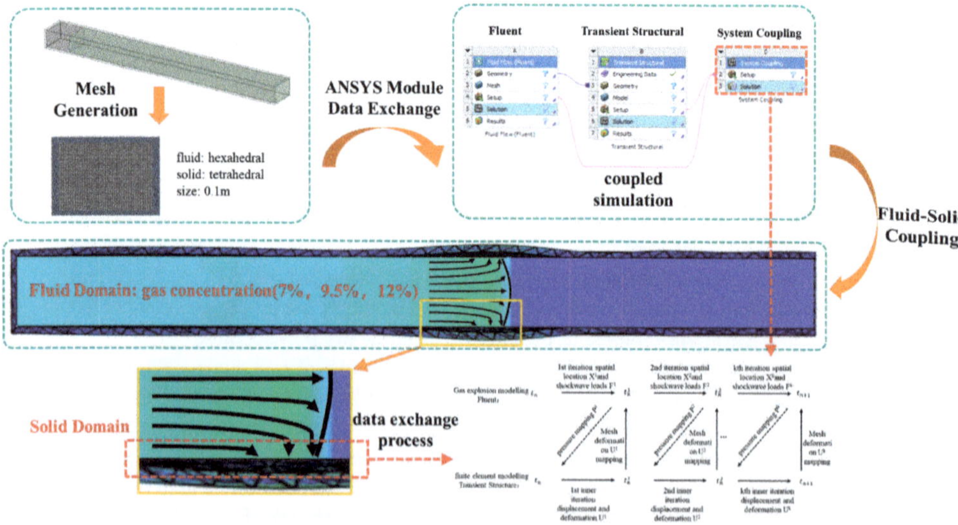


**Figure 1.** Technical route and data interaction mechanism of the coupled numerical simulation of multi-physics fields.

*2.1. Governing Equations for Methane-Air Mixture Explosion Simulations*

The explosion process of methane-air premixed gas is essentially a rapid combustion reaction process in a confined space, and its numerical simulation requires the establishment of a system of conservation equations for the coupled chemical reaction, including the mass conservation law, momentum conservation law (Navier-Stokes equations), energy conservation equations, and species transport equations.

A detonation is a combustion induced by an excitation wave that propagates in a steady manner. Because of the intense chemical reaction of this combustion, it has a very high flame propagation rate and its products have very high temperatures and pressures. After weak ignition, a laminar flame is generated, which is destabilized and accelerated by boundary action. The accelerated flame and the wall have a positive feedback effect, so that the flame continues to accelerate, is caused by sudden turbulence, the formation of turbulent flame, resulting in further enhancement of the combustion rate; combustion products of the expansion of the flame front of the medium have a compression effect, the formation of a series of compression waves, and ultimately develop into a surge. The induced surge, with the continuous acceleration of the flame, continues to enhance the flame; the accelerated flame formation of the induced surge is strong enough, and the surge pre-compression zone may be a local explosion, the formation of a bombardment, so that the combustion suddenly transforms into a strong bombardment. Strong bombardment is unstable, gradually decaying into a stable propagation of CJ bombardment. In the deflagration-to-detonation process, the flame acceleration and the enhancement of the surge are closely related to turbulent combustion. Deflagration to detonation is a more complex physicochemical process, including turbulent combustion, compression wave reflection, and other phenomena. The process is not only strongly nonlinear, but there is also a wide range of spatial and temporal characteristics of the scale, so the use of numerical calculations needs to be a suitable model of turbulent combustion [18]. The $k - \varepsilon$ model quantifies the turbulent mixing effect through turbulent kinetic energy ($k$) and dissipation rate ($\varepsilon$), which can significantly enhance the flame face stretching and local combustion rate, and also affects the momentum equation by varying the effective Reynolds stress to capture the interaction of the shock wave with turbulent eddies such as pressure oscillations, and wavefront surface distortions, which are crucial for accurately modeling the reflection and superposition of the pressure wave in a confined space [19,20]. The theoretical framework constructed in this study is based on the following set of fundamental control equations:

Mass Conservation Equation:

$$\frac{\partial \rho}{\partial t} + \nabla \cdot (\rho v) = 0 \tag{1}$$

Ensures the conservation of mass within the fluid domain. The temporal change in density ($\partial\rho/\partial t$) balances the spatial divergence of mass flux ($\nabla\cdot(\rho v)$).

Energy Conservation Equation:

$$\frac{\partial}{\partial t}(\rho h) + \nabla \cdot (\rho h v) = \frac{\partial P}{\partial t} + (\nabla \cdot v)P + \nabla \cdot \tau \tag{2}$$

Accounts for enthalpy ($h$) transport, pressure work, and thermal conduction ($\tau$).

Momentum Conservation Equation:

$$\frac{\partial}{\partial t}(\rho v) + \nabla \cdot (\rho v v) + \nabla P = \nabla \cdot \tau + S \tag{3}$$

Describes the balance between inertial forces, pressure gradients ($\nabla P$), viscous stresses ($\tau$), and external forces ($S$).

Species Transport Equation:

$$\frac{\partial}{\partial t}\left(\rho V_{fv}\right) + \nabla \cdot \left(\rho v V_{fv} - \frac{v_e}{G_{fv}} \nabla V_{fv}\right) = R_{fv} \tag{4}$$

In this equation, ρ denotes the density (kg/m$^3$),t represents time (s), and v is the velocity vector (m/s), h is the specific enthalpy, P denotes the fluid pressure (Pa), and τ represents the shear stress tensor. $V_{fv}$ represents the mass fraction of the flammable component (%), $v_e$ is the effective viscosity coefficient, $G_{fv}$ denotes the effective thermal conductivity, and $R_{fv}$ corresponds to the premixed gas explosion rate (m/s).

The standard model resolves turbulent kinetic energy and its dissipation rate to characterize turbulence effects, which is critical for capturing stagnation zones, recirculation, and flame wrinkling in explosion simulations. Turbulence increases the rate of combustion by increasing the surface area of the flame, which is necessary for flame acceleration, and neglecting turbulence will make it impossible to accurately model the explosion evolution.

Turbulent kinetic energy transport modeling:

$$\frac{\partial \rho}{\partial t}(\rho \mathrm{k}) + \frac{\partial}{\partial x_i}(\rho k a_i) = \frac{\partial}{\partial x_j}\left[\left(\mu + \frac{\mu_t}{\delta_k}\right)\frac{\partial k}{\partial x_j}\right] + G_k + G_b - \rho\varepsilon - Y_m + S_k \tag{5}$$

Turbulent dissipation rate transport equations:

$$\frac{\partial}{\partial t}(\rho \mathrm{k}) + \frac{\partial}{\partial x_i}(\rho k u_i) = \frac{\partial}{\partial x_j}\left[\left(\mu + \frac{\mu_t}{\sigma_\varepsilon}\right)\frac{\partial k}{\partial x_j}\right] + c_{1\varepsilon}\frac{\varepsilon}{k}(G_k + C_{3\varepsilon}G_b) - C_{2\varepsilon}\rho\frac{\varepsilon^2}{k} + S_\varepsilon \tag{6}$$

Turbulent viscosity calculation equation:

$$\mu_{\mathrm{t}} = \rho C_\mu \frac{\mathrm{k}_{\mathrm{t}}^2}{\varepsilon} \tag{7}$$

In this equation, $G_k$ is the turbulent kinetic energy produced by the mean velocity gradient; $G_b$ is the turbulent kinetic energy generated for buoyancy; $Y_m$ is the effect of fluctuating expansion on turbulent dissipation; $C_\mu = 0.09$; $c_{1\varepsilon} = 1.44$; $C_{2\varepsilon} = 1.92$; $C_{3\varepsilon} = 1.2$; $\delta_k = 1.0$; and $\sigma_\varepsilon = 1.3$.

Based on the structural dynamics analysis framework of nonlinear finite-element theory, the continuous medium structure is decomposed into a finite number of units by spatial discretization and the unit-level dynamics control equations are established [21]. The time evolution characteristics of the displacement, velocity, and acceleration fields of the structural system under external load excitation can be obtained by numerically solving this set of coupled equations. The system kinematic governing equations of this theoretical framework can be formulated as

$$Ma + Cv + Kd = F(t) \tag{8}$$

In this equation, $M$, $C$, and $K$ are the mass, damping, and stiffness matrices, respectively; $a$, $v$, and $d$ denote the acceleration, velocity, and displacement vectors; and $F(t)$ represents the external force vector.

*2.2. Bi-Directional Fluid-Structure Interaction Method for Gas Explosion Simulations*

The bi-directional fluid-solid coupled numerical simulation method provides an effective computational framework for solving fluid-structure interaction dynamics problems. The method realizes the coupling solution between the fluid and solid domains through a two-way physical quantity transfer mechanism, which strictly follows the dynamic equilibrium principle at the interface. For the strong nonlinear coupling problem of the interaction between the tunnel-enclosing rock and the methane-air premixed gas explosion shock wave, in view of the strong nonlinear characteristics of this physical process, it is necessary to adopt a multi-physics field strong coupling iterative algorithm to achieve the convergent transfer of the interface parameters at each time step in order to ensure computational accuracy [22–24]. This study is based on the Ansys Workbench co-simulation platform, integrating the Fluent fluid dynamics solver and the Transient Structure nonlinear structural analysis module, realizing real-time interaction between pressure load and structural displacement through the System Coupling system interface, and constructing a complete two-domain two-way coupling numerical model. The fluid-structure coupling interface needs to satisfy the following basic conservation laws:

$$\begin{gathered} \tau_f \cdot n_f = \tau_s \cdot n_s \\ d_f = d_s \end{gathered} \tag{9}$$

In this equation, $\tau_f$, $\tau_s$ is Fluid and solid domain stresses (MPa), $n_f$, $n_s$ is the unit normal vectors of fluid and solid boundaries, and $d_f$, $d_s$ is the displacement vectors ($m$) at the interface.

Within the Ansys co-simulation framework, the roadway rock-gas contact interface is defined as a fluid-solid coupling interface. Based on the principle of energy conservation, the real-time bi-directional transfer of the pressure field in the fluid domain and the displacement field in the solid domain is realized by an implicit coupling algorithm. In each iteration step, the Fluent transient fluid solver and the Transient Structure nonlinear structure solver complete the convergent data exchange of dynamic loads (shock wave pressure, shear stress) and structural response (displacement field, deformation tensor) through the System Coupling module (see Figure 2 for details of the process). This study mainly focuses on the role of the explosion shock wave on the surrounding rock, and the explosion time scale is very short. The surrounding rock heat absorption can be ignored, so the data exchange process in the System Coupling module ignores heat transfer. The coupling system solution result contains the following: (1) the fluid domain pressure field resolution (composed of the dynamic pressure component and the wall shear stress component), and the boundary load is transferred to the surrounding rock surface through the spatial mapping algorithm; (2) the solid domain displacement field is solved by the nonlinear finite-element method, and then it is fed back to the fluid domain through the interface coordination condition. At each iteration, the fluid domain topology is updated using dynamic mesh technology, and the shock wave propagation path and load distribution are modified simultaneously. Through multiple iteration cycles, the peak shock wave pressure, the plastic deformation of surrounding rock, and the interface shear stress gradually reach the convergence threshold, and finally the stable flow field distribution and structural dynamic response characteristics that meet the residual requirements are obtained at the end of the time step.

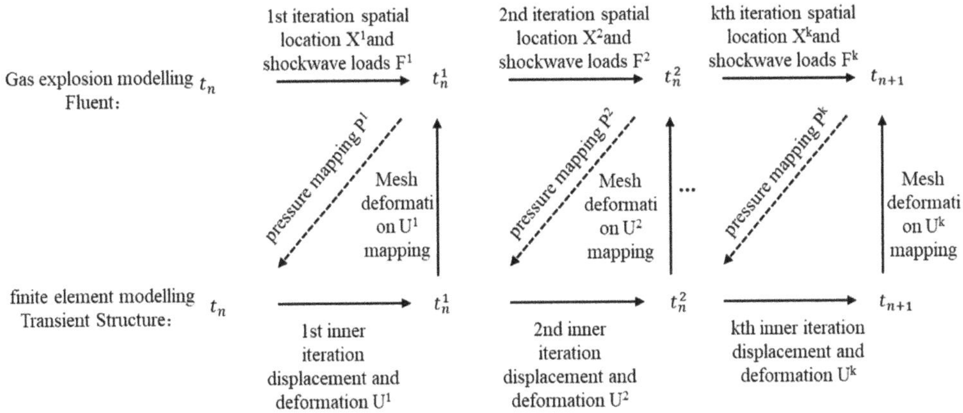


**Figure 2.** Bi-directionally coupled iterative time-step data exchange diagrams.

### 2.3. Numerical Modeling of Fluid-Solid Coupling

A three-dimensional simplified computational model was constructed based on a typical tunnel engineering prototype, which is characterized by a semi-enclosed structure (axial length of 50 m, cross-section size of 4.5 m × 3 m). The methane pre-mixing zone is defined as a restricted spatial domain 0 − 5 m from the closed end, to study the propagation of shock waves in the air zone after forming a complete wavefront in the premixed zone. In order to make the damage region more significant, its boundary constraint layer is set to be 0.3 m thick and is defined as an impermeable wall, where there are no cracks or pores during the simulation time, and the gas cannot pass through the computational domain. The ignition source is positioned at the geometric center axis 1.0 m from the closed end. The computational domain was discretized using a hybrid mesh discretization strategy: isotropic discretization was achieved by a structured hexahedral mesh for the fluid domain, and an unstructured tetrahedral mesh was used for morphologically adapted discretization in the solid domain, and mesh suitability was verified by orthogonal mass [25]. Three sets of methane volume fraction working conditions were constructed within the methane pre-mixing zone (7% stoichiometric, 9.5% optimal, and 12% rich mixture). The volume fraction of $O_2$ in the whole model area is 21%, $CO_2$ and $H_2O$ volume fractions are 0.01% and 0.03%, respectively, and the rest of the components are $N_2$. The monitoring system was arranged with monitoring arrays along the axial direction of the roadway wall according to the distance parameter (see Figure 3 for details of the spatial distribution of the projected grid).

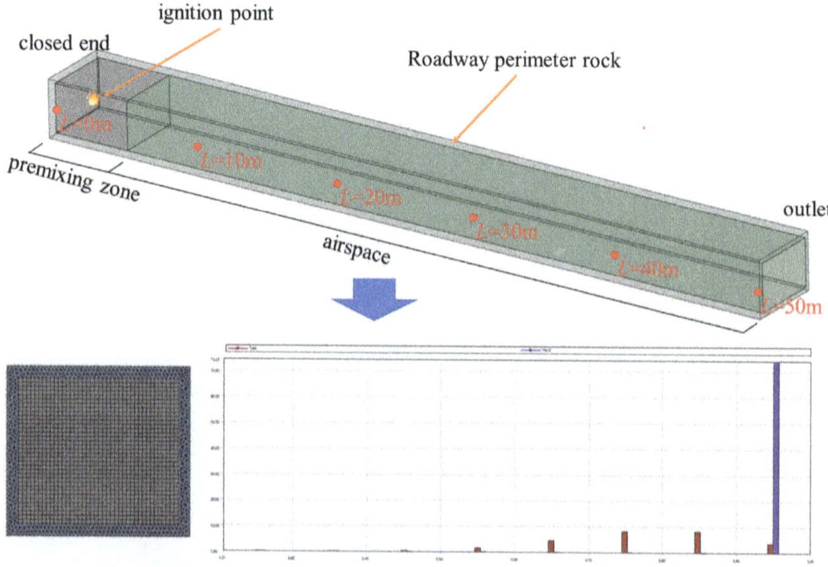


**Figure 3.** Distribution of monitoring points and grid quality.

The fast chemical reaction kinetics of methane-air premixed gas explosion is characterized by a finite-rate/eddy-dissipation (FR/ED) volumetric reaction model. The model combines the Arrhenius formula and the eddy-dissipation model, which can take into account both kinetic and turbulence factors, and is suitable for the local premixed combustion case, where the reaction rate is taken as the smaller value of the Arrhenius rate and the eddy-dissipation rate where the Arrhenius formula is

$$k = AT^{\beta}exp(-E_a/RT) \tag{10}$$

In this equation, $k$ is the rate of reaction; $R$ is the molar gas constant with a value of 8.314 J/(mol-K); $T$ is the thermodynamic temperature (K); $E_a$ is the apparent activation energy (J/mol) which has a value of 2.027 × 108; and $A$ is the pre-exponential factor, which has a value of 2.119 × 1011.

The eddy-dissipation model reaction rate is calculated as

$$R_{i,r} = v'_{i,r} M_{w,i} AB\rho \frac{\varepsilon_1}{k_1} \frac{\sum_p Y_p}{\sum_j^N v'_{i,r} M_{w,j}} \tag{11}$$

In this equation, $Y_p$ is the mass fraction of the product; $A$ and $B$ are empirical constants, $A = 4$ and $B = 0.5$; $v'_{i,r}$ is the stoichiometric number of product $i$ in reaction $r$; and chemical reaction rates are controlled by large eddy mixing timescales $\frac{\varepsilon_1}{k_1}$.

The net reaction rate is taken to be the smallest of the above two rates. Finite-rate kinetics prevent the reaction from occurring before the flame stabilizer. Once the flame is ignited, the vortex dissipation rate is usually less than the Arrhenius rate, and the reaction is limited by mixing.

This is constructed on the basis of the following basic assumptions: (1) the fluid medium follows the ideal gas state equation; (2) the thermodynamic system is adiabatic and the effects of bulk forces are neglected; (3) the single-step irreversible lumped reaction mechanism ($CH_4 + 2O_2 \rightarrow CO_2 + 2H_2O$); and (4) the ignition mechanism is realized through the spark ignition setting in the component transport model, and different spark ignition parameters have certain effects on the peak value of the shock wave and the propagation speed. In this paper, the ignition energy is 1 mJ, the initial radius is 0.002 m, and the duration is 1 ms., with the initial conditions set at ambient temperature (298 K) and standard atmospheric pressure (101.325 kPa), and with a uniform distribution of the concentration and temperature fields.

Full displacement constraint boundary conditions are used at the closed end of the tunnel to maintain numerical stability; the top and bottom plates were used as fixed supports at the junction of the top and bottom plates with the two gangs of surrounding rock to maintain numerical stability, while a pressure of 1 MPa was applied to the surface of the surrounding rock to simulate the initial ground stress, and the surrounding rock medium is idealized by an equivalent isotropic linear elasticity constitutive model. The key material constitutive parameters (including Young's modulus, Poisson's ratio, density, and kinetic parameter) were set according to the experimental measurements shown in Table 1, where the explosion rate transport coefficient of the explosive mixture is obtained by calculating the kinetic theory Equation (4).

**Table 1.** Enclosed rock model material parameters.

| Parameter Type | $\rho$/kg/m$^3$ | $E$/MPa | $v$ | $Rm$/MPa | $\sigma_c$/MPa | $K$/MPa |
|---|---|---|---|---|---|---|
| parameter value | 2350 | 2500 | 0.25 | 1.5 | 24 | 920 |

In Table 1, $\rho$ is the density; $E$ is Young's modulus; $v$ is Poisson's ratio; $Rm$ is the tensile ultimate strength; $\sigma_c$ is the ultimate compressive strength; and $K$ is the strength factor.

In this study, an implicit strongly coupled iterative algorithm is used to implement the fluid-structure coupling solution, and the numerical framework is set as follows: a single bi-directional data transfer is executed within each global time step ($\Delta t = 1 \times 10^{-5}$ s), and the physical field coupling is achieved through a multi-field convergence mechanism. In the specific implementation, the fluid solver performs 20 sub-iterative computations within each time step to ensure convergence of the Navier-Stokes equations, which is determined by dynamic optimization of the mesh eigenvelocity according to the Courant–Friedrichs–Lewy condition. The time discretization strategy of the coupled system adopts a fixed time-step scheme ($\Delta t = 1 \times 10^{-5}$ s), and this parameter setting is verified by the leading mesh independence analysis and the time-step sensitivity to achieve optimal computational efficiency while ensuring the computational accuracy. The data convergence criterion is based on the triple constraints of interfacial load residual (pressure residual $< 1\%$), displacement continuity error ($L^2 < 0.5\%$), and energy conservation threshold ($\frac{\Delta E}{E_0} < 5‰$).

## 3. Analysis of the Evolutionary Pattern of Wall Loads in a Gas Explosion

### 3.1. Numerical Model Validation

In order to verify the accuracy of the parameter settings and boundary conditions used in this paper, the numerical simulation of the gas explosion pipe is compared with the experimental results with reference to the existing studies [26]. The inner diameter of the circular pipe is 18 cm, and the data of 0.23 m$^3$ of gas volume in case 1 is selected for simulation and comparison. The experimental schematic and the distribution of measurement points are shown in Figure 4.

According to the experimental results, the numerical simulation results are plotted and the peak values of overpressure at different measurement points are shown in Table 2. Relative error is calculated by averaging the experimental results with the numerical simulation results over three experimental results. From the results, it can be seen that the

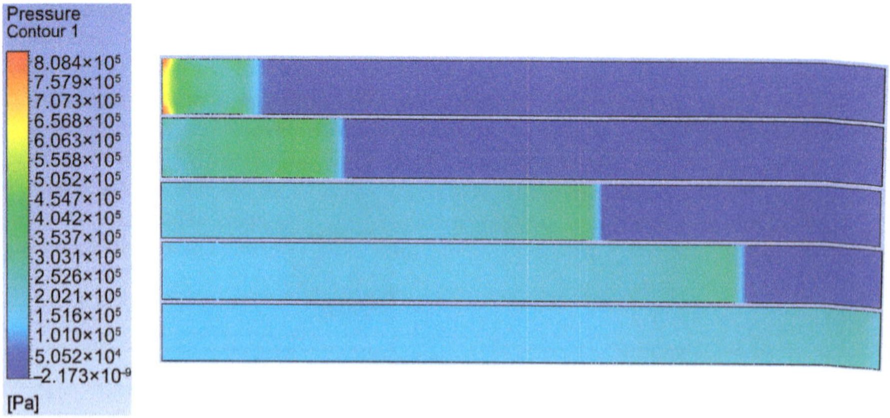


(**a**)

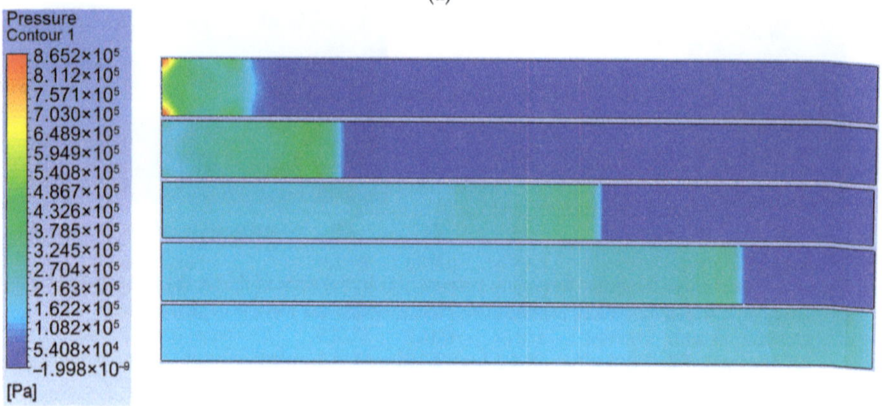


(**b**)

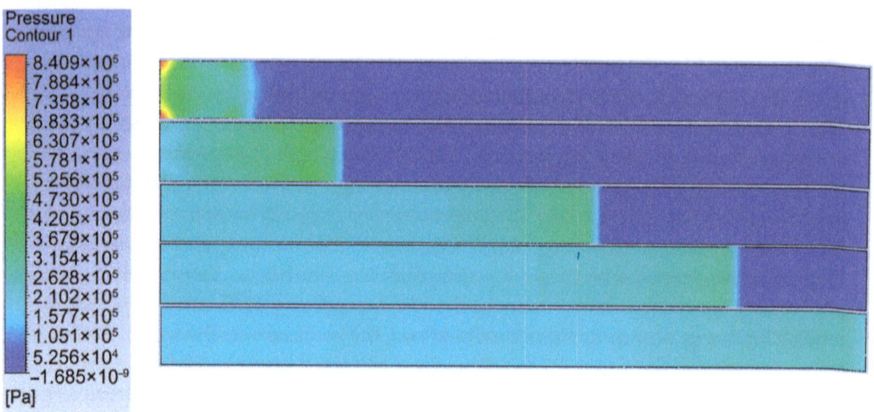


(**c**)

**Figure 6.** Pressure contour of the location of the point where the blast wave reached the measurement point. (**a**) 7% gas concentration explosion contour; (**b**) 9.5% gas concentration explosion contour; (**c**) 12% gas concentration explosion contour.

Figure 7 reveals the regulatory mechanism of methane volume fraction on the temporal evolution of explosion overpressure. Moreover, 7% (black curve), 9.5% (red curve), and 12% (blue curve) conditions all show characteristic bimodal pressure evolution. The first peak corresponds to the initial positive impact between the shock wave and the surrounding rock of the tunnel, and the second peak is dominated by the superposition effect of the compression wave formed by the reflection of the shock wave in the closed end face. At this stage, shock waves dominate energy release. Detonation formation occurs primarily in three phases, as illustrated by the pressure curve at 9.5% concentration. Before 11.8 ms, shock-flame interactions induce Richtmyer–Meshkov instability, causing flame surface destabilization. Turbulent flames significantly increase combustion surface area, enhancing energy release and generating high-intensity pressure waves. During 11.8–21.3 ms, the shock wave contacts the confined space wall, generating reflected shocks that heat unburned gases. This causes shock bifurcation, forming recirculation zones that entrain flames, leading to Hot Spots consistently appearing in unburned regions. At 21.3 ms, driven by the Zeldovich Gradient Mechanism, spontaneous reaction waves propagate along gradients and transition to detonation [27]. Moreover, 12% rich-fire conditions and 9.5% equivalent conditions meet the CJ blast condition (Chapman–Jouguet velocity theory $v = 682\ \mathrm{m/s}$ [28]), and the actual propagation velocity of the shock wave reaches 512 m/s and 618 m/s, respectively, and this propagation characteristic makes the first peak appear 1.8 ms earlier ($\frac{\Delta t}{t_1} = 14.6\%$) than the lean-burn condition (7%). In the lean condition (7%), the combustion efficiency decreases to 73.2% (calculated by integrating the heat release rate) due to fuel limitation, resulting in a 9% decrease in the first peak pressure amplitude ($\Delta P_{max} = 0.19$ MPa) compared to the equivalent condition, with a significant phase delay.

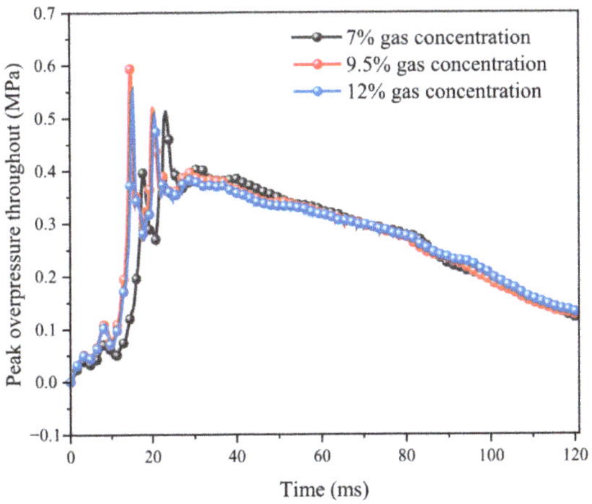


**Figure 7.** Peak explosion overpressure for different gas concentrations.

### 3.3. *Laws of Shock Wave Propagation in Gas Explosions*

After ignition triggering, the time-course evolution of the pressure on the wall surface of the tunnel enclosure showed significant nonlinear dynamics (Figure 8a–c), and the physical process can be deconstructed into three characteristic phases. The stage of detonation and turbulence development ($t_1 = 0 - 18$ ms): the initial compression wave system and the wall surface undergo multimodal reflection coupling, which induces the instability of the shear layer to form a Kelvin–Helmholtz vortex structure [29], leading to periodic pressure oscillations in the near-field area ($L = 0$ m). In this stage, the pressure rise rate reaches an extreme value, and the pressure pulsation shows multi-peak characteristics. Quasi-steady propagation phase ($t_2 = 18 - 82$ ms): a self-similar propagation mode is formed under the constraints of the channel geometry, and the energy dissipation mechanism is dominated by viscous dissipation together with turbulent kinetic energy transport. The along-track decay of the pressure peak is consistent with the $P(x) = P_0 e^{-\alpha x}$ model, where $P_0$ is the initial pressure, $\alpha$ is the pressure decay rate constant, and $x$ is the propagation distance, and the time delay of the pressure rise at the neighboring monitoring point ($\Delta L = 10$ m) is linear. Expansion dissipation phase ($t_3 = 82 - 120$ ms): the open-end boundary induces a Prandtl–Meyer expansion wave system, forming a characteristic pressure plunge zone.

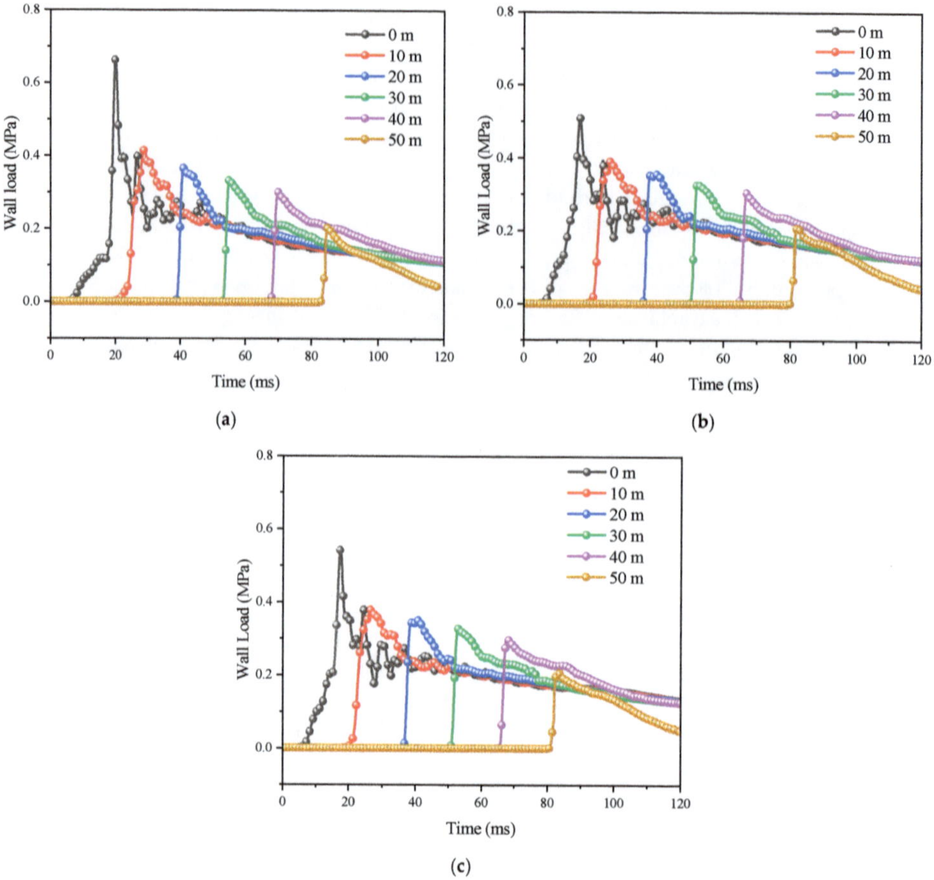


**Figure 8.** Variation of perimeter rock loads at different measurement points. (**a**) Notably, 7% gas concentration; (**b**) 9.5% gas concentration; (**c**) 12% gas concentration.

Figure 9 quantitatively reveals the modulation mechanism of methane volume fraction on the spatial decay pattern of explosion overpressure. The data from the near-field monitoring point ($L = 0$) showed that the initial overpressure peak was 0.51 MPa (the lowest for all conditions) for the equivalence ratio condition (9.5% $CH_4$), while that for the lean-burn condition (7%) reached 0.67 MPa (the highest for all conditions), an anomaly closely related to the quadratic polar characteristic of the laminar combustion rate of the gas mixture. The time-course evolution analysis shows that the pressure decay rate constants $\alpha = 0.309$ m$^{-1}$ and $\alpha = 0.203$ m$^{-1}$ for the lean combustion condition are increased by

67.9% and 10.3% compared with those of the equivalent condition ($\alpha = 0.184\ m^{-1}$), and the residual pressure decreases to 0.207 MPa at the exit of the roadway ($L = 50$) (decaying by 69.3% from the initial peak value). The equivalent condition benefits from a near-theoretical optimal concentration, with a combustion efficiency of $\eta_c = 98.2\%$ (validated by the Arrhenius combustion model [30]), resulting in an energy density of 23.7 $MJ/m^3$, which is 28.8% and 41.1% higher than the lean (18.4 $MJ/m^3$) and rich (16.8 $MJ/m^3$) conditions, respectively. This energy advantage results in a shock wave front velocity of 618 m/s for the equivalent case. Although the lean-burn condition generates transient high-pressure pulses due to the rapid consumption of finite fuel, its low energy deposition efficiency leads to rapid dissipation of kinetic energy in the late stage of shock wave propagation. Laws of shock wave propagation in gas explosions.

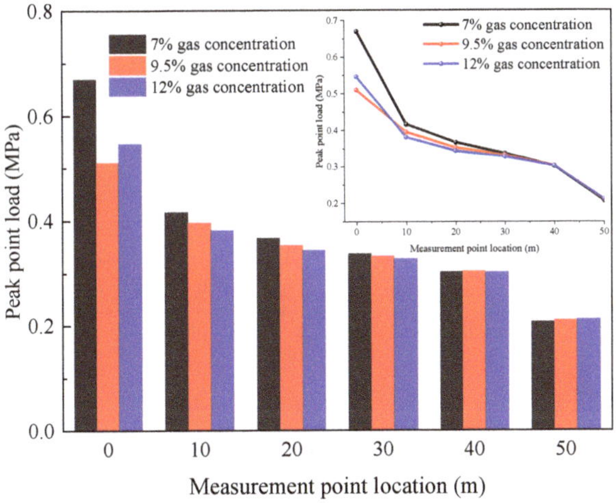


**Figure 9.** Plot of peak loads at each monitoring point for different gas concentrations.

## 4. Stress and Strain Analysis of Tunnel Wall Under Flow-Solid Coupling

### *4.1. Stress Analysis of Tunnel Wall Under Fluid-Solid Coupling*

The dynamic mapping of the fluid pressure field to the solid mechanics field is realized by the System Coupling module, and the temporal and spatial evolution characteristics of the von Mises stress field in the surrounding rock at different measurement points are successfully reconstructed (Figure 10a–c). The numerical results show that the stress evolution patterns of different equivalent ratio working conditions have universal characteristics, so a typical equivalent working condition (9.5% $CH_4$) is selected for the mechanistic analysis:

Initial impact stage ($t = 0\sim20$ ms): the monitoring point at the confined end is affected by the interference effect of the multiple reflected waves, the stress appears to be a significant energy aggregation, and the maximum von Mises stress is 2.09 (the lowest for the full working condition). This phenomenon is in accordance with the Saint-Venant principle, but induces local stress tensor anisotropy. Steady-state propagation stage ($t = 20\sim80$ ms): the stress wave shows a characteristic hyperbolic decay pattern, and the peak stress of 10.02 MPa is reached at the position of $L = 10$. The speed of the stress wave remains constant during the propagation process, and the stress phase difference between the neighboring monitoring points ($L = 20$) is converged from 3.2 ms to 0.8 ms, which indicates an increase in the rate of energy accumulation. Boundary dissipation phase ($t > 80$ ms): the monitoring points at the open end are affected by the turbulence-structure coupling resonance effect, and the amplitude of stress fluctuation increases. At this time, the stress energy flow density decreases compared to the initial value.

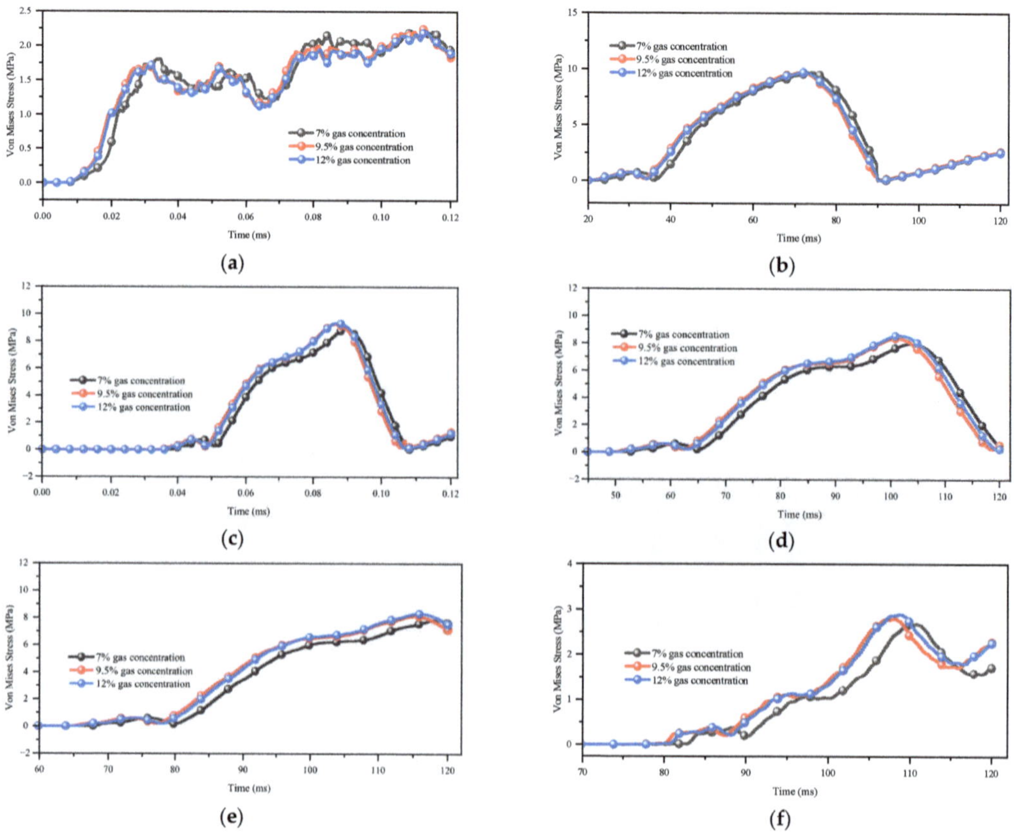


**Figure 10.** Equivalent force variation curves at different measurement points. (**a**) Notably, 0 m measurement point; (**b**) 10 m measurement point; (**c**) 20 m measurement point; (**d**) 30 m measurement point; (**e**) 40 m measurement point; (**f**) 50 m measurement point.

The numerical solution shows that there is a phase lag of 18 ms in the stress response relative to the pressure loading, and this phase lag phenomenon originates from the continuous loading in the high-pressure region behind the shock wave. The geometrical constraints at the closed end dominate the effect of the equivalence ratio, while the turbulence effect at the open end increases the sensitivity of the equivalence ratio. In particular, although the depleted combustion case (7%) has a maximum pressure pulse, its short-term loading characteristics do not match the dynamic response characteristics of the rate-dependent plasticity ontological model of the enclosing rock, resulting in only a small portion of the shock energy being converted into effective stress work, which is lower than that of the equivalent case.

### *4.2. Characterization of Damage Analysis of Roadway Perimeter Rock Under Fluid-Solid Coupling*

The analysis of the explosion impact dynamics based on the equivalent ratio working condition (9.5% $CH_4$) shows (Figure 11) that the spatial and temporal evolution of the strain field of the tunnel enclosure follows a three-phase dynamic pattern:

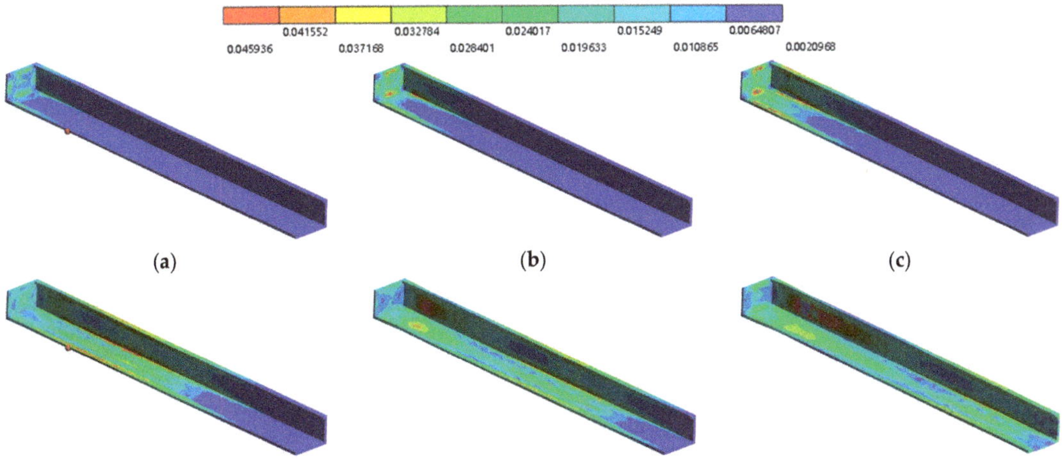


(**a**) (**b**) (**c**)

(**d**) (**e**) (**f**)

**Figure 11.** Strain contour of the roadway perimeter rock. (**a**) T = 15 ms; (**b**) T = 35 ms; (**c**) T = 55 ms; (**d**) T = 75 ms; (**e**) T = 95 ms; (**f**) T = 115 ms.

Initial strain-inducing phase: when the shock wave array first contacts the enclosing rock interface, an elastic deformation field with a maximum principal strain of $\varepsilon_1 = 0.01 \pm 0.002$ mm/mm is induced. The closed end is constrained by the Saint-Venant boundary effect to form a low-strain gradient zone, while the wall pinch point produces a strain concentration due to the stress three-dimensionality coefficient, forming a high-strain core with a diameter of $d = 0.8 \pm 0.1$ m. Strain accumulation and propagation stage: the rear high-pressure zone drives plastic strain accumulation at a rate of $\varepsilon = 2.1 \times 10^{-3}\ \text{s}^{-1}$. The formation of composite strain bands in the near field of the burst source. The strain field is redirected to migrate through the Hill anisotropy model, resulting in a characteristic bulging deformation in the center of the wall. Residual strain stabilization phase: the open end of the expansion wave system is triggered by the reverse propagation of the strain decay front, and the explosion source area due to the accumulation of equivalent plastic strain forms a permanent deformation zone. At this point, the energy dissipation rate returns to the initial value of the enhancement.

The high stiffness property of the closed end limits the strain energy density to $U = 0.18$ MJ/m$^3$ (67% lower than the open end). The high energy density of the equivalent case drives a two-mechanism deformation: plastic flow dominated by dislocation slip in the early stage ($T < 35$ ms), and grain boundary migration dominated by the diffusion creep mechanism in the late stage ($T > 35$ ms), with the dynamic equilibrium point of the two occurring at $T = 35 \pm 1.2$ ms. Strain-phase analyses show that the equivalent plastic strains lag behind the von Mises stresses.

Figure 12 shows the motion vector diagram of the surrounding rock during the propagation of the shock wave, and it can be seen that at the initial stage of the motion T = 53.9 ms. At this time, the motion direction of the wall surface of the closed end of the surrounding rock is facing outward, while the junction of the surrounding rock is moving inward. With the propagation of displacement, at T = 74.1 ms, the bottom plate motion vector is still outward and the range of displacement becomes larger but the speed decreases, while the two gang areas change to inward motion, and the density behind the shock wave decreases at this time. At the same time, comparing different time vector diagrams, it can be seen that when the two gangs' displacement direction is towards the more dramatic internal region, the size of the bottom plate velocity is lower; on the contrary, when the bottom plate velocity is larger, the two gangs' inward velocity is smaller. Reflected and tensile waves influenced by shock waves are particularly important for rock fracture.

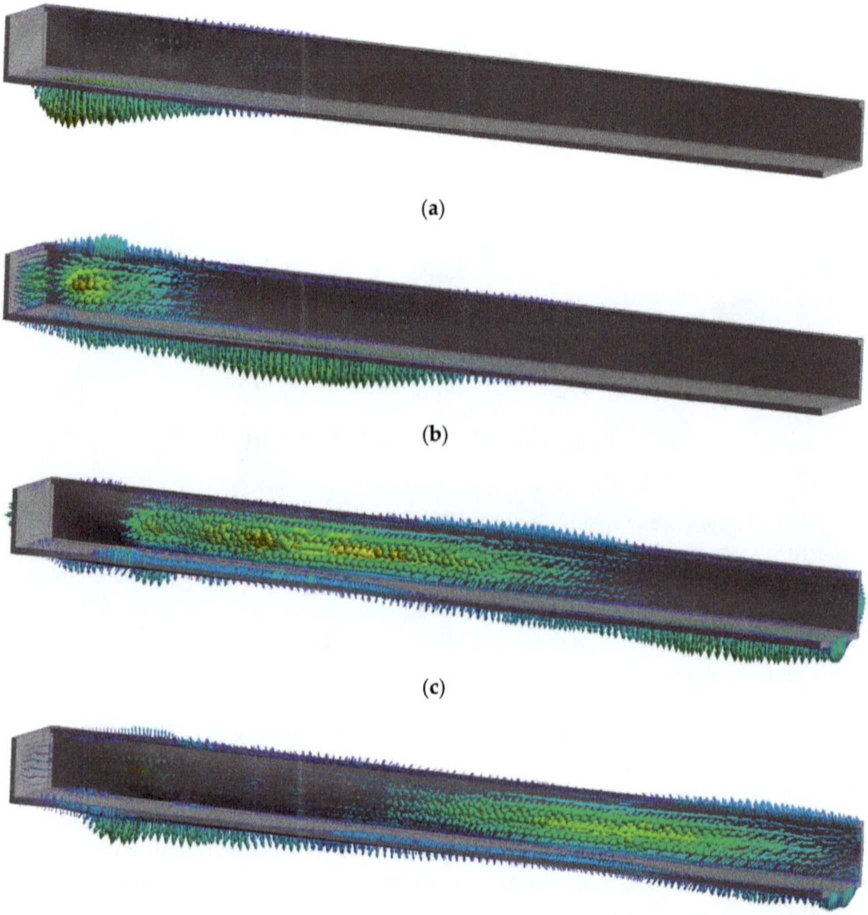

(a)

(b)

(c)

(d)

**Figure 12.** Vector diagram of perimeter rock movement. (**a**) T = 53.9 ms; (**b**) T = 74.1 ms; (**c**) T = 107.8 ms; (**d**) T = 128 ms. The size and color of the arrow simultaneously represent the magnitude of speed. Colors closer to red indicate higher speeds, while colors closer to blue indicate lower speeds.

The kinetic coupling between the methane-air premixed gas explosion shock wave and the surrounding rock of the roadway can induce the coal rock medium to produce multi-scale damage evolution. The degree of damage is closely related to the shock wave pressure amplitude and action time. The crack initiation threshold of coal rock can be characterized by the following critical strain criterion [31–33].

Using the equivalent strain, the statistical damage expression for the strength of uniaxial coal rock under impact loading is obtained as

$$R_m = E\varepsilon_{\text{threshold}}(1-D) = E\varepsilon_{\text{threshold}}e^{\left[-\left(\frac{F}{F_0}\right)^m\right]} \quad (12)$$

$$F_0 = \left(\beta + \frac{1}{\sqrt{3}}\right) E\varepsilon_m m^{\frac{1}{m}} \tag{13}$$

$$m = -\frac{1}{\ln[\sigma_{max}/(E\varepsilon_m)]} \tag{14}$$

In this equation, $\varepsilon_{threshold}$ is the strain threshold for crack generation; $\alpha$ is the crack generation coefficient of the coal rock; $R_m$ is the ultimate tensile strength, MPa, and $E$ is the modulus of elasticity of the coal rock, MPa; $F$ is the distribution variable for the strength of the micromeres; and $F_0$ and m is the Weibull distribution parameter, which reflects the mechanical properties of the coal rock material.

Through the uniaxial impact test results of different coal samples, the extreme value method was used to determine with, and the values are shown in Table 3.

**Table 3.** Table of solid model parameters.

| $\dot{\varepsilon}/s^{-1}$ | $F_0$/MPa | m |
|---|---|---|
| 24.8 | 12.8 | 2.95 |
| 33.8 | 14.7 | 1.63 |
| 42.3 | 23.2 | 1.41 |
| 46.2 | 23.4 | 1.74 |
| 53.8 | 32.3 | 1.20 |

The $\dot{\varepsilon}$ versus $F_0$ and m relationship is obtained by fitting Table 3's data and simplifying it by substituting it into Equation (16):

$$\varepsilon_{threshold} = \frac{\alpha R_m}{E} \tag{15}$$

$$\alpha_t = \frac{1}{-e^{\left(\frac{F}{F_0}\right)^m}} \tag{16}$$

In this equation, $\alpha_t$ is the crack generation coefficient of the coal rock.

Based on the basic assumptions of linear elastic fracture mechanics (tensile strength $R_m = 1.5$ MPa, modulus of elasticity $E = 2500$ MPa). It is known from previous studies that the usual value range is between 0.3 and 1.2, and according to the nature of coal rock, this paper takes the value of 0.7 [34], and the critical damage strain threshold is obtained by calculation as $\varepsilon_{threshold} = 0.042$ mm/mm. The damage evolution tracking algorithm is used to establish the following failure criterion in the Transient Structure solver: when the equivalent plastic strain at the integration point of the unit is greater than the threshold value, the unit failure marker parameter is activated. When the equivalent plastic strain at the integration point of the unit is larger than the threshold value, the unit failure marking parameter is activated. The spatial distribution characteristics of the damage field of the surrounding rock after the explosion impact were successfully reconstructed by this intrinsic model (Figure 13).

Based on the quantitative strain field analysis of the damage characteristics of the surrounding rock during the shock wave propagation stage in Figure 10a, the closed end is constrained by the high stiffness boundary condition ($\nu = 0.25$, $E = 2500$ MPa), and its maximum principal strain is limited to the range of $\varepsilon_1 < 0.03 \pm 0.002$ mm/mm, which does not reach the threshold value of crack initiation ($\varepsilon_{threshold} = 0.042$ mm/mm). The wall-top plate angle zone is subjected to bi-directional stress constraints, and the decrease in strain concentration factor leads to a decrease in damage density. The center zone of the surrounding rock is subject to elastic strain energy release, dominated by the shrinkage of the damage area compared to the pinch point zone. The strain migration effect induces local strain intensification in the central zone, but it is still below the critical threshold. The damage mechanism at the closed end revealed by the damage characteristics of the surrounding rock in the shock wave release stage in Figure 8b shows that the superposition of the reflected shock waves causes the dynamic strain energy density in the central zone to break through the threshold to form a radial branching fissure system. The damage density is suppressed in the sidewall region due to the higher lateral confinement strength than the top plate. The central area of the roof plate is governed by the tensile-shear composite stress state, which induces a band-shaped damage zone.

**Figure 13.** Characteristics of damage to the surrounding rock in the roadway of a gas explosion. (**a**) Damage conditions during shock wave propagation; (**b**) Final form of the damage.

### 4.3. Strain Thresholding Results for Thicker Perimeter Rock Models

The damage distribution under the thicker surrounding rock conditions is verified by thickening the thickness of the surrounding rock to 3 m, 5 m, and 10 m. Figure 14a–c The left figure shows the model before the damage, and in order to more intuitively observe the distribution of the damaged area in the thicker surrounding rock, the strain greater than 0.042 is retained in the area, and the right figure shows the damaged area of the surrounding rock under the thicker surrounding rock conditions. It can be seen that in the area near the closed end, the two groups of surrounding rock are more seriously damaged, while the top and bottom plates are basically not damaged. As the surrounding rock is more stable away from the closed end, the damage area starts to expand after the shock wave forms a stable wave surface. At the thicknesses of 3 m and 5 m, in the middle of the roadway near the exit area, the top slab was damaged near the outside of the model, but not inside. In all the models, the surrounding rock near the exit position is damaged internally while the two sides of the wall are more intact. Figure 9 shows that the top plate is more affected by the reflected wave of the shock wave, and the surrounding rock is mostly extended outward, so the external strain is greater than the internal.

After the gas explosion, the support method can significantly change the surrounding rock response. For the central area of the roof plate, surface shotcrete can reduce the surface tensile stress of the surrounding rock and inhibit the spalling damage induced by the explosion shock wave, but it may increase the local compressive stress concentration; anchor support can deform through the pre-tensioning force and the surrounding rock synergistically, to improve the uniformity of the stress field and reduce the scope of the plastic zone; through Flexible arch support, the explosion energy can be dissipated, but Flexible arch support may cause secondary collapse. The use of high-strength prestressing anchors and silicate-based grouting materials for coordinated support can significantly reduce the equivalent plastic strain; the closed end of the roadway is equipped with a polyurethane-rubber composite buffer layer, which can reduce the depth of damage to the surrounding rock through energy dissipation and pressure wave attenuation.

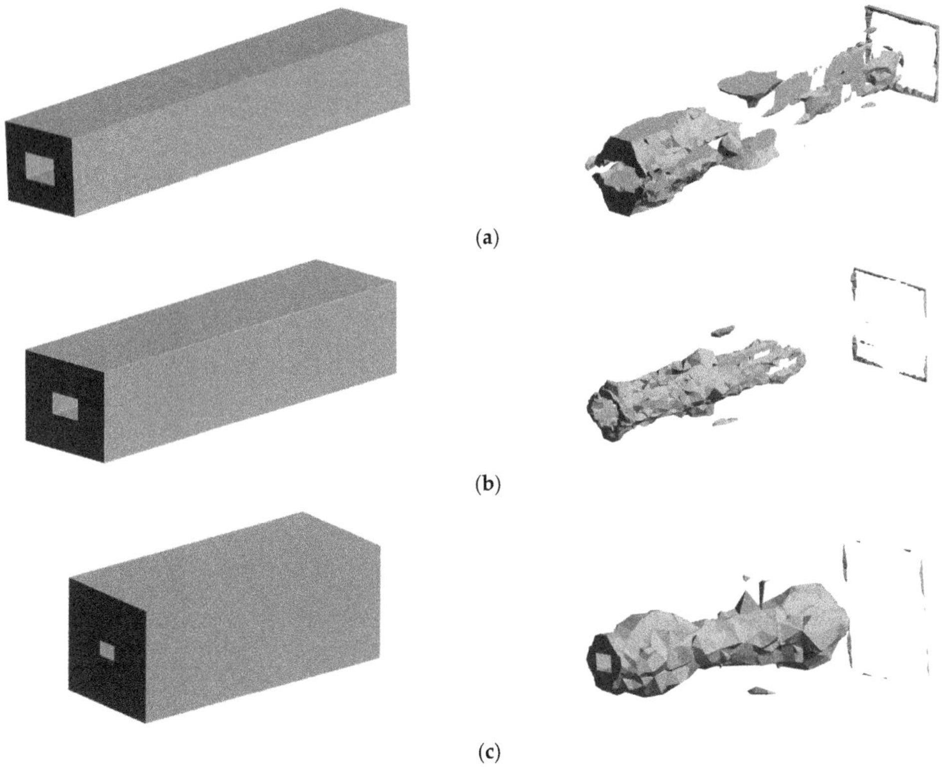

Figure 14. Verification Results for Thicker Envelope Conditions. (a) Schematic of modeled damage to 3 m thick perimeter rock; (b) Schematic of modeled damage to 5 m thick perimeter rock; (c) Schematic of modeled damage to 10 m thick perimeter rock.

## 5. Conclusions

Based on multi-physics field-coupled numerical simulation and fracture mechanics theory, this study systematically reveals the coupling mechanism between the methane-air premixed gas explosion shock wave and the dynamic response of the tunnel's surrounding rock. The following innovative conclusions are obtained through the establishment of a two-way fluid-solid coupling numerical model and damage evolution constitutive equation:

(1) The methane volume fraction and shock wave overpressure show a significant quadratic relationship, and the equivalent condition (9.5%) produces the maximum overpressure peaks of 31% and 9% compared with those of the lean combustion/rich combustion condition, and its pressure decay rate constant is reduced by 67.9% and 10.3% compared with that of the other conditions. This phenomenon is highly consistent with the laminar flow combustion velocity extreme value characteristics and the CJ burst theory.

(2) The dynamic evolution of the surrounding rock stress field is revealed through the numerical simulation of fluid-solid coupling. The initial shock stage at the closed end is interfered with by multiple reflected waves to produce stress aggregation, and the maximum von Mises stress is 2.09 MPa, which is in accordance with the Saint-Venant principle; the steady-state propagation stage shows hyperbolic decay characteristics

of the stress wave, and the phase difference between the neighboring monitoring points is converged; and the open end of the boundary dissipation stage is aggravated by the stress fluctuation due to the turbulence-coupled resonance.

(3) The initial elastic deformation forms a high-strain kernel; the plastic accumulation stage is caused by the redirected migration of the strain field, which triggers the central bulge; and the residual stage is caused by the permanent deformation of the open end by the reverse attenuation. There is a significant difference in the displacement between the bottom plate of the wall and the two gangs. The initial displacement is more concentrated. With the propagation of the shock wave, the displacement area increases but the displacement velocity decreases. By increasing the thickness of the perimeter rock, the movement of the base plate perimeter rock is oriented towards the outside, resulting in less strain near the roadway than on the side away from the roadway, leading to more internal damage to the perimeter rock. Tension-shear stress in the center of the roof plate formed band damage, and the anchor support and buffer layer effectively suppressed the damage expansion.

The results of this research provide the following theoretical support for the anti-explosive design of coal mine roadways: (1) the quantitative relationship of explosion equivalence ratio-damage evolution is established; (2) a two-way coupled simulation framework is used; and (3) a dynamic damage model is proposed. Future research will integrate the coupling mechanism of geological tectonic effects and multi-component gas, and verify the engineering applicability of the model through on-site distributed fiber-optic monitoring.

**Author Contributions:** Conceptualization, Q.Z. and Z.Y.; methodology, P.D.; software, Q.Z. and P.D.; validation, Q.Z., Z.Y., Y.Z. and J.Z.; formal analysis, P.D.; investigation, Q.Z.; resources, J.Z.; data curation, P.D.; writing—original draft preparation, Y.Z.; writing—review and editing, P.D.; visualization, Q.Z.; supervision, Y.Z.; project administration, Z.Y.; funding acquisition, Q.Z. All authors have read and agreed to the published version of the manuscript.

**Funding:** This research was funded by the Key Research and Development Program of Shaanxi Province, grant number 2022GY-150.

**Institutional Review Board Statement:** Not applicable.

**Informed Consent Statement:** Not applicable.

**Data Availability Statement:** The data are not publicly available due to commercial confidentiality, as they contain information that could compromise the privacy of research participants.

**Conflicts of Interest:** The authors declare no conflicts of interest.

## References

1. Wang, Y.; Fu, G.; Lyu, Q.; Li, X.; Chen, Y.; Wu, Y.; Xie, X. Modelling and analysis of unsafe acts in coal mine gas explosion accidents based on network theory. *Process Saf. Environ.* **2023**, *170*, 28–44.
2. Wang, B.; Lv, Y.; Liu, C. Research on fire early warning index system of coal mine goaf based on multi-parameter fusion. *Sci. Rep.* **2024**, *14*, 485. [CrossRef]
3. Kang, Y.; Wu, Z.; Ma, S.; Zhao, M.; Li, W. CFD-based assessment and visualization of the failure consequences of LPG tankers. *J. Loss Prev. Process Ind.* **2023**, *82*, 105008. [CrossRef]
4. Xu, Y.; Huang, Y.; Ma, G. CFD-based AHP-TOPSIS decision-making model for gas explosion overpressure mitigation measures in underground utility tunnels. *J. Loss Prev. Process Ind.* **2024**, *89*, 105299. [CrossRef]
5. Mandal, J.; Goel, M.; Agarwal, A. Underground structures subjected to various blast loading scenarios: A scoping review. *Arch. Comput. Methods Eng.* **2022**, *29*, 2491–2512. [CrossRef]
6. Jia, Q.; Xue, S.; Si, X.; Si, R.; Wang, L.; Liu, X.; Li, Z. Dynamics of gas composition and thermal behavior post explosions: An experimental investigation across varied initial concentrations. *Case Stud. Therm. Eng.* **2024**, *59*, 104564. [CrossRef]

7. Gao, K.; Liu, Z.; Wu, C.; Li, J.; Liu, K.; Liu, Y.; Li, S. Effect of low gas concentration in underground return tunnels on characteristics of gas explosions. *Process Saf. Environ.* **2021**, *152*, 679–691. [CrossRef]
8. Li, R.; Si, R.; Wang, L. Propagation of gas explosions of different volumes in a large test tunnel. *Energy Sources Part A Recovery Util. Environ. Eff.* **2025**, *47*, 3222–3234. [CrossRef]
9. Yang, K.; Si, R.; Huang, Z.; Wang, L.; Zhao, J.; Li, Z. Experimental study on the deflagration to detonation transition process of gas explosions in engineering-scale pipelines. *Fuel* **2025**, *384*, 133995. [CrossRef]
10. Tan, B.; Liu, Y.; Liu, H.; Wang, H.; Li, T. Research on size effect of gas explosion in the roadway. *Tunn. Undergr. Space Technol.* **2021**, *112*, 103921. [CrossRef]
11. Zhao, Y.; Wu, J.; Zhou, R.; Cai, J.; Bai, Y.; Pang, L. Effects of the length and pressure relief conditions on propagation characteristics of natural gas explosion in utility tunnels. *J. Loss Prev. Process Ind.* **2022**, *75*, 104679. [CrossRef]
12. Wang, S.; Li, Z.; Fang, Q.; Yan, H.; Chen, L. Performance of utility tunnels under gas explosion loads. *Tunn. Undergr. Space Technol.* **2021**, *109*, 103762. [CrossRef]
13. Liu, W.; Mu, C.; Li, Z. Influence of cavity structure on gas explosion characteristics in coal mine. *Powder Technol.* **2022**, *398*, 117084. [CrossRef]
14. Yuan, M.; Hu, Q.; Huang, Z.; Shen, X.; Qian, X.; Yang, H.; Qi, S.; Jiang, J.; Xia, D. Gas explosion impact behavior and disaster analysis based on structural failure: Numerical modeling. *J. Loss Prev. Process Ind.* **2024**, *87*, 105234. [CrossRef]
15. Cheng, J.; Fu, Q.; Wang, Z.; Qin, Y.; Jiang, H.; Gao, K.; Zhang, L. Dynamic response characteristics and structural optimization of isolation seal under the impact of gas explosion. *J. Loss Prev. Process Ind.* **2023**, *84*, 105095. [CrossRef]
16. Cheng, C.; Si, R.; Wang, L.; Jia, Q.; Xin, C.; Chen, X. Experimental study on the effect of initial accumulation pattern on gas explosion and explosion suppression in a real roadway. *Case Stud. Therm. Eng.* **2023**, *51*, 103544. [CrossRef]
17. Cheng, C.; Si, R.; Wang, L.; Jia, Q.; Xin, C. Explosion and explosion suppression of gas/deposited coal dust in a realistic environment. *Fuel* **2024**, *357*, 129710. [CrossRef]
18. Gui, M.; Zhang, L.; Cui, H.; Zhang, H. Turbulent combustion related with detoantion. *Acta Aerodyn. Sin.* **2019**, *38*, 515–531.
19. Maremonti, M.; Russo, G.; Salzano, E.; Tufano, V. Numerical simulation of gas explosions in linked vessels. *J. Loss Prev. Process Ind.* **1999**, *12*, 189–194. [CrossRef]
20. Du, J.; Chen, J.; Zhu, L.; Guo, L.; Wang, F.; Hu, X. Numerical study on the dynamic response of gas explosion in uneven coal mine tunnels using CESE reaction dynamics model. *Appl. Sci.* **2024**, *14*, 2372. [CrossRef]
21. Mo, L.; Wang, R.; Yang, H.; Yang, Y.; Wu, X.; Jia, W.; Li, C.; Chen, C. Dynamic response of spherical tanks subjected to the explosion of hydrogen-blended natural gas. *Fuel* **2024**, *377*, 132834. [CrossRef]
22. Zheng, G.; Shi, J.; Li, L.; Li, Q.; Gu, Z.; Xu, W.; Lu, B.; Wang, C. Fluid-solid coupling-based vibration generation mechanism of the multiphase vortex. *Processes* **2023**, *11*, 568. [CrossRef]
23. Zhang, D.; Wu, C.; Shi, Z.; Li, Y.; Zhao, Y.; Wu, X. A fluid–solid coupling model for hydraulic fracture of deep coal seam based on finite element method. *Phys. Fluids* **2024**, *36*, 066609. [CrossRef]
24. Tan, W.; Chen, Z.; Li, Z.; Yan, H. Thermal-fluid-solid coupling simulation and oil groove structure optimization of wet friction clutch for high-speed helicopter. *Machines* **2023**, *11*, 296. [CrossRef]
25. Shamsuddin, D.S.N.A.; Fekeri, A.F.M.; Muchtar, A.; Khan, F.; Khor, B.C.; Lim, B.H.; Rosli, M.I.; Takriff, M.S. Computational fluid dynamics modelling approaches of gas explosion in the chemical process industry: A review. *Process Saf. Environ.* **2023**, *170*, 112–138. [CrossRef]
26. Qiu, J. Study on Propagation Characteristics of Shock Wave in Pipe Network and Simulation of Catastrophic Process for Gas Explosion. Ph.D. Thesis, Anhui University of Science and Technology, Huainan, China, 2018.
27. Oran, E.S.; Gamezo, V.N. Origins of the deflagration-to-detonation transition in gas-phase combustion. *Combust. Flame* **2007**, *148*, 4–47. [CrossRef]
28. Saif, M.; Wang, W.; Pekalski, A.; Levin, M.; Radulescu, M.I. Chapman–Jouguet deflagrations and their transition to detonation. *Proc. Combust. Inst.* **2017**, *36*, 2771–2779. [CrossRef]
29. Kumar, K.; Bandyopadhyay, P.; Singh, S.; Dharodi, V.S.; Sen, A. Kelvin-Helmholtz instability in a compressible dust fluid flow. *Sci. Rep.* **2023**, *13*, 3979. [CrossRef] [PubMed]
30. Nagy, T.; Valkó, É.; Sedyó, I.; Zsély, I.G.; Pilling, M.J.; Turányi, T. Uncertainty of the rate parameters of several important elementary reactions of the H2 and syngas combustion systems. *Combust. Flame* **2015**, *162*, 2059–2076. [CrossRef]
31. Li, S. Stress or strain? *Proc. R. Soc. A* **2024**, *480*, 20240269. [CrossRef]
32. Gong, B.; Wan, X.; Huo, X.; Liu, Y.; Deng, C.; Wang, D.; Dai, L. Development of a strain-based fracture assessment approach for X80 steel pipe welded girth by modified Mohr-Coulomb model. *Int. J. Press. Vessel. Pip.* **2023**, *206*, 105056. [CrossRef]

33. Zhang, J.-Z.; Zhou, X.-P. Integrated acoustic-optic-mechanics (AOM) multi-physics field characterization methods for a crack: Tension vs. shear. *Eng. Fract. Mech.* **2023**, *287*, 109339. [CrossRef]
34. Yu, Z.; Haoran, S.; Xiaohui, L.; Wenju, Z. Failure characteristics and constitutive model of coal rock at different strain rates. *Explos. Shock Waves* **2021**, *41*, 053103.

applied sciences

MDPI

*Article*

# Analysis of Internal Explosion Vibration Characteristics of Explosion-Proof Equipment in Coal Mines Using Laser Doppler

**Xusheng Xue [1,2,3], Junbiao Qiu [1,2,3,*], Hongkui Zhang [4], Wenjuan Yang [1,2,3], Huahao Wan [1,2,3] and Fandong Chen [4]**

1 School of Mechanical Engineering, Xi'an University of Science and Technology, Xi'an 710054, China; xuexsh@xust.edu.cn (X.X.); yangwj@xust.edu.cn (W.Y.); whh15152206115@163.com (H.W.)
2 Shaanxi Key Laboratory of Mine Electromechanical Equipment Intelligent Detection and Control, Xi'an 710054, China
3 Shaanxi 'Four Subjects and One Union' Coal Mine Intelligent Management and Control System and Disaster Prevention Equipment School-Enterprise Joint Research Center, Xi'an 710054, China
4 CCTEG Shenyang Research Institute, Shenyang 113122, China; 18940021505@126.com (H.Z.); cfd1987@163.com (F.C.)
* Correspondence: 13585355811@163.com

**Abstract:** Currently, there is a lack of methods for detecting the mechanism of gas explosion propagation within flameproof enclosures and the dynamic behavior of flameproof enclosures under explosion impact. Therefore, this paper studies a method for detecting the vibration characteristics of coal mine explosion-proof equipment under internal gas explosions using laser Doppler. First, a model of gas explosion propagation and explosion transmission response in flameproof enclosures is established to reveal the mechanism of gas explosion transmission inside coal mine flameproof enclosures. Second, a laser Doppler measurement method for coal mine flameproof enclosures is proposed, along with a step-by-step progressive vibration characteristic analysis method. This begins with a single-frequency dimension analysis using the Fourier transform (FFT), extends to time–frequency joint analysis using the short-time Fourier transform (STFT) to incorporate a time scale, and then advances to a three-dimensional linkage of scale, time, and frequency using the wavelet transform (DWT) to solve the limitation of the fixed window length of the STFT, thereby achieving a dynamic characterization of the detonation response characteristics. Finally, a non-symmetric Gaussian impact load inversion model is constructed to validate the overall scheme. The experimental results show that the FFT analysis identified a 2000 Hz main frequency, along with the global frequency components of the flameproof enclosure vibration signal, the STFT analysis revealed the dynamic evolution of the 2000 Hz main frequency and global frequency over time, and the wavelet transform achieved higher accuracy positioning of the frequency amplitude in the time domain, with better time resolution. Finally, the experimental platform showed an error of less than 5% compared with the actual measured impact load, and the error between the inverted impact load and the actual load was less than 15%. The experimental platform is feasible, and the inversion model has good accuracy. The laser Doppler measurement method has significant advantages over traditional coal mine flameproof equipment measurement and analysis methods and can provide further failure analysis and prevention, design optimization, and safety performance evaluation of flameproof enclosures in the future.

**Keywords:** flameproof enclosures; laser Doppler; gas explosion; vibration characteristics; time–frequency-domain analysis

 https://doi.org/10.3390/app15179255

## 1. Introduction

Coal accounts for more than 90% of China's fossil energy resources and is the most stable, economical, and independently secure energy source. China's coal mining geological conditions are complex and characterized by high levels of flammable and explosive gases, and failures of key components of coal mine equipment, such as flameproof enclosures, can cause gas explosions and other disasters. The safety of explosion-proof equipment in coal mines directly impacts the smooth and safe operation of coal mine production. The frequent occurrence of methane explosions and other accidents have placed higher demands on the performance of explosion-proof equipment [1]. On 19 July 2022, the National Mine Safety Administration issued the "14th Five-Year Plan for Mine Safety Production", explicitly stating the need to enhance the safety and reliability of equipment and eliminate production safety accidents caused by equipment failures. Currently, the performance of flameproof enclosures is mainly judged based on the requirements of GB3836.2 [2] and past experience. Pressure resistance tests are used to verify the strength of flameproof enclosures, but this method cannot reveal the explosion process inside the enclosure, nor can it obtain rich dynamic data, and it is even more difficult to measure the dynamic response of the enclosure for mechanical performance analysis. Therefore, it is of great significance to study the propagation and action mechanism of gas explosions inside flameproof enclosures, to improve the testing methods and analysis techniques for flameproof enclosures, to prevent safety accidents caused by the failure of flameproof enclosures, to optimize explosion-proof designs, and to reduce testing costs.

As an important part of coal mine equipment, the explosion-proof performance of flameproof enclosures directly determines the safety of equipment in explosive environments. Among these, explosion-proof type "d" is the most reliable and widely used type of electrical explosion protection [3]. Its principle mainly relies on the explosion resistance of the outer shell, requiring the flameproof enclosure to have sufficient strength to withstand the explosion pressure of explosive gases inside the outer shell. Currently, research on the mechanisms of methane explosions is relatively well developed. Early studies [4] examined the generation of methane explosion flames and pressure waves based on chemical reactions, while Kundu [5] and others investigated combustion, explosion, deflagration, and detonation phenomena related to pressure waves. Concurrently, scholars have also studied various explosion theory models, including general models, isothermal models, and isochoric adiabatic models [6]. Many scholars [7,8] have also conducted research on gas explosion pressure and the mechanical behavior of flameproof enclosures based on the above models, but there is a lack of research on the explosion transmission mechanism of flameproof enclosures under the action of gas explosions. At present, the measurement system for coal mine flameproof enclosures is relatively basic. At this stage, pressure sensors are used to collect pressure data through pressure resistance tests [9] on flameproof enclosures, which serve as the evaluation criteria for product explosion-protection qualification. Therefore, current research on testing methods for flameproof enclosures is limited to pressure sensor measurement research [10,11], which can only obtain the impact load of gas explosions and cannot determine the dynamic response of the enclosure. Laser Doppler measurement technology has the advantages of high accuracy and high time resolution in explosion enclosure measurement in other fields [12] and can fully obtain the dynamic response of the enclosure surface, providing a new technical reference for the measurement of coal mine flameproof enclosures. The explosion-proof testing and qualification assessment of coal mine equipment primarily rely on the aforementioned type tests. While this method can, to some extent, verify the explosion-proof qualification of the equipment, pressure sensors cannot capture the dynamic response of the equipment in a real-world explosion environment, nor can they analyze mechanical behavior for quantitative assessment, and they cannot optimize designs at weak points [13]. Therefore, dynamic response analysis methods for flameproof enclosures are currently lacking, whereas laser Doppler measurement can achieve response characteristics through time–frequency-domain analysis of the collected signals [14–16]. Therefore, laser Doppler technology can be used to analyze the vibration response of coal mine flameproof enclosures, conduct in-depth research on the mechanical characteristics of flameproof enclosures, and establish scientific analysis methods, which can provide more effective data support and more scientific guidance for improving coal mine safety production, analyzing and preventing the failure of explosion-proof equipment, and designing and manufacturing equipment.

In summary, this paper establishes a model for gas explosion propagation and explosion response in flameproof enclosures. The model reveals the propagation process from the start of the gas explosion inside the enclosure to the enclosure wall, as well as the response pattern after the explosion. A laser Doppler measurement method for flameproof enclosures is proposed, which solves the problems of relying on single pressure measurement data and difficult pressure sensor installation in traditional methods. It proposes a method for analyzing the vibration characteristics of flameproof enclosures by combining one-dimensional FFT frequency domain to STFT time–frequency and DWT scale–time–frequency three-dimensional linkage. This method gradually reveals the vibration characteristics of flameproof enclosures in multiple dimensions and scales, and has significant advantages over traditional coal mine flameproof equipment explosion-proof characteristic testing and analysis methods.

## 2. Principles of Measurement and Analysis of Vibration Characteristics of Coal Mine Flameproof Enclosures

This study aims to systematically analyze the vibration response characteristics of explosion-proof equipment in coal mines under internal explosion conditions. It uses laser Doppler technology to replace traditional pressure contact measurement methods and establishes a complete technical path from theoretical analysis, detection methods, and vibration characteristic analysis. The overall research process is shown in the Figure 1 below and mainly includes three core links:

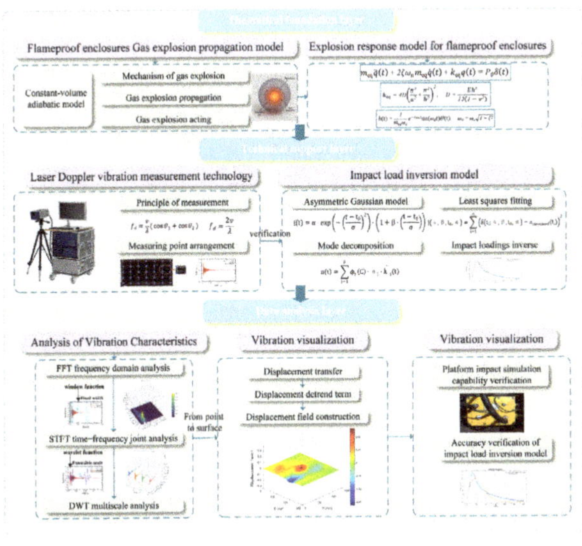


**Figure 1.** Principle of vibration characteristic measurement and analysis of explosion-proof equipment in coal mines.

(1) Gas explosion propagation and detonation transmission response model for flameproof enclosures

In the explosion propagation and response model section, the basic mechanism of gas explosions and their propagation characteristics in flameproof enclosures are discussed. By establishing a gas explosion propagation model inside flameproof enclosures and an explosion transmission response model for the enclosures, a theoretical basis is provided for subsequent laser Doppler measurement layout, impact simulation experiment design, and physical interpretation of vibration response data.

(2) Laser Doppler measurement method for flameproof enclosures

In the vibration measurement of coal mine flameproof enclosures, the laser Doppler measurement method (LDV) is proposed, which can accurately capture the dynamic response of flameproof enclosures under the impact of explosions, avoiding the signal interference and installation complexity of contact sensors during explosions. The accuracy and reliability of vibration measurement are ensured by using the direct measurement method and the global measurement point layout.

(3) Method for analyzing vibration characteristics of flameproof enclosures

We propose a step-by-step progressive method for analyzing the vibration characteristics of coal mine flameproof enclosures, starting from a single-frequency dimension analysis using FFT, to a two-dimensional time–frequency joint analysis using STFT, and finally to a three-dimensional scale–time–frequency joint analysis using DWT. This method enables a step-by-step progressive analysis of vibration characteristics, revealing the frequency and energy distribution patterns of vibration signals during the propagation of explosion shock waves. By constructing a vibration displacement field through time-domain alignment and the integration of vibration measurements obtained using laser Doppler technology, this method enables a comprehensive understanding of vibration characteristics from point to surface.

(4) Overall verification

Finally, an impact load inversion model was constructed based on an asymmetric Gaussian model, combining the least squares method and modal decomposition to invert the time history of the impact load from laser Doppler data, confirming the feasibility of the laser Doppler measurement method and its potential to replace traditional methods.

## 3. Coal Mine Explosion-Prevention Equipment: Methane Explosion Transmission and Detonation Response Model

### 3.1. Gas Explosion Propagation Model

In the explosion test, the gas explosion inside the flameproof enclosure was achieved by igniting a mixture of methane and oxygen at a standard concentration. The explosion process released a large amount of heat and generated an impact force, with the shock wave propagating outward in a spherical direction [17], as shown in Figure 2a, including a precursor wavefront and a flame wavefront, whose propagation can be expressed by the following equation:

$$R(t) = \xi \left( \frac{E}{\rho_0} \right)^{1/5} t^{2/5} \tag{1}$$

where $R(t)$ is the shock wave radius (m); $E$ and $\rho_0$ are the total energy released by the explosion (J) and the medium density (kg/m$^3$), respectively; and $\xi$ is a dimensionless constant related to the specific heat ratio $\gamma$.

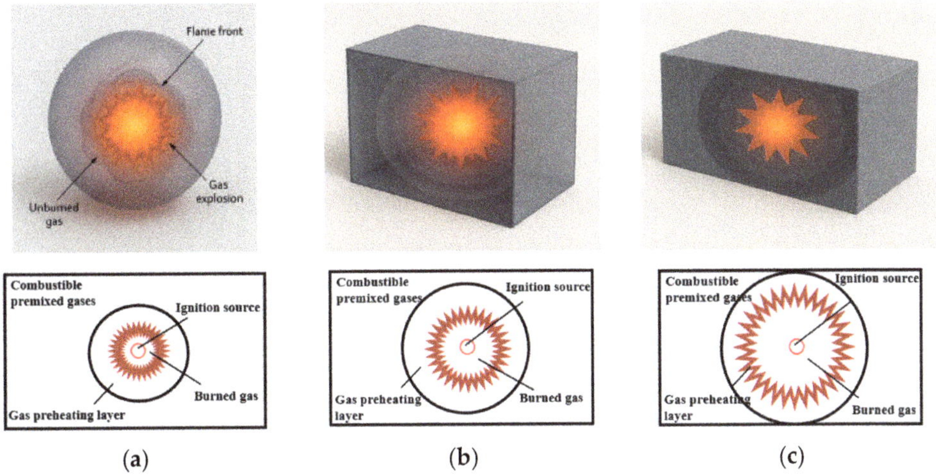


**Figure 2.** Shock wave propagation diagram of internal gas explosions in an explosion-proof shell. (**a**) Initial state. (**b**) Propagation state. (**c**) Interaction state.

Currently, explosion models include isothermal models, adiabatic models, and general models. Explosions inside flameproof enclosures are classified as closed container explosions, with no heat exchange with the outside environment, and are suitable for constant-volume adiabatic models [18]. It is assumed that the temperature of unreacted gas $T_u$ and the temperature of reacted gas $T_b$ increase as the pressure P in the container increases:

$$T_u = T_0\left(\frac{P}{P_0}\right)^{1-\frac{1}{Y_u}} \quad T_b = T_m\left(\frac{P}{P_m}\right)^{1-\frac{1}{Y_b}} \tag{2}$$

where $P_0$ is the initial pressure, $T_0$ is the initial temperature, $Y_u$ and $Y_b$ represent the specific heat ratio of the unreacted gas and the specific heat ratio of the reacted gas. The fixed-volume explosion pressure can be obtained using the ideal gas equation of state:

$$P_n = \frac{n_m T_m}{n_0 T_0} P_0 \tag{3}$$

where $P_0$ is the initial pressure, $n_m$ and $T_m$ are the molar quantities and temperatures of the reaction products, and $n_0$ and $T_0$ are the molar quantities and temperatures of the reactants.

In the aforementioned model, within a sealed container, the chemical reaction rate is extremely high, causing the temperature to rise sharply and the pressure to increase rapidly, leading to an explosion. Finally, the explosion shock wave is transmitted to the shell wall, as shown in Figure 2b,c,with the center of the wall being the first to be affected, as it is the earliest and strongest point of impact of the explosion shock wave. This can be regarded as the center of the wall array being affected, and subsequent research is based on this theory. When the shock wave reaches the wall, the shell will produce a strong response, which initially manifests as a vibration response.

### 3.2. Detonation Transmission Response Model for Flameproof Enclosures

After an explosion occurs during an explosion test, the gas inside the flameproof enclosure burns instantaneously, releasing heat and expanding rapidly, forming a high-amplitude, short-duration instantaneous impact load. On a time scale, this load can be approximated as a pulse excitation. The pulse excitation characteristics of a gas explosion determine how it affects the structure. This pulse excitation can be expressed by the following equation:

$$F(t) = F_0\delta(t) \tag{4}$$

where $F_0$ represents the intensity of the pulse impact, and $\delta(t)$ represents the instantaneous impact force.

The response of flameproof enclosures to this pulse impact can be modeled using a single-degree-of-freedom system, whose system dynamics equation is as follows:

$$m\ddot{q}(t) + c\dot{q}(t) + kq(t) = P_0\delta(t) \tag{5}$$

where $q(t)$ represents the displacement of a point, m is the mass, c is the damping coefficient, and k is the stiffness. The impulse response function is obtained as follows:

$$\mathrm{h(t)} = \frac{1}{\mathrm{m}\omega_\mathrm{d}}\mathrm{e}^{-\xi\omega_0\mathrm{t}}\sin(\omega_\mathrm{d}\mathrm{t}) \tag{6}$$

where $\omega_0$ represents the natural frequency of the shell without damping, and $\omega_\mathrm{d}$ represents the natural frequency of the shell with damping.

To more accurately reflect the true dynamic characteristics of thin plate shells under explosive impact, this paper introduces a multi-modal vibration model based on the material properties of Q235 steel on the basis of the traditional single-degree-of-freedom system. This model comprehensively considers the natural modal distribution of rectangular thin plates under four-sided fixed boundary conditions, modal coupling effects, and additional damping caused by welds and fixed constraints, making the calculation results closer to the actual structural response. By applying corrections based on equivalent mass, equivalent damping, and equivalent stiffness, complex multi-modal responses can be mapped to an equivalent single-degree-of-freedom system. Equivalent mass is used to reflect the contribution of each modal order to the overall inertial characteristics, equivalent damping represents the combined effect of material internal dissipation and boundary energy dissipation, and equivalent stiffness combines the bending stiffness of thin plates and boundary constraint stiffness to perform equivalent processing on the overall stiffness of the system.

(1) Structural damping and stiffness of materials

In order to more accurately describe the dynamic response of coal mine flameproof enclosures, considering the internal damping of Q235 steel and the additional damping caused by welds and fixed boundaries, the vibration energy will rapidly decay in a short period of time. Therefore, we introduce equivalent mass, equivalent damping coefficients, and stiffness coefficients, and the dynamic equation is improved as follows:

$$m_{eq}\ddot{q}(t) + c_{ef}\dot{q}(t) + k_{ef}q(t) = P_0\delta(t) \tag{7}$$

where $m_{eq}$ is to correct the mass coefficient, $c_{eq}$ is to correct the damping coefficient, and $k_{eq}$ is to correct stiffness coefficient.

(2) Material and structural parameter expression

The equivalent stiffness of the shell is derived from thin plate theory as follows:

$$k_{eq} = 4D\left(\frac{\pi^2}{a^2} + \frac{\pi^2}{b^2}\right)^2, D = \frac{Eh^3}{12(1-\nu^2)} \tag{8}$$

The equivalent mass is the effective mass under the first-order mode of the structure:

$$m_{eq} \approx 0.236\rho hab \tag{9}$$

where $E$ is Young's modulus, h is the shell thickness, ab is the panel size, $\nu$ is Poisson's ratio, and $D$ is the bending stiffness; all parameters can be determined through shell design parameters.

(3) Improved response function

Combining (7), (9), and the initial response function (6), we obtain the improved response function for a unit pulse as follows:

$$h(t) = \frac{1}{m_{eq}\omega_d} e^{-\zeta\omega_n t} sin(\omega_d t) H(t),\ \omega_d = \omega_n\sqrt{1-\zeta^2} \tag{10}$$

In summary, by modeling the propagation process of gas explosions and their impact response on flameproof enclosures, the spatial distribution of explosion excitation and the response characteristics of the enclosures were clarified, providing a theoretical basis for subsequent laser Doppler measurement layout, impact simulation experiment design, and physical interpretation of vibration response data. Based on this propagation and response model, this paper further constructs a measurement and analysis system to carry out the dynamic analysis of the vibration behavior of the enclosures and the identification of impact loads.

## 4. Principle of Explosion Shock Vibration Measurement Using Flameproof Enclosures with Laser Doppler

Laser Doppler measurement flameproof enclosures technology is based on the laser Doppler frequency shift effect [19], using a single-frequency coherent light beam emitted by a laser, which is separated into a reference beam and a measurement beam by a beam splitter. The measurement beam is irradiated onto the surface of the flameproof enclosure, and the reflected light carries the instantaneous velocity information of the target surface.

The principle is shown in Figure 3 below.

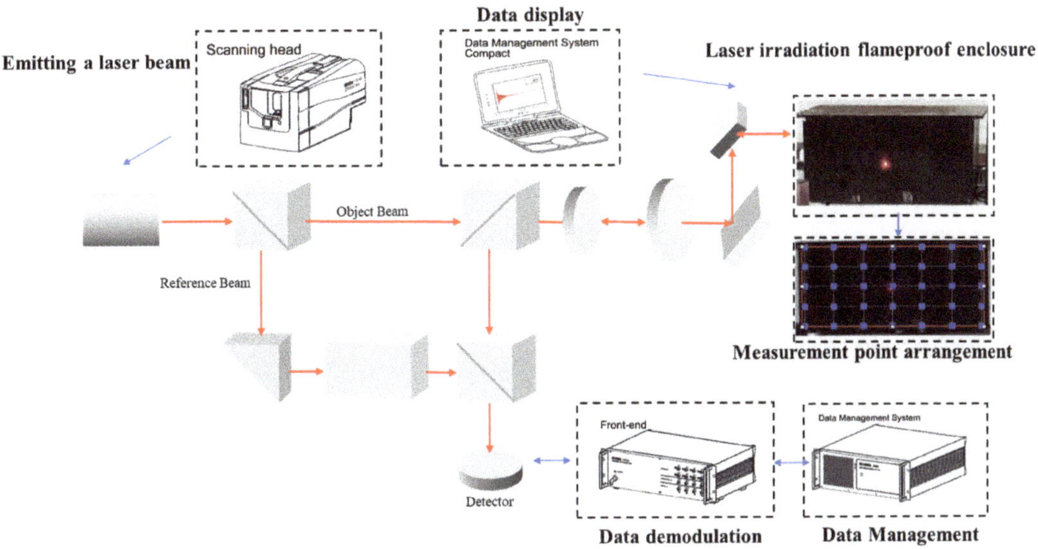


**Figure 3.** Schematic diagram of laser Doppler measurement in flameproof enclosures.

The following measurement methods are specified for flameproof enclosures:

(1) Direct measurement method

The amount of laser frequency shift reflected back is directly proportional to the object's velocity $v$. By demodulating the frequency shift signal based on the laser Doppler effect, the target's vibration velocity can be determined. The detected Doppler frequency shift $f_d$ has the following relationship with the measured object's velocity $v$:

$$f_d = \frac{v}{\lambda}(cos\theta_1 + cos\theta_2) \tag{11}$$

where $\lambda$ is the wavelength, $\theta_1$ and $\theta_2$ are the angles between the incident light direction, the scattered light direction, and the object's motion direction, respectively.

In order to simplify the expression of Doppler frequency shift and improve measurement accuracy, this paper adopts a direct measurement method, where the incident direction of the laser beam is parallel to the vibration direction of the target, and the incident angle and reflection angle are both zero ($\theta_1 = \theta_2 = 0°$). The expression for calculating the object's velocity using Doppler frequency shift is

$$f_d = \frac{2v}{\lambda} \tag{12}$$

where $f_d$ is the Doppler shift; $v$ is the object's velocity; and $\lambda$ is the wavelength of light.

By calculating the time-varying frequency difference $f_d$ between the transmitted wave and the reflected wave from a single acquisition, the vibration velocity curve of a single point on the shell surface can be obtained.

(2) Measurement point layout

To comprehensively measure the vibration characteristics of the shell structure, a two-dimensional rectangular array grid is arranged on the surface of the test area, as shown in Figure 4. There are a total of 35 measurement points arranged in seven columns and five rows, with a spacing of 5 cm in the horizontal direction and 3.5 cm in the vertical direction to ensure that the spatial sampling meets the structural modal resolution requirements.

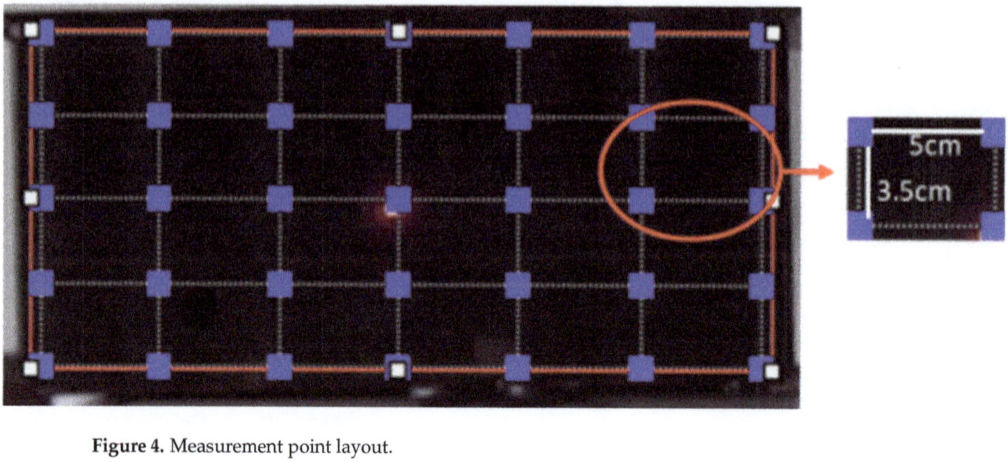


**Figure 4.** Measurement point layout.

Repeat experiments were conducted at each measurement point to ensure that the excitation remained consistent during each collection. The laser beam was vertically incident on the center of the measurement point during each collection to achieve "direct measurement," thereby maximizing the signal-to-noise ratio and simplifying Doppler shift calculations. This arrangement is suitable for subsequent parameter extraction and structural response visualization processing, providing high-resolution data support for the study of shell dynamic behavior. The single-point data collected is shown in Figure 5.

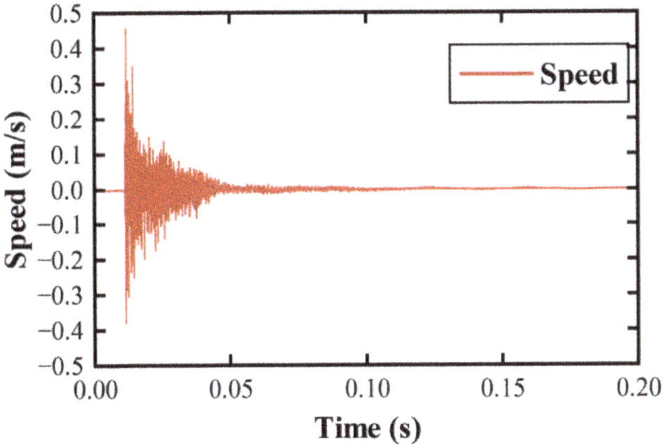


**Figure 5.** Data example.

## 5. Method for Analyzing Vibration Characteristics of Flameproof Enclosures

Vibration signals from flameproof enclosures have obvious transient characteristics, non-stationarity, and complex time-varying characteristics, such as multi-frequency component superposition. A comprehensive analysis of their time–frequency characteristics is required. Therefore, a time–frequency analysis method of vibration characteristics of coal mine explosion-proof shell is proposed. The method progresses from the single-frequency dimension analysis using Fourier transform (FFT) to two-dimensional time–frequency joint analysis with the short-time Fourier transform (STFT), and finally to three-dimensional scale–time–frequency linkage through the wavelet transform (DWT).

The specific method is shown in Figure 6. The FFT is used to obtain the frequency distribution of the explosion vibration signal, but this method lacks time-varying frequency information. Therefore, the window function introduced by the STFT is used to obtain the signal frequency distribution in each time window, but its fixed window width cannot take into account both time and frequency resolution. Therefore, DWT is further applied to perform multi-level filtering and down-sampling of the signal, analyzing the signal in layers based on different frequency bands. The scalability of the wavelet basis functions within the time window achieves high frequency resolution in the high-frequency band and high time resolution in the low-frequency band, optimizing the resolution balance issue of STFT. The analytical methods not only gradually expand in processing dimensions but also progressively enhance their ability to reveal features. Finally, a displacement field is constructed based on the single-sided measurement point data of the flameproof enclosures, achieving a comprehensive understanding of the vibration distribution of the flameproof enclosures, from point to surface and from local features to the whole.

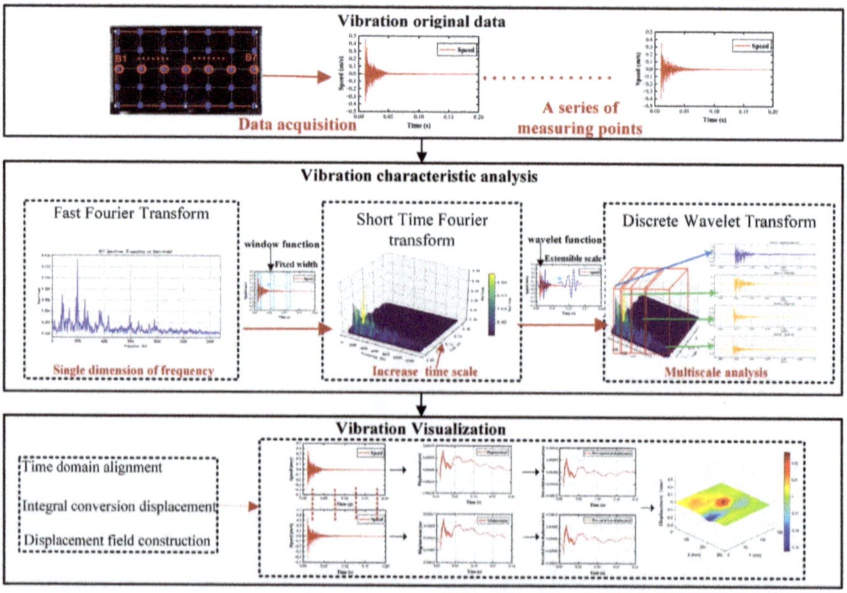


**Figure 6.** Method for analyzing the vibration characteristics of flameproof enclosures.

### 5.1. Time–Frequency Analysis Method for Vibration Characteristics of Flameproof Enclosures

(1) Fourier transform method for analyzing vibration characteristics

The FFT method maps the time-domain signal of flameproof enclosures to the frequency domain, reflecting the global distribution of each frequency component through the spectrum [20]. The conversion formula for the FFT transformation is as follows:

$$X[k] = \sum_{n=0}^{N-1} x[n] \cdot e^{-j2\pi k/n} \tag{13}$$

where $x[n]$ is the nth sample value of the input signal, $N$ is the total number of samples, and $e^{-j2\pi k/n}$ is the Fourier basis function.

FFT can only obtain the frequency components contained in the vibration signals of flameproof enclosures. The spectrum obtained is global, and it is impossible to obtain the dynamic changes in the frequencies and corresponding vibration energy of the explosion vibration signals at different times.

(2) Vibration characteristic analysis method of short-time Fourier transform

The STFT method extends the analysis dimension of FFT, which can only be observed in the frequency domain, to a combined time–frequency representation [21]. It uses a window function to perform a Fourier transform on each small segment of the time domain to analyze the frequency transient changes and time-varying characteristics of explosive vibration non-stationary signals. The formula is as follows:

$$X(m,\omega) = \sum_{n=0}^{N-1} x[n] \cdot w[n-m] e^{-j2\pi k/n} \tag{14}$$

where $x[n]$ is the original time-domain discrete vibration signal, $w[n-m]$ is the window function, and n is the sampling point time index.

The width of the STFT window function results in a trade-off between time and frequency resolution, as it is not possible to achieve both simultaneously. A narrow window provides higher time accuracy but limits the ability to judge signal periodicity, resulting in lower frequency resolution. A wide window has the opposite effect. Therefore, STFT cannot achieve high-precision time–frequency analysis across the entire frequency band.

(3) Vibration characteristic analysis method of wavelet transform

The wavelet transform (CWT) replaces the window function in STFT with a wavelet function, which can be scaled arbitrarily to solve the STFT resolution trade-off problem [22]. Vibration signals are discrete data, and discrete wavelet transform (DWT) is used to decompose the signal into different frequency bands through multiscale filtering, thereby achieving multiscale analysis of the vibration characteristics of flameproof enclosures. The wavelet function, decomposition formula, and principle are as follows:

$$\psi_{j,k}(t) = 2^{j/2}\psi\left(2^{j}t - k\right) \tag{15}$$

$$a[k] = \sum_{n=0}^{n=k-1} x[n]\cdot h[2k-n] \quad d[k] = \sum_{n=0}^{k-1} x[n]\cdot g[2k-n] \tag{16}$$

where $j$ is the scale parameter, $k$ is the translation parameter, $x[n]$ is the shell vibration signal, $d[k]$ is the high-frequency filtered signal, and $a[k]$ is the low-frequency filtered signal.

Figure 7 shows that each layer of the vibration signal $x[n]$ extracts detailed information $d[k]$ through a high-pass filter $g[n]$, extracts approximate information $a[k]$ through a low-pass filter $h[n]$, and recursively decomposes the approximate part to ultimately obtain a multi-scale representation of different frequency bands. The signals in each frequency band have good time localization capabilities, which optimizes the resolution trade-off problem of STFT under fixed window function constraints.

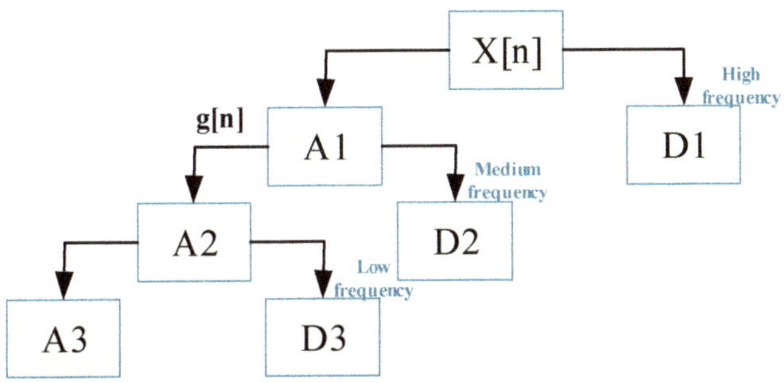


**Figure 7.** DWT decomposition principle.

## 5.2. Displacement Field Construction

The above method for analyzing the vibration characteristics of flameproof enclosures targets individual measurement points and analyzes the time–frequency characteristics of the vibration signals of flameproof enclosures. It further constructs the enclosure displacement field based on the time-series measurement point data to visualize the vibration response in the spatial dimension, thereby progressing from a description of local point characteristics to a description of the overall deformation mode.

The velocity signals measured at single points on the flameproof enclosures are integrated and converted into displacement signals, which are then aligned on a time axis to construct multi-point synchronous time series data, thereby forming a high-resolution displacement field of the single-sided structure. This displacement field can show the overall dynamic deformation profile of the flameproof enclosures under explosion excitation, thereby realizing the visualization of the vibration displacement field.

## 6. Impact Load Inversion Model

Laser Doppler replaces pressure sensors. Since the pressure distribution of how gas explosion shock waves act on flameproof enclosures is unknown, a model of the relationship between the vibration response of the enclosure surface and the internal wall impact pressure can be constructed to reconstruct the gas explosion impact pressure. Generally, the contour of any shock time history can be approximated as a Gaussian-like function [23], while methane explosion shock loads are closer to a non-symmetric Gaussian model, with the mathematical formula as follows:

$$f(t) = \alpha \cdot \exp\left(-\left(\frac{t-t_0}{\sigma}\right)^2\right) \cdot \left(1+\beta \cdot \left(\frac{t-t_0}{\sigma}\right)\right) \tag{17}$$

where $\alpha$ is the intensity factor of the impact force, $t_0$ is the start time, $\sigma$ is the standard deviation, which controls the width of the force, $\beta$ is the parameter controlling asymmetry.

The acceleration response $a(t)$ of flameproof enclosures under impact loads can be expressed as a linear combination of a set of modes, as shown in the following formula:

$$a(t) = \sum_{j=1}^{p} \phi_j(c) \cdot a_j \cdot \ddot{h}_j(t) \tag{18}$$

where $\phi_j(c)$ is the displacement shape of the $j$th mode, $a_j$ is the modal coefficient, and $\ddot{h}_j(t)$ is the acceleration response of the $j$th mode.

We can fit the acceleration response using experimental acceleration data $a_m$ and obtain $j$ and other modal parameters. The difference between the acceleration response $a_1$ of the impact load model fitted by the least squares method and the experimentally measured acceleration response $a_2$ is expressed by the objective function:

$$J(\alpha, \beta, t_0, \sigma) = \sum_{i=1}^{n} \left(\hat{a}(t_i; \alpha, \beta, t_0, \sigma) - a_m(t_i)\right)^2 \tag{19}$$

where $\hat{a}(t_i; \alpha, \beta, t_0, \sigma)$ is the acceleration response predicted by fitting the impact force history, and $a_m(t_i)$ is the actual measured acceleration data. By minimizing the objective function, we can obtain the optimal parameters $\alpha, \beta, t_0,$ and $\sigma$.

Once the least squares fitting is completed, and the optimal parameter is obtained, we can calculate the peak load $f_{max}$ based on the fitted impact force model, thereby reconstructing the impact load history $f(t)$.

## 7. Analysis

### *7.1. Experimental Design for Vibration Measurement of Flameproof Enclosures*

Based on the aforementioned explosion model and propagation mechanism, it is evident that the shock wave expands spherically in all directions during an explosion. Therefore, theoretically, the four central points are the first points to be subjected to force, and the initial impact is the most significant factor influencing the shell's response. Thus, the impact model is simplified to a central impact at the four corners of the shell. Therefore, this paper simulates the process of an explosion-proof enclosure being subjected to shock waves using mechanical impact simulation. This configuration employs a magnetic impactor to form a two-dimensional impact array, simulating the scenario of multi-point synchronous impact in a real explosion, and possesses good scalability and reproducibility.

The laboratory simulation test used a flameproof enclosure model with dimensions of 321 mm × 321 mm × 160 mm and a thickness of 3 mm. The impact force of a single magnetic impactor was fixed at 60 N, which was calculated to generate a pressure of approximately 0.36 MPa, consistent with the pressure generated by a gas explosion in the explosion test described in the literature [24]. The complete test device is shown in Figure 8:

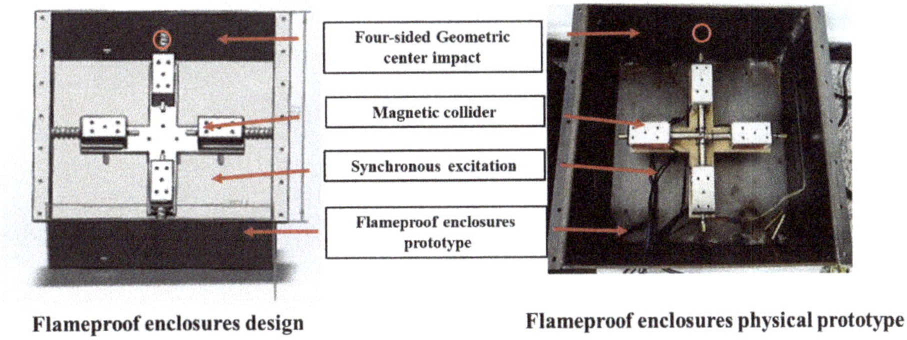


**Figure 8.** Experimental subjects.

The experimental platform for collecting surface vibrations at the moment of explosion in coal mine flameproof enclosures was constructed as shown in Figure 9. The laser detection point of the laser vibration meter was placed on the front of the flameproof enclosure, and the measurement parameters are shown in Table 1. The sampling frequency was set to 25,600 Hz. Due to the characteristics of the helium–neon laser as the light source, its measurement distance has a periodic effect on the signal-to-noise ratio. Specifically, the optimal measurement distance is S = 141 mm + n × 204 mm. A measurement distance of 2181 mm was taken to ensure that the laser head camera fully covered the flameproof enclosure for the arrangement of measurement points. Perform 2D alignment on one side of the flameproof enclosure to ensure the accuracy of the measurement point layout. According to the method described in 3.1 above, use the rectangular single side of the enclosure as the measurement area, set up a 7 × 5 rectangular grid layout for the measurement points, and arrange the measurement point spacing as shown in Figure 10. During measurement, ensure that the lens is aligned with the shell surface, and the measurement distance remains constant. Measure the 35 measurement points sequentially from top to bottom and from left to right. The excitation device is synchronously triggered at each point to achieve real-time measurement of the vibration response.

**Figure 9.** Experimental platform.

Table 1. Laser Doppler vibrometer parameters.

| Measurement Range | Resolution | Sampling Frequency | Speed Range | Laser Wavelength | Number of Channels |
|---|---|---|---|---|---|
| 0–15 m | μm level | 0–50 MHz | ±10 m/s | 633 nm He-Ne | single channel |

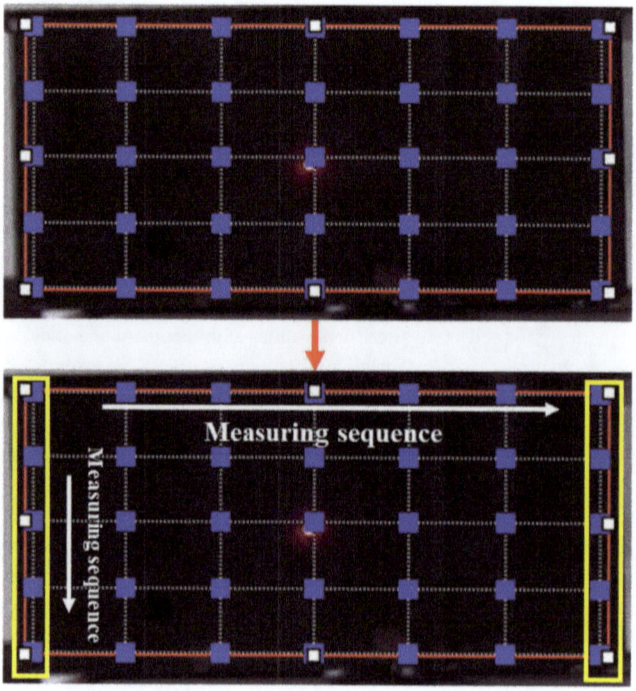


Figure 10. Measuring point arrangement and measuring sequence.

*7.2. Vibration Characteristics Analysis of Flameproof Enclosures*

7.2.1. Time–Frequency Domain Analysis of Flameproof Enclosures Under Vibration

Laser Doppler was used to collect the vibration velocity signals of the outer surface of the explosion-proof enclosure at the moment of impact. Through equivalent repeated experiments, the above three calibrated measurement points were collected in sequence. The results are shown in the middle row, as shown in Figure 11. The row data vibration time-domain diagrams correspond to the measurement points (B1–B7) in order from left to right.

Based on the maximum vibration amplitude and vibration waveform diagram of the middle row, it can be observed that the curve exhibits a relatively large initial vibration amplitude, which gradually decreases over time. This is due to the free vibration generated by the explosion-proof housing after being subjected to an instantaneous impact load gradually decaying, with the vibration amplitude stabilizing around 0.1 s. Additionally, under conditions where the impact value remains constant, as the distance from the impact source increases, the maximum vibration velocity and vibration amplitude exhibit a gradual decreasing trend. That is, the maximum vibration velocity and vibration amplitude are inversely proportional to distance, and the intensity of the shock wave gradually decays as the propagation distance increases.

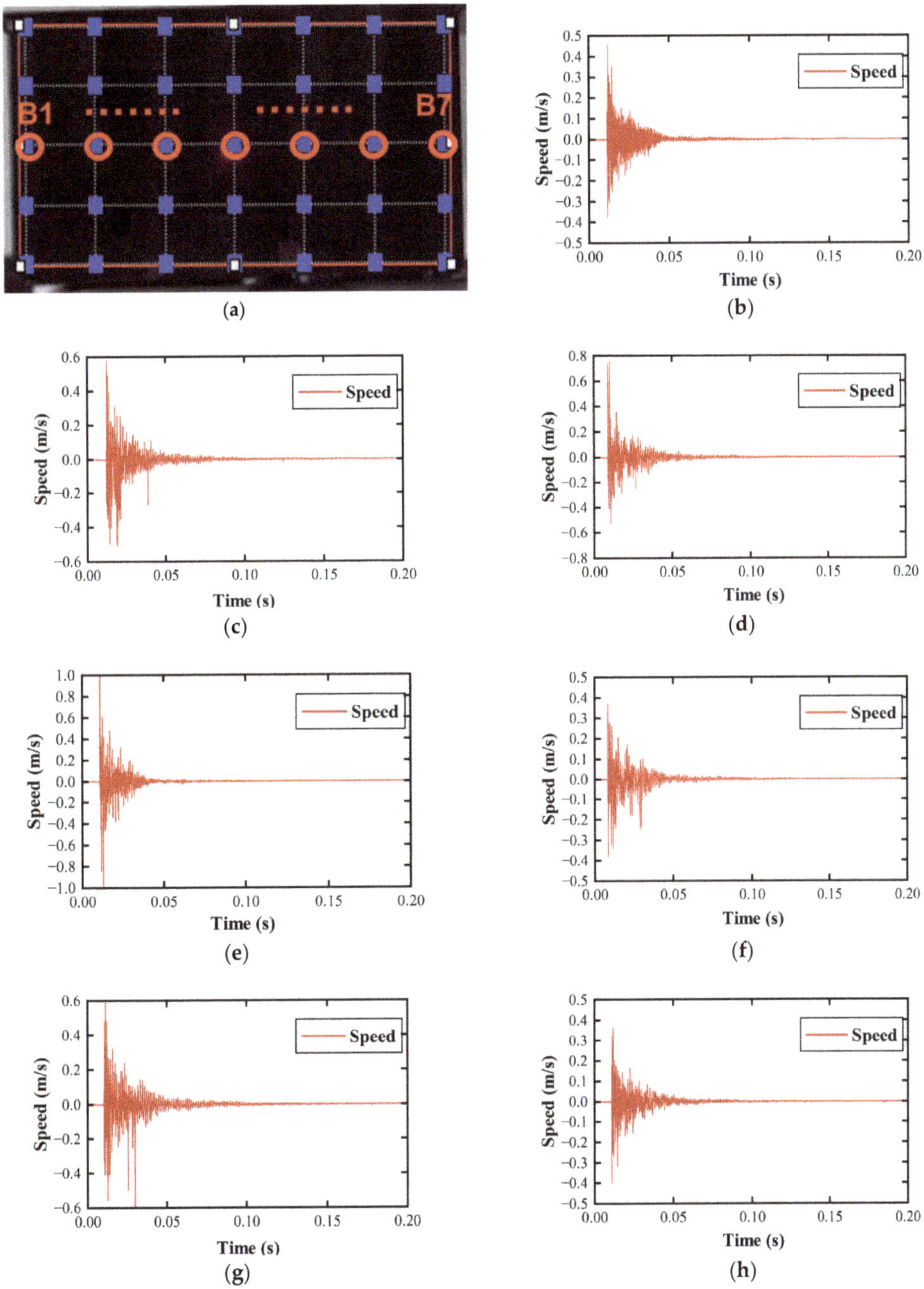


**Figure 11.** Display of row data results. (**a**) Arrangement of data points in the middle row. (**b**) B1 measurement point. (**c**) B2 measurement point. (**d**) B3 measurement point. (**e**) B4 measurement point. (**f**) B5 measurement point. (**g**) B6 measurement point. (**h**) B7 measurement point.

The four points B1–B4 mentioned above have a larger data span and are more representative, so they were selected as the vibration analysis objects. Initially, Fourier transform (FFT) was used to convert the time-domain signal to the frequency domain. The results are shown in Figure 12, where the right column shows the frequency-domain distribution obtained by FFT for measurement points B1–B4.

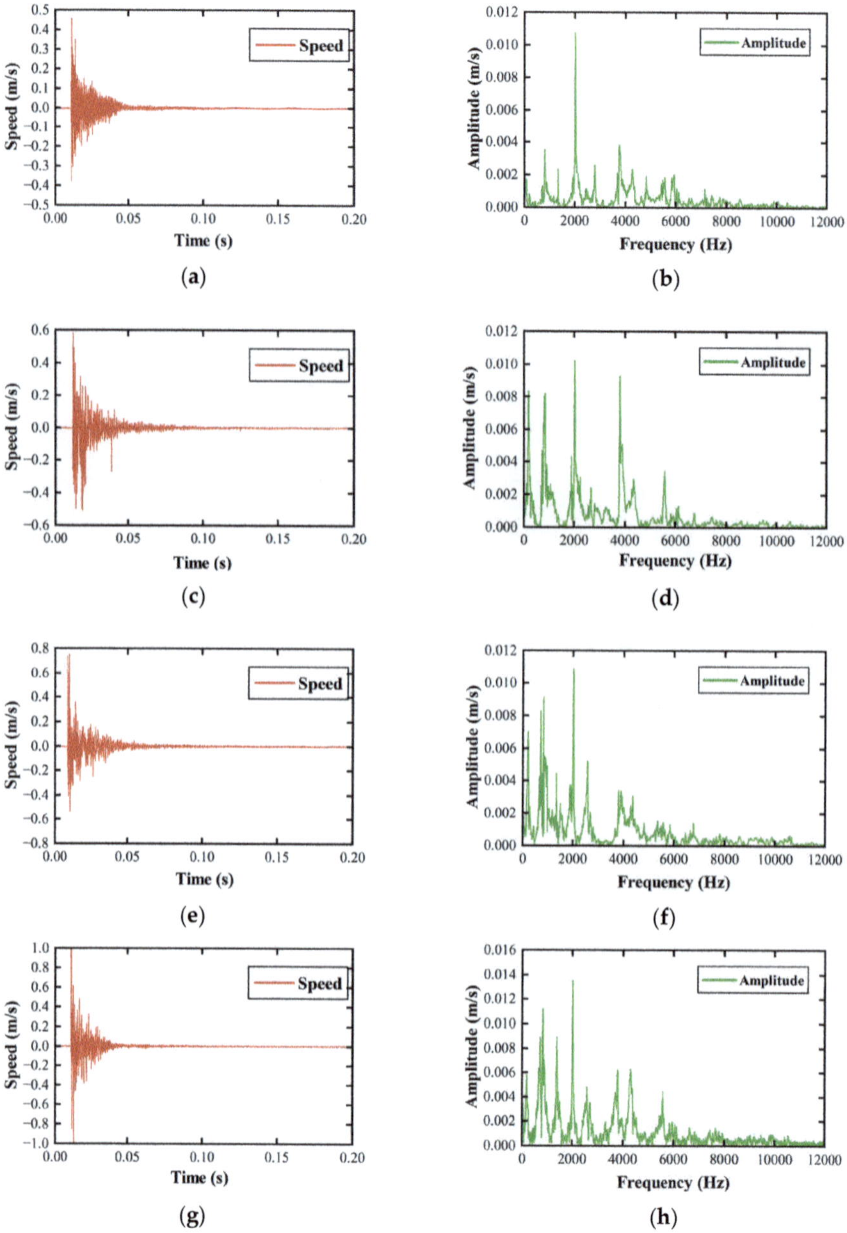


**Figure 12.** Time-domain and frequency-domain diagrams for measurement points B1–B4. (**a**) Time-domain diagram of measurement point B1. (**b**) Frequency-domain diagram of measurement point B1. (**c**) Time-domain diagram of measurement point B2. (**d**) Frequency-domain diagram of measurement point B2. (**e**) Time-domain diagram of measurement point B3. (**f**) Frequency-domain diagram of measurement point B3. (**g**) Time-domain diagram of measurement point B3. (**h**) Frequency-domain diagram of measurement point B3.

B1, being the farthest edge point from the impact location, shows a time-domain plot where the vibration amplitude first increases and then decreases. The peak value of 0.46 m/s is reached at 0.001 s, with the frequency-domain peak corresponding to a frequency of 2000 Hz. This indicates that the corresponding time at this frequency is 0.001 s, and it is the dominant frequency. B2 has a time-domain peak of 0.6 m/s, slightly higher than B1, and its corresponding frequency-domain dominant frequency is also 2000 Hz. B3 has a time-domain peak of 0.78 m/s, slightly higher than B2, and its corresponding dominant frequency is also 2000 Hz. B4, being the closest point to the impact center, has a time-domain amplitude of 1 m/s, which is the highest peak value among the four measurement points, corresponding to the main frequency of 2000 Hz. It can be seen that the main peak values at different positions increase as the distance from the impact center decreases, indicating that the explosive energy is primarily concentrated in the central region of the shell, with higher energy levels closer to the impact center. In addition, the frequency distribution of each measurement point is relatively broad, mainly distributed between 500 Hz and 8000 Hz. Except for the main frequency at 2000 Hz, each measurement point also shows relatively obvious sub-frequencies at 800 Hz, 1300 Hz, 4000 Hz, and 6000 Hz, indicating that the overall vibration distribution at each position on the single side of the shell is relatively consistent.

Based on the above FFT analysis, STFT analysis was performed, and the results are shown in Figure 13. The left figure shows the two-dimensional time–frequency results of measurement points B1 to B4 after STFT. In addition to obtaining the frequency distribution of the FFT results, the changes in each frequency and corresponding amplitude over time can be clearly seen. The right figure shows the three-dimensional time–frequency distribution of each measurement point after STFT, which more clearly displays the amplitudes corresponding to different frequencies.

The two-dimensional time–frequency plot of B1 is most prominent at 2000 Hz, consistent with the 2000 Hz dominant frequency obtained from the FFT. It reaches a peak of 0.051 m/s within the range of 0.014 s to 0.016 s. Subsequently, the amplitude of this dominant frequency gradually decays to zero, corresponding to the decay of the shell impact and the disappearance of vibration, with a duration of approximately 0.05 s. The corresponding three-dimensional plot clearly shows a trend of first increasing and then decreasing. The dominant frequency obtained for B2 is also consistent with the FFT results, reaching a peak of 0.055 m/s within the range of 0.014 s to 0.016 s, with a motion pattern consistent with B1; B3 reaches its peak of 0.071 m/s between 0.014 s and 0.016 s; B4, being the closest point to the impact, also reaches its peak between 0.014 s and 0.016 s, at 0.086 m/s. It can be clearly seen that the dominant frequencies at each measurement point are generally consistent with the FFT results, and the corresponding energy changes over time also follow a trend of first increasing and then decreasing. This temporal variation is an advantage that the FFT does not possess. Additionally, the four measurement points can clearly distinguish the sub-frequencies at 800 Hz, 1300 Hz, 4000 Hz, and 6000 Hz, and the energy corresponding to these frequencies also follows a trend of first increasing and then decreasing.

STFT adds a time scale to FFT, allowing for a clear view of the dynamic changes in different frequencies and corresponding energies over time, resulting in a qualitative improvement in the dynamic analysis and multi-dimensional analysis of vibration characteristics.

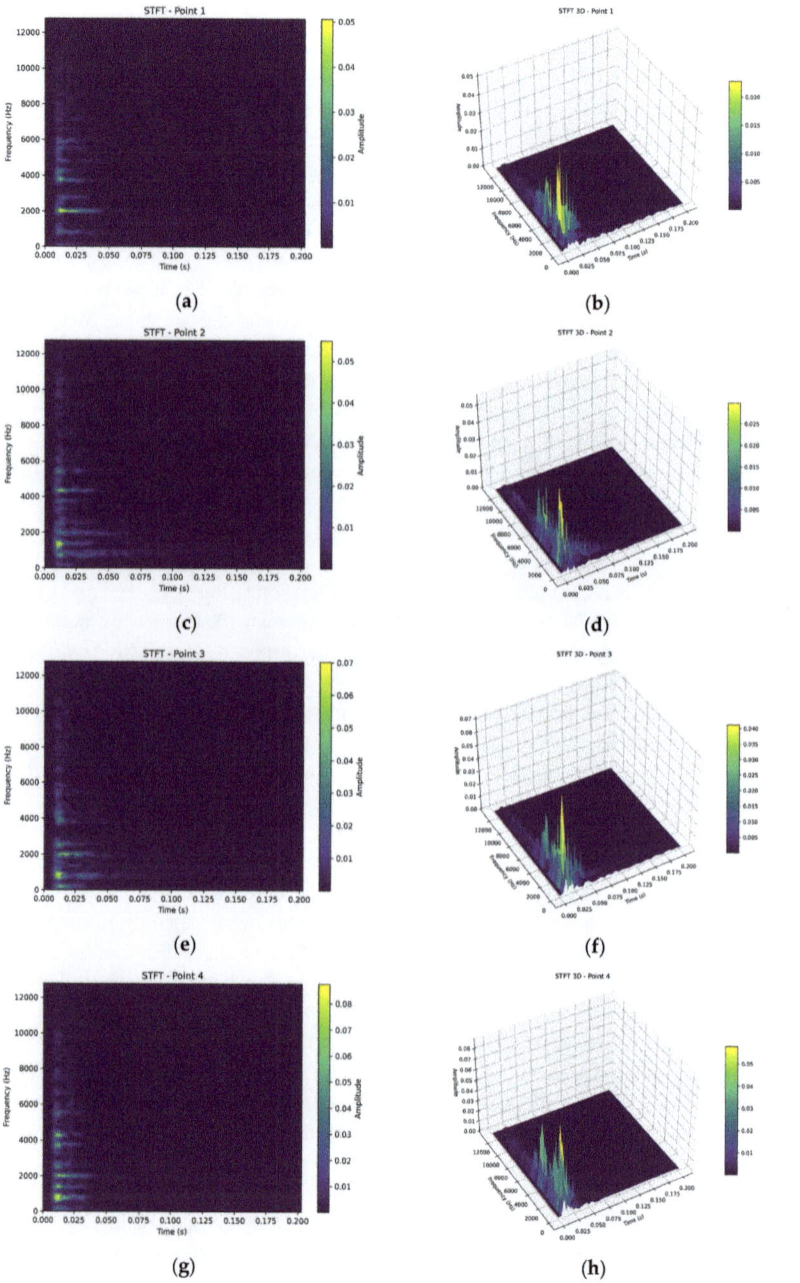


**Figure 13.** Time–frequency diagram of measurement points B1–B4 after STFT. (**a**) Two-dimensional time–frequency plot of STFT at measurement point B1. (**b**) Three-dimensional time–frequency diagram of STFT at measurement point B1. (**c**) Two-dimensional time–frequency plot of STFT at measurement point B2. (**d**) Three-dimensional time–frequency diagram of STFT at measurement point B2. (**e**) Two-dimensional time–frequency plot of STFT at measurement point B3. (**f**) Three-dimensional time–frequency diagram of STFT at measurement point B3. (**g**) Two-dimensional time–frequency plot of STFT at measurement point B4. (**h**) Three-dimensional time–frequency diagram of STFT at measurement point B4.

Further analysis of the four points of the flameproof enclosure using wavelet transform, three-layer wavelet decomposition of the flameproof enclosure vibration signal, and frequency information for each layer is shown in Table 2. The wavelet transform results are shown in Figure 14. The left figure displays the wavelet transform results for B1 to B4 in sequence, while the right figure shows the specific information for each scale of the corresponding wavelet transform. Each measurement point can precisely divide the scale of each layer according to the desired frequency range, enabling the precise observation of the decomposed signals at each scale.

**Table 2.** Hierarchical information in wavelet transforms.

| Frequency Band | D1 | D2 | D3 | A3 |
|---|---|---|---|---|
| Frequency | $\left[\frac{f_s}{4}, \frac{f_s}{2}\right] = [6400, 12800]$ | $\left[\frac{f_s}{8}, \frac{f_s}{4}\right] = [3200, 6400]$ | $\left[\frac{f_s}{16}, \frac{f_s}{8}\right] = [1600, 3200]$ | $\left[\frac{f_s}{0}, \frac{f_s}{16}\right] = [0, 1600]$ |
| Explanation | High-frequency details | Mid–high frequency details | Midrange details | Low-frequency trend |

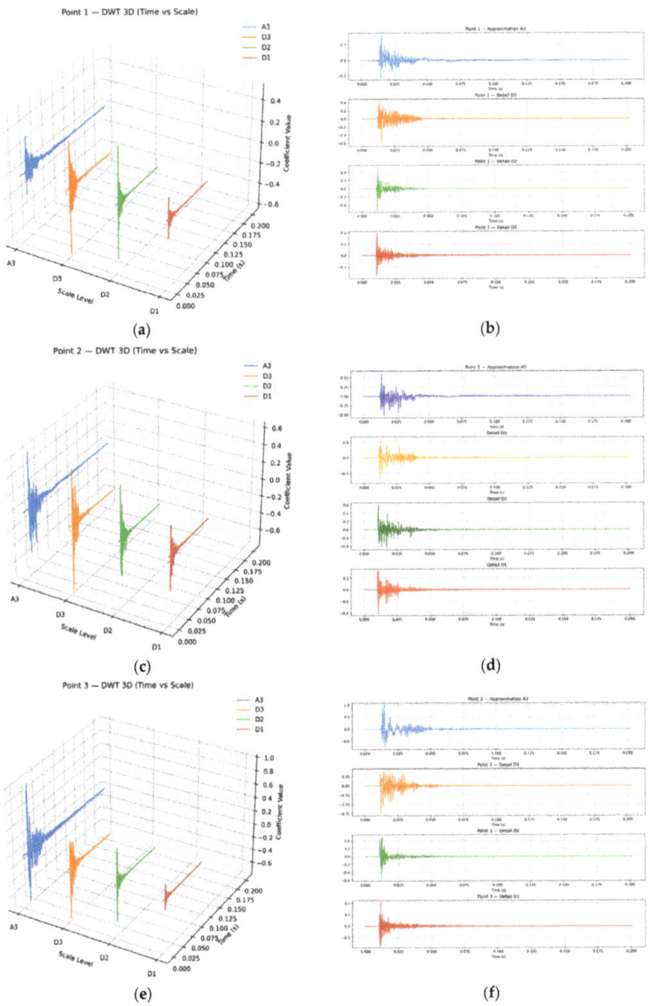


(a) (b)

(c) (d)

(e) (f)

**Figure 14.** *Cont.*

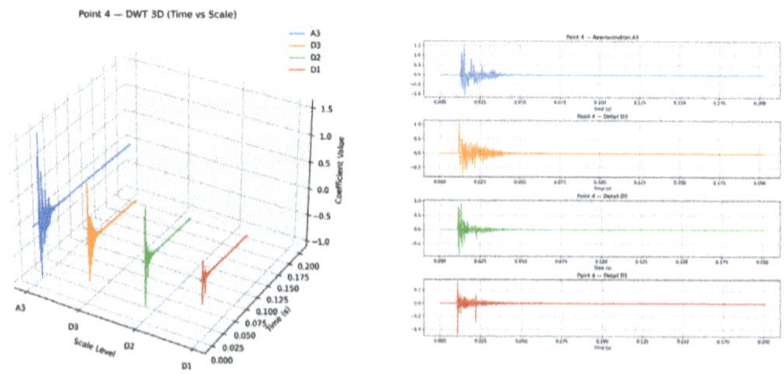


(g) (h)

**Figure 14.** Wavelet transform results for measurement points B1–B4. (**a**) Wavelet transform results for measurement point B1. (**b**) Wavelet transform of B1 measurement point at various scales. (**c**) Wavelet transform results for measurement point B2. (**d**) Wavelet transform of B2 measurement point at various scales. (**e**) Wavelet transform results for measurement point B3. (**f**) Wavelet transform of B3 measurement point at various scales. (**g**) Wavelet transform results for measurement point B1. (**h**) Wavelet transform of B1 measurement point at various scales.

The aforementioned B1 point in the frequency range includes a main frequency of 2000 Hz. It can be clearly seen that the signal reaches its peak at 0.015 s, while the peak position obtained from the STFT is located within a short time interval, with its accuracy influenced by the window width. It is evident that the time localization accuracy of the wavelet transform is significantly higher than that of the STFT. The peaks of B1 to B4 in this frequency band are 0.41, 0.58, 0.53, and 1.12, respectively, and the amplitude changes over time can be precisely determined at specific time points. The wavelet transform demonstrates superior time resolution. Additionally, although each layer of the wavelet transform represents the results of a specific frequency range, each scale exhibits a clearer overall trend. For example, at the high-frequency scale of measurement point B1, the vibration amplitude gradually increases from the 0 s mark, reaching a peak of 0.29 at the 0.015 s mark, and then gradually decreases. It decays to 0 at 0.075 s. The vibration amplitude values of measurement points B2 to B4 also follow this pattern. It can be seen that although STFT has advantages in continuous spectrum expression, wavelet transform has multi-scale time–frequency resolution capabilities and performs better in identifying non-steady sudden changes and high-frequency impact responses, such as explosion vibrations in flameproof enclosures, demonstrating its unique advantages for flameproof enclosure vibration analysis.

### 7.2.2. Visualization of Vibration in Flameproof Enclosures

The vibration velocity signal is converted into a displacement signal using numerical integration methods, as shown in Figure 15a. However, during the integration process, there is interference from trend terms. The least squares method is used to remove the trend terms. The signal changes exhibit more symmetrical characteristics, thereby more accurately reflecting the true displacement characteristics of the explosion-proof enclosure during the vibration process, as shown in Figure 15b.

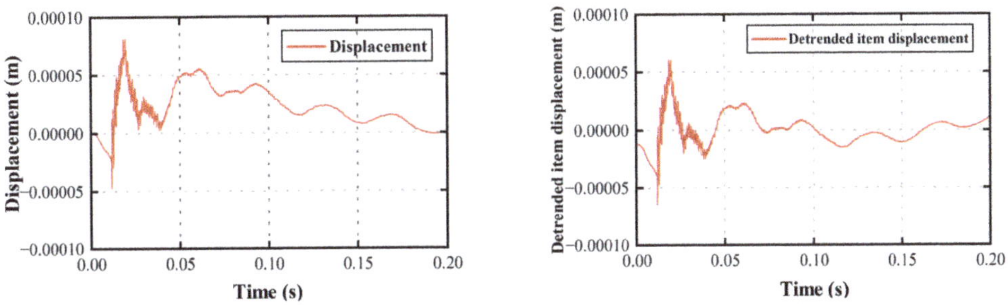


(a) (b)

**Figure 15.** Displacement curve of measurement points on flameproof enclosures. (**a**) Displacement after integration. (**b**) Elimination of trend displacement.

By aligning the time domain of each measurement point, a single-sided displacement field of the enclosure was constructed to achieve preliminary visualization of the vibration. The displacement fields at four different times are shown in Figure 16.

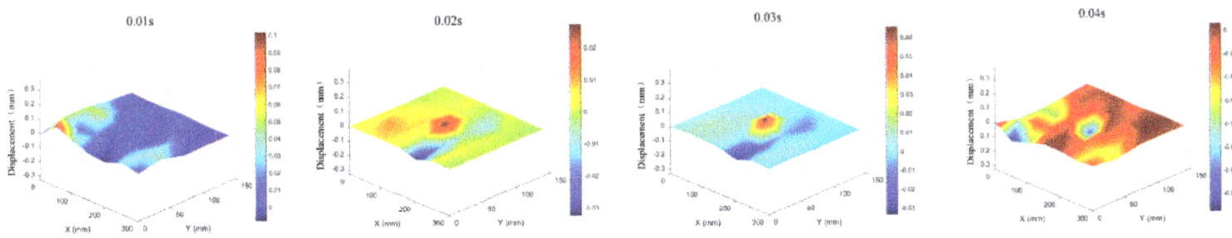


(**a**) 0.01 s (**b**) 0.02 s (**c**) 0.03 s (**d**) 0.04 s

**Figure 16.** Visualization of displacement fields at different times.

Figure 16a shows the displacement field of the shell at 0.01 s after impact, which is the state just before large-scale deformation occurs. Figure 16b shows the displacement field of the shell surface at 0.02 s after impact, at which point the shell has already undergone significant deformation. Figure 16c shows the deformation state of the shell at 0.03 s, with larger displacement than at 0.02 s. Figure 16d shows the deformation state of the shell surface at 0.04 s. Since the shell deformation is in a back-and-forth oscillating state, the displacement field at this time is opposite to that at 0.03 s and more intense. Subsequently, the displacement field oscillates and decreases with time until it reaches zero.

The dynamic changes in the displacement field provided a comprehensive understanding of the vibration characteristics of coal mine flameproof enclosures from point to surface.

### *7.3. Impact Load Inversion Verification*

To verify the feasibility of the experimental platform and the accuracy of the impact load inversion model, the following dual verification scheme was designed:

(1) Platform impact simulation capability verification

The center impact measurement point B4 was selected, and the pressure sensor measured the impact load time history of the experimental platform. The results were compared with the gas explosion impact pressure data from the actual explosion test. The actual explosion test platform is shown in Figure 17a. The comparison of the results is shown in Figure 17b. The various errors are shown in Table 3. All errors are within 5%, indicating that the experimental platform can simulate a gas explosion impact.

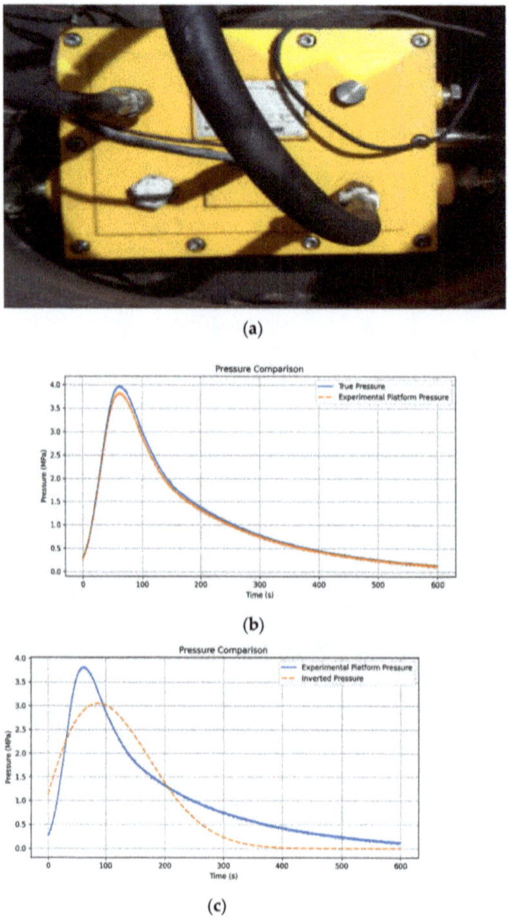


**Figure 17.** Experimental verification. (**a**) Explosion test platform. (**b**) Comparison of actual pressure and platform pressure. (**c**) Comparison of platform pressure and inversion pressure.

**Table 3.** Platform measurement of impact load and error in actual pressure resistance test load.

| Indicator Name | MSE | Peak Error | Ascent Rate Error | Attenuation Rate Error |
|---|---|---|---|---|
| Error | 0.039 | 1.5% | 4.1% | 3.9% |

(2) Verification of the accuracy of the impact load inversion model

Based on the platform feasibility verification, in order to further evaluate the accuracy of the inversion model, the inversion model obtained the impact force inversion results according to the vibration response signals of the flameproof enclosures measured by laser Doppler. Figure 17c above shows the inverted impact pressure and the measured impact pressure curve of the experimental platform. The error between the inversion impact force and the known actual impact force is shown in the Table 4 below. All errors are within 15%, and the model has good inversion accuracy, indicating that the laser Doppler measurement method can replace traditional pressure measurement.

**Table 4.** Inverse impact load and platform measurement load error.

| Indicator Name | MSE | Peak Error | Ascent Rate Error | Attenuation Rate Error |
|---|---|---|---|---|
| Error | 0.183 | 11.33% | 13.12% | 13.89% |

By comparing the above 7.2 laser Doppler analysis data with the actual explosion test pressure data in Figure 14, it can be seen that traditional pressure measurement can only obtain the time history of the impact force but cannot determine the dynamic response of the shell. The laser Doppler measurement method not only provides richer data, but also characterizes the dynamic response characteristics of the surface of flameproof enclosures, providing new technical means for measuring coal mine flameproof enclosures and providing data support for the failure mode and analysis, failure prevention, design and safety performance evaluation of flameproof enclosures in the future.

## 8. Conclusions

In summary, a gas explosion propagation and flameproof enclosure detonation transmission model was established, and a laser Doppler measurement method for flameproof enclosures was proposed to address issues with traditional pressure measurement methods. Through stepwise vibration analysis, the frequency and energy distribution patterns of explosion shock waves were revealed, offering enhanced data richness and substitutability. The main conclusions are as follows:

(1) A model for gas explosion propagation inside flameproof enclosures was established. The single pulse response function was improved by introducing the material and structural characteristics of flameproof enclosures, and a detonation transmission response model for flameproof enclosures was established, providing a theoretical basis for the propagation laws and detonation transmission characteristics of gas explosion shock waves in coal mine flameproof equipment.

(2) Laser Doppler replaces traditional explosion test pressure measurement, using direct measurement and global measurement point layout to accurately capture the dynamic response of flameproof enclosures, solving the installation difficulties of traditional contact measurement.

(3) FFT, STFT, and DWT were used to analyze the vibration characteristics of the surface of flameproof enclosures. The experimental results show that the FFT method can be used to obtain the global time-domain distribution of vibrations, while the STFT can be used to further analyze the changes in the main frequency and other significant frequencies over time. The DWT can be used to further analyze each frequency band's amplitude to a specific time through multi-scale analysis, gradually revealing the vibration characteristics of flameproof enclosures in multiple dimensions and scales.

(4) The measured values of the experimental platform and the actual explosion test impact pressure error index are both less than 5%, and the inversion load errors are all within 15%, which verifies the feasibility of the experimental platform, the accuracy of the model inversion, and the good substitutability of the laser Doppler measurement method.

(5) Vibration analysis and comparison with traditional explosion test data solved the problem of traditional pressure measurement data being single and unable to characterize the dynamic response of the enclosure. The laser Doppler measurement method provided new technical support for measuring and analyzing flameproof enclosure failure modes.

**Author Contributions:** X.X.: writing—review and editing, validation, software, resources, formal analysis. J.Q.: writing—original draft, methodology, investigation, formal analysis, data curation. H.Z.: writing—review and editing, supervision, resources, project administration, conceptualization. W.Y.: validation, software, resources. H.W.: software, resources, methodology, investigation. F.C.: project administration and conceptualization. All authors have read and agreed to the published version of the manuscript.

**Funding:** This research was funded by the National Key Research Development Program Young Scientists Project of China, 2022YFF0605300, and by the, Shaanxi Provincial Department of Education serves local special—industrialization cultivation project, 23JC048.

**Institutional Review Board Statement:** Not applicable.

**Informed Consent Statement:** Not applicable.

**Data Availability Statement:** The supported anchor-rod dataset presented in this article is not publicly available due to the data being part of ongoing research.

**Conflicts of Interest:** The authors declare no conflict of interest.

## References

1. Liu, H. Laying a solid foundation and taking multiple measures to shape a new trend of safe development in mines. *China Coal Ind.* **2025**, 56–57.
2. Zhang, Y. Analysis of Explosion Pressure and Its Influencing Factors of Flameproof Shell. *Colliery Mech. Electr. Technol.* **2023**, *44*, 67–70.
3. Chen, F. Influence Factors of Reference Pressure Measurement in Pressure Test of Flameproof Shell. *Saf. Coal Mines* **2019**, 232–234. [CrossRef]
4. Deffet, L.; Brinkley, S.R.; Laderman, A.J.; Urtiew, P.A.; Oppenheim, A.K. On the Generation of a Shock Wave by Flame in an Explosive Gas. In *Symposium (International) on Combustion*; Elsevier: Amsterdam, The Netherlands, 1963; pp. 265–274.
5. Kundu, S.; Zanganeh, J.; Moghtaderi, B. A review on understanding explosions from methane–air mixture. *J. Loss Prev. Process Ind.* **2016**, *40*, 507–523. [CrossRef]
6. Ogle, R.A. Explosion hazard analysis for an enclosure partially filled with a flammable gas. *Process Saf. Prog.* **1999**, *18*, 170–177. [CrossRef]
7. Ma, H.; Long, Y.; Li, X.; Zhong, M.; Wu, J.; Zhou, Y. Study on Vibration Characteristics of Natural Gas Pipeline Explosion Based on Improved MP-WVD Algorithm. *Shock. Vib.* **2018**, *2018*, 8969675. [CrossRef]
8. Liang, Y.; Wang, L.; Luo, H.; Fen, W.; Tian, F. Computational model of reaction kinetic for gas explosion in constant volume combustion reactor. *J. China Coal Soc.* **2015**, *40*, 1853–1858.
9. Li, H. Pressure Test System of Flameproof Enclosure and Analysis of Influencing Factors. *Electr. Explos. Prot.* **2020**, 36–38. [CrossRef]
10. Zhou, W. Method to Test Explosion Pressure of Flameproof Enclosure and Analysis on Influence Factors. *Coal Sci. Technol.* **2012**, *40*, 91–94.
11. An, P. Pressure Testing method on Flameproof Enclosure of Electrical Equipment. *Electr. Explos. Prot.* **2010**, 38–40. [CrossRef]
12. Liu, H.L.; Shao, L.; Zuo, H.B.; Liu, N. Application of Laser Doppler Technique in Measurement of Explosive Container Based on Frequency Correction. *Appl. Mech. Mater.* **2013**, *391*, 497–500. [CrossRef]
13. Li, M. Discussion on the Main Problems and Precautions in the Processing of Mining Flameproof Enclosure. *Electr. Explos.* **2025**, 20–23. [CrossRef]
14. Xiao, T.; Shi, F.; Yi, X.; Chen, X. Research on vibration characteristics of electric spindle of milling machine based on laser vibration measurement technology. *Chin. J. Constr. Mach.* **2024**, *22*, 191–195.
15. Bai, X.; Guo, P.; Chen, S.; Zhang, Y.; Chen, H.; Bu, Z. Simulation in the Time Domain and Time-Frequency Analysis for Coherent Doppler Wind Lidar. *Chin. J. Laser* **2015**, *42*, 0114003.
16. Labelle, L.; Roozen, N.B.; Vandenbroeck, J.; Akasaka, S.; Glorieux, C. Elastic characterization of polymer fibers by laser Doppler vibrometry. *Opt. Lasers Eng.* **2017**, *99*, 88–97. [CrossRef]
17. Taylor, G.I. The formation of a blast wave by a very intense explosion I. Theoretical discussion. *Proc. R. Soc. Lond. A* **1950**, *201*, 159–174.
18. Zhang, W.; Zou, Y.; Jia, Y. Influence of filling ratio on gas explosion in flameproof chamber. *Saf. Coal Mines* **2023**, *54*, 229–233.
19. Briggs, M.E.; Hill, L.; Hull, L.; Shimas, M. *Application and Principles of Photon-Doppler Velocimetry for Explosives Testing (No. LA-UR-10-01427; LA-UR-10-1427)*; Los Alamos National Laboratory (LANL): Los Alamos, NM, USA, 2010.

20. Nian, G.; Gu, X.; Cheng, F.; Zeng, X. The Research on Vibration Characteristic of Cylindrical Explosive Vessel with Flat End Plat. *J. Proj. Rocket. Missiles Guid.* **2011**, *31*, 105–108.
21. Wang, K.; Pan, H.; Zhang, T.; Wang, H. Experimental study on the radial vibration characteristics of a coal briquette in each stage of its life cycle under the action of $CO_2$ gas explosion. *Fuel* **2022**, *320*, 123922. [CrossRef]
22. Huang, D.; Cui, S.; Li, X. Wavelet packet analysis of blasting vibration signal of mountain tunnel. *Soil Dyn. Earthq. Eng.* **2019**, *117*, 72–80. [CrossRef]
23. Goutaudier, D.; Gendre, D.; Kehr-Candille, V.; Ohayon, R. Single-sensor approach for impact localization and force reconstruction by using discriminating vibration modes. *Mech. Syst. Signal Process.* **2020**, *138*, 106534. [CrossRef]
24. Li, S. On Explosion Pressure in Explosion-Proof Enclosures and Its Influencing Factors. *Saf. Coal Mines* **2003**, 101–103. [CrossRef]

applied sciences

MDPI

Article

# Application of UAVs to Support Blast Design for Flyrock Mitigation: A Case Study from a Basalt Quarry

**Józef Pyra [1,*] and Tomasz Żołądek [2]**

1 Faculty of Civil Engineering and Resource Management, AGH University of Krakow, 30-059 Kraków, Poland
2 Holcim Kruszywa Sp. z o.o., 02-305 Warszawa, Poland; tomasz.zoladek@holcim.com
* Correspondence: pyra@agh.edu.pl

**Abstract:** Blasting operations in surface mining pose a risk of flyrock, which is a critical safety concern for both personnel and infrastructure. This study presents the use of unmanned aerial vehicles (UAVs) and photogrammetric techniques to improve the accuracy of blast design, particularly in relation to controlling burden values and reducing flyrock. The research was conducted in a basalt quarry in Lower Silesia, where high rock fracturing complicated conventional blast planning. A DJI Mavic 3 Enterprise UAV was used to capture high-resolution aerial imagery, and 3D models were created using Strayos software. These models enabled precise analysis of bench face geometry and burden distribution with centimeter-level accuracy. The results showed a significant improvement in identifying zones with improper burden values and allowed for real-time corrections in blasthole design. Despite a ten-fold reduction in the number of images used, no loss in model quality was observed. UAV-based surveys followed software-recommended flight paths, and the application of this methodology reduced the flyrock range by an average of 42% near sensitive areas. This approach demonstrates the operational benefits and enhanced safety potential of integrating UAV-based photogrammetry into blasting design workflows.

**Keywords:** UAV; flyrock; blast design; photogrammetry; burden optimization; surface mining

## 1. Introduction

The use of explosives for rock fragmentation remains the primary method of extraction in surface mining operations for rock raw materials. The application of blasting agents allows for effective resource exploitation; however, it also poses significant environmental and safety risks. One such hazard is flyrock, which presents a danger to mine personnel, equipment, and surrounding infrastructure. Although this risk is critical, it remains difficult to accurately assess.

During blasting operations, flyrock can be ejected over considerable distances. As a result of the detonation of explosives within blastholes, rock fragments are displaced, typically causing horizontal movement of the entire blasted rock mass. When the blasting sequence is carefully designed and properly executed, the process should not pose substantial risks. Nevertheless, in addition to the expected displacement, flyrock may be expelled unpredictably, creating safety hazards for personnel, equipment, and structures within its range. Flyrock refers to any rock fragment projected into an unintended area [1].

The extent of the flyrock zone depends on a number of factors that can be quantitatively determined under controlled conditions through measurements. Identifying and monitoring these parameters enables the implementation of measures to mitigate flyrock hazards. The key contributing factors include [2–14]:

- Burden and spacing of blastholes,
- Specific charge (powder factor),
- Inadequately characterized geological structure,
- Blasthole deviation,
- Length and quality of stemming,
- Presence of back fractures,
- Initiation sequence,
- Human error.

In most cases, these factors are interrelated; for instance, blasthole deviation affects burden and spacing, which in turn can locally increase the powder factor. For this reason, the literature offers a range of empirical methods and formulas for estimating the maximum flyrock range, such as the Swedish method [15,16], the American method (Roth) [17], Pokrowski's method [18–21], and those based on explosive energy indices. Traditional empirical formulas and statistical regression models often proved inadequate for accurate flyrock prediction, showing low predictive capacity. In the past decade, a wide range of artificial intelligence (AI) and machine learning (ML) techniques have been applied to this problem, generally achieving much higher accuracy than empirical approaches. This section reviews these techniques, grouped by model type, and discusses their predictive performance, practical field applications, and noted limitations (e.g., data scope and input sensitivity).

### 1.1. *Artificial Neural Network (ANN) Approaches*

Artificial neural networks have been extensively used to model blast-induced flyrock. Early studies demonstrated that ANN models can significantly outperform conventional statistical methods. For example, Trivedi et al. [22] applied back-propagation neural networks (BPNNs) to predict flyrock distance in Indian limestone quarries and found the ANN's predictions to be far more accurate than multivariate regression analysis (MVRA) on the same data. In a follow-up study, Trivedi et al. [22] simultaneously predicted flyrock distance and rock fragmentation using a BPNN. The network yielded substantially lower root-mean-square error (RMSE) and mean absolute error (MAE) than MVRA, confirming the ANN's superior accuracy. Monjezi et al. [23] trained a three-layer ANN on 192 blast records from Sangan iron mine (Iran) and likewise observed reliable predictions, identifying key influential inputs, such as powder factor, stemming length, and charge-per-delay. Notably, by using the ANN's insights to adjust blast design (e.g., reducing the charge per delay), the mine was able to reduce observed flyrock distances from 165 m to 25 m, illustrating a direct practical benefit of ANN modeling in the field. Overall, these studies established that well-trained ANNs can capture the highly nonlinear relationships governing flyrock and serve as a better predictive tool than regression models [22,23], with typical reported coefficients of determination ($R^2$) above 0.90 in training and validation phases.

Researchers have also explored optimizations and variants of ANN to further improve accuracy. Monjezi et al. [24] introduced a neuro-genetic model, using a genetic algorithm (GA) to optimize an ANN's architecture and parameters. This hybrid ANN-GA model (sometimes called a genetic algorithm optimized neural network) achieved higher predictive efficiency than a standard neural net on the same dataset. Similarly, other works integrated evolutionary algorithms with ANN training (see the Hybrid Models Section below). In general, ANNs have proven highly effective for flyrock prediction, consistently outperforming empirical equations and providing actionable insights (e.g., sensitivity analysis from ANN weights often highlights the most critical factors controlling flyrock throw). However, one limitation noted is that an ANN's performance and derived insights are only as good as the training data—if the data come from a single site or do not span a wide range of conditions, the network may not generalize well to other mines (a point revisited later). To mitigate this, some studies trained ANNs on multi-mine datasets: for instance, Trivedi et al. [25] pooled data from four different limestone mines (125 blast events) to train an ANN and still achieved a high prediction accuracy (the ANN's outputs correlated much better with observed flyrock distances than any regression fit). This indicates that with sufficient and diverse data, ANN models can be robust tools for flyrock forecasting in practice.

### 1.2. Support Vector Machines and SVR Models

Support vector machine (SVM) algorithms have also been successfully applied to flyrock prediction, especially in cases with complex, nonlinear parameter interactions. SVM-based models generally outperform linear regression and even perform favorably against ANNs in certain cases. Amini et al. [26] tested SVM for predicting flyrock at the Sungun copper mine and found it "faster and more precise" than an ANN model on the same data. Khandelwal and Monjezi [27] developed an SVM regression model for flyrock prediction and compared it to MVRA. The SVM achieved a coefficient of determination $R^2 \approx 0.95$, dramatically higher than the 0.44 achieved by multiple regression, and reduced prediction error (MAE) to about 3.1 m versus 7.7 m for MVRA [27]. These results demonstrate SVM's ability to capture nonlinear patterns that simple regressions miss, yielding much improved accuracy [28]. The success of SVM is further reflected in its variants like support vector regression (SVR). Guo et al. [29] proposed an ensemble of SVR models (with different kernels) combined via a regularized linear meta-learner. This stacked SVR model achieved excellent performance (test $R^2 = 0.993$ and RMSE ~ 3.74 m), significantly better than any single SVR (whose $R^2$ ranged from 0.92 to 0.97). Likewise, Li et al. [30] utilized a hybrid SVR optimized by the Harris Hawks algorithm, which attained test $R^2 \approx 0.97$ and RMSE ~9.7 m, outperforming not only a standard SVR but also other machine learners like ANN and extreme learning machine (ELM) on the same dataset. These studies confirm that SVM-based approaches are highly capable for flyrock modeling, often matching or exceeding ANN accuracy. A practical consideration with SVM/SVR models is the need to tune kernel and regularization parameters. Advanced optimization techniques (genetic algorithms, particle swarm, etc.) have been applied to automate this tuning with great success in flyrock problems (e.g., yielding the $R^2$ ~ 0.97 noted above for hybrid SVR [31]). As with ANNs, SVM models benefit from adequate training data—in fact, Khandelwal and Monjezi's SVM was trained on the same small dataset (from one mine) as their regression yet generalized far better [27]. This robustness suggests SVMs can be a powerful tool even when data are limited or noisy, provided the model parameters are properly optimized.

### 1.3. Fuzzy Logic and Neuro-Fuzzy Systems

Fuzzy inference systems provide an alternative AI approach that has been applied to handle the uncertainty and nonlinearity in flyrock phenomena. In a pure fuzzy logic model, expert knowledge or data-driven rules map input parameters to flyrock distance using linguistic fuzzy rules. Rezaei et al. [32] developed a Mamdani-type fuzzy model for an iron mine in Iran and reported its performance was "much better" than a conventional statistical regression on the same 490-blast dataset. The fuzzy model yielded more accurate flyrock predictions and was able to incorporate vague relationships between inputs and outputs that crisp equations could not capture. Similarly, Ghasemi et al. [33] compared an ANN and a fuzzy inference system for flyrock prediction in the Sungun mine—both models proved useful, but the fuzzy model slightly outperformed the ANN in predictive performance. These outcomes suggest that fuzzy systems, which do not require an explicit functional form, can effectively model the complex interplay of blasting factors. A noted advantage is that fuzzy models provide interpretable if–then rules, which can be meaningful for engineers (e.g., a rule might relate "high powder factor and short stemming" to "long flyrock distance" in linguistic terms).

To leverage the strengths of both neural networks and fuzzy logic, researchers have employed the Adaptive Neuro-Fuzzy Inference System (ANFIS) for flyrock prediction. ANFIS is essentially a neural network that implements fuzzy inference rules and can learn rule parameters from data. Hudaverdi and Agan [34] applied ANFIS (alongside a standard ANN and Gaussian process regression) to data from a Turkish quarry, using important input features selected via statistical analysis. The ANFIS model proved most accurate, with a mean absolute error (MAE) of only ~5.36 m—outperforming the back-propagation ANN in that study. In fact, the intelligent models (ANN, ANFIS, and GPR) all predicted flyrock with <6 m error (mean percentage errors < 10%), but ANFIS slightly edged out the others [35]. In a related study, Hudaverdi [36] demonstrated that simplifying the input set can improve neuro-fuzzy models: using variable selection procedures, he developed two ANFIS models with only 3–4 inputs each, which achieved median errors < 5 m and MAPE < 8%. These ANFIS models were not only accurate but also less complex, underlining the benefit of focusing on the most predictive parameters [34]. Another advantage of neuro-fuzzy approaches is their ability to highlight influential inputs. For instance, Hudaverdi's studies identified the burden-to-hole diameter ratio and specific charge (powder factor) as consistently significant features for flyrock prediction.

Recent research has pushed the performance of ANFIS even further by integrating metaheuristic optimizers (essentially creating hybrid neuro-fuzzy models). Nguyen et al. [37] introduced an ANFIS enhanced by a Lévy-flight Jaya algorithm (a modern swarm intelligence optimizer) to predict flyrock. This ANFIS–LJ model achieved $R^2 \approx 0.981$ on the test set (with MAE ~ 1.42 m), and when validated on 13 new production blasts it reached $R^2 = 0.988$ with only ~1.3 m MAE. Such high accuracy (within ~1–2 m of observed values) is exemplary for flyrock studies and was obtained using 5-fold cross-validation to ensure the model generalizes well. In a complementary study, another team led by Nguyen [38] tested ANFIS models optimized by different algorithms (adaptive differential evolution, genetic algorithm, fireworks, and bee colony). They found the JADE-ANFIS (ANFIS optimized by JADE) to be the best, with an overall prediction accuracy of ~97.8% and very low error (MAPE = 1.1%). Competing models in that comparison achieved 89–96% accuracy (MAPE up to 3%), so the JADE-ANFIS was clearly superior [37,38]. Interestingly, the sensitivity analyses from these neuro-fuzzy models often concur with earlier findings. For example, the JADE-ANFIS study noted that stemming length was the most critical factor influencing flyrock distance, which aligns with other works that frequently identify stemming and charge-related variables as key drivers (these factors essentially control the confinement of explosive energy). In summary, fuzzy and neuro-fuzzy approaches (especially when boosted with learning algorithms) have attained excellent predictive performance in flyrock modeling. They combine accuracy with interpretability, making them attractive for practical implementation—though like all data-driven models, their rules are valid within the range of training data and should be applied cautiously beyond that range [37,38].

### *1.4. Hybrid Metaheuristic and Advanced Machine Learning Models*

Given the inherent complexity of blasting processes, many researchers have turned to hybrid models that integrate machine learning with evolutionary or swarm optimization techniques. The motivation is two-fold: (1) to optimize model parameters/architecture for better accuracy, and (2) to possibly find optimal blasting designs that minimize flyrock. A common strategy is coupling an ANN or other learner with an optimizer to fine-tune its weights or hyperparameters. For example, Armaghani et al. [39] combined particle swarm optimization with an ANN (PSO-ANN) to improve convergence and avoid local minima in training. Applied to 44 blast cases in Malaysian quarries, the PSO-ANN was able to predict flyrock distance with "a high degree of accuracy", better than a standard back-propagation ANN, and the authors reported that powder factor and charge per delay were the most influential inputs on the model. Marto et al. [40] used a different evolutionary algorithm (imperialist competitive algorithm, ICA) to optimize an ANN and showed that this ICA-ANN hybrid outperformed a conventional ANN as well as two newly developed empirical equations on a dataset of 113 blasts. Likewise, Monjezi et al. [24] had earlier demonstrated a GA-optimized ANN (neuro-genetic model) with superior performance to an unoptimized network, underscoring the value of such hybridization.

Beyond ANN enhancements, researchers have also explored hybrid models with alternative learning algorithms. Extreme learning machine (ELM) is a fast neural network variant, and Murlidhar et al. [41] optimized it using biogeography-based optimization (BBO). Their BBO-ELM achieved a testing $R^2 \approx 0.94$, compared to $R^2 \approx 0.79$ for a plain ELM on the same data (262 blast samples). In fact, the hybrid BBO-ELM slightly outperformed a PSO-tuned ELM as well (which had $R^2 \approx 0.93$), making BBO-ELM the top model in that study [40]. A recent study by Bhatawdekar et al. [42] introduced an equilibrium optimizer coupled ELM (EO-ELM) and benchmarked it against PSO-ELM and PSO-ANN models. Using 114 blast records for training/testing, the EO-ELM attained an impressive $R^2 \approx 0.97$ (RMSE ~ 32 m) on the test set, whereas the PSO-ANN yielded $R^2 \sim 0.87$ (RMSE ~ 64 m) and PSO-ELM ~ 0.88 (RMSE ~ 49 m). This huge accuracy gain (EO-ELM reducing error by ~50% relative to the ANN model) demonstrates how modern metaheuristic optimizers can dramatically improve model performance [41]. It is worth noting that such hybrids also often include a sensitivity analysis component; for instance, Bhatawdekar et al. [42] identified powder factor and blastability index as the most sensitive inputs (each with a normalized sensitivity of 0.98) in their EO-ELM model.

Another category of hybrid models aims to combine predictive modeling with optimization of blasting parameters. Zhou et al. [43] took a two-stage approach, first training an ANN to predict flyrock, then using a PSO algorithm to search for optimal blast design parameters that minimize the predicted flyrock. In a case study, they showed that by adjusting the pattern (within practical limits), the anticipated flyrock distance could be reduced to ~34 m under ideal conditions (or ~109 m under typical field conditions), whereas the baseline scenario had much greater flyrock throw. This highlights the potential of AI not just to predict outcomes, but to actively improve blasting practices. Similarly, Saghatforoush et al. [44] combined an ANN with an ant colony optimization (ACO) procedure. Their trained network predicted flyrock and back-break with high accuracy (for the ANN alone, Ea $\approx$ 0.014 and RMSE $\approx$ 0.063 in scaled units), and then the ACO algorithm was used to optimize the blast layout. By implementing the ACO-derived design changes, the mine could potentially achieve a 61% reduction in flyrock distance (and 58% reduction in back-break) according to the model's estimations. These studies demonstrate a practical extension of predictive models into prescriptive tools for blast optimization.

### 1.5. Tree-Based Ensembles and Other Methods

In addition to ANN and SVM, various tree-based machine learning methods have been applied to flyrock prediction, often yielding performance on par with the best ANN/SVM models. Decision tree ensembles like random forests and gradient boosting machines can naturally handle nonlinear interactions and have the advantage of built-in variable importance analysis. Yari et al. [45] evaluated four tree-based models—decision tree (DT), random forest (RF), extreme gradient boosting (XGBoost), and adaptive boosting (AdaBoost)—for a mine dataset. All four techniques provided accurate predictions ($R^2 > 0.95$), with AdaBoost being the most precise ($R^2 \approx 0.99$ for both training and testing). Such near-perfect fits indicate that the models captured almost all variance in the data, though the authors cautioned that burden and spacing had the least impact on the model output while powder factor had the highest influence. Hasanipanah et al. [46] similarly found that a regression tree model outperformed a multiple linear regression for flyrock distance prediction in a Malaysian quarry, based on metrics like median absolute error and $R^2$. The regression tree was more precise, and a sensitivity check again pointed to powder factor as the dominant factor affecting flyrock throw. These results reinforce that ensemble or rule-based tree models can be highly effective. One advantage is that they are relatively straightforward to apply and interpret; for example, a regression tree yields if–then rules similar to expert reasoning (e.g., "if stemming < X and charge > Y, expect flyrock > Z"). Indeed, Hudaverdi and Akyildiz [35] proposed a novel classification approach using multiple discriminant analysis (MDA) to categorize flyrock throw severity in a quarry. Instead of predicting an exact distance, their model classifies blasts into risk categories (no flyrock, moderate flyrock, and severe flyrock) based on blast parameters, aligned with regulatory safety criteria. The MDA-based model successfully classified flyrock outcomes for test blasts and has the benefit of simplicity—it does not require specialized software or extensive training to use. The authors even produced a territorial map for the quarry so that engineers can easily determine the expected flyrock severity class from blast design parameters [35]. This approach, while not yielding a numeric prediction, is very practical for field use as a quick risk assessment tool.

Ensemble learning has also been used in creative ways. Guo et al. [29] (as mentioned earlier) combined multiple SVRs and a GLM-net aggregator to exploit diverse models' strengths, achieving $R^2 = 0.99$-level accuracy. Nguyen et al. [47] took an ensemble of ANNs approach: they trained five different ANN models and then ensembled their predictions to train a meta-ANN (referred to as EANNs). This stacked model attained $R^2 \approx 0.986$ (with very low error, e.g., RMSE ~4.3 m) on the flyrock dataset, outperforming any single ANN and even a benchmark ANN of identical structure. Such ensemble methods benefit from averaging out individual models' errors, often yielding more stable and accurate predictions. However, a potential downside is the complexity of implementation and the need for more data to train multiple models. In practice, these approaches show that combining models can push the predictive accuracy to extremely high levels—albeit with the caveat that models trained and tested on the same site's data (especially if limited in size) might overfit to that context. For example, an AdaBoost or EANN model achieving ~99% accuracy on a given quarry's blasts [45,47] may not maintain that level on a different mine without retraining. Recognizing this, some recent studies explicitly tested generalization: Zhang et al. [48] optimized a random forest using two novel algorithms (light spectrum optimizer and puma optimizer) and evaluated it on two separate quarry datasets (one from Malaysia and one from Istanbul). Both optimized RF models performed very well (test $R^2 \approx 0.96$–$0.97$ and RMSE on the order of 6–16 m) and significantly outperformed an unoptimized RF. More importantly, when the model trained on one site was validated on a completely different site (an "engineering case" not seen in training), it still achieved a respectable $R^2$ ~0.88. This indicates strong generalization capacity, likely due to the robustness conferred by optimization and the use of diverse training data. Moreover, by using SHAP (Shapley Additive Explanations) analysis, Zhang et al. [48] identified the critical factors for each site's model—for instance, hole diameter and maximum charge per delay were most important in the Malaysian quarry, while specific charge and the spacing-to-burden ratio mattered most in the Istanbul site. Such insights are valuable for practitioners to understand regional differences and to focus on controlling the right parameters in each operation.

### 1.6. Practical Applications and Common Findings

The breadth of studies above shows a consensus that AI/ML techniques can predict blast-induced flyrock with far greater accuracy than older empirical methods. Models such as ANN, SVM, tree ensembles, and hybrids routinely achieve $R^2$ values in the 0.90–0.99 range, with RMSE on the order of only a few meters to a few tens of meters (depending on dataset scale)—a substantial improvement over classical formulas that often had $R^2$ values well below 0.5 in comparative tests [27]. Notably, many different model types (from simple fuzzy systems [32] to deep ensembles [29]) have proven successful, suggesting that as long as the major influencing factors are captured, a variety of AI models can map them to flyrock outcomes effectively. Across the literature, there is strong agreement on which input parameters are most influential in flyrock distance. Nearly all studies highlight some measure of explosive charge and rock confinement as key drivers, e.g., powder factor (explosive per rock volume) emerges as a top factor in numerous sensitivity analyses, while maximum charge per delay and stemming length are also repeatedly identified as critical inputs. In contrast, parameters like burden, spacing, or rock density often rank lower in influence, although their effects are not negligible and can be site-specific. The general finding is that flyrock risk is most governed by how much explosive energy is released and how well that energy is confined—which aligns with practical blasting experience. This gives confidence that AI models are picking up meaningful physical relationships, not just statistical coincidences.

From a practical standpoint, several research works have moved beyond pure prediction into field implementation and validation. For instance, Monjezi et al. [23] used their ANN model's recommendations to adjust blasting in an active mine, achieving drastic flyrock reduction (as noted, from 165 m down to 25 m). Armaghani et al. [28] built a regression-based model and then performed a Monte Carlo simulation to establish a probabilistic safe distance for flyrock; interestingly, the simulated mean flyrock (236.3 m) closely matched the observed mean (238.6 m), giving credence to using such models for defining blast clearance zones. However, they cautioned that models should be used only under conditions similar to those they were developed for and directly applying them elsewhere is not recommended. This caution about generalizability is echoed by many authors. Flyrock is a highly site-specific phenomenon, influenced by local geology and blast practices, so an AI model trained on one mine's data may need retraining or recalibration for another site. The risk of overfitting looms, especially when models boast extremely high $R^2$ on limited datasets (e.g., an AdaBoost or ensemble ANN that fits one mine's blasts almost perfectly [41,45]). Without rigorous cross-validation or external testing, such performance might be optimistic. To address this, recent studies increasingly adopt techniques like k-fold cross-validation (as in [37]) and multi-site data fusion (as in 48]) to ensure models generalize. Some works also explicitly exclude less relevant inputs to reduce noise—e.g., using random forest or statistical selection to prune variables—which can improve model robustness when moving to new conditions [36,49,50].

Another important practical aspect is data acquisition for model development and use. Traditional methods of measuring flyrock (manual spotters and video analysis from a single camera) can be error-prone (e.g., due to parallax) and dangerous. The adoption of modern monitoring technologies like unmanned aerial vehicles (UAVs) and high-speed videography is making it easier to collect accurate flyrock data for both model training and operational prediction. Lawal et al. [51] performed a comprehensive bibliometric review and noted that conventional means of locating flyrock fragments suffer from biases and errors, and they demonstrated a field trial where UAV-mounted cameras captured flyrock trajectories more reliably. The UAV-based method showed clear advantages in capturing the full flight of rock fragments, which were then analyzed with motion analysis software, and a soft computing model was used to predict the recorded flyrock distances [51]. Similarly, Mishra et al. [52] used high-speed cameras onsite (including UAV-mounted) to record flyrock from blasts at five mines, generating a rich dataset of launch velocities and distances. They then trained several ML models, deploying the best one (an extremely randomized trees regressor) in a prototype software tool to simulate flyrock trajectories for new blasts. This integration of AI models into user-friendly interfaces allows engineers to input planned blast parameters and visualize the predicted flyrock hazard zone, greatly aiding in blast design and risk mitigation. More generally, UAV photogrammetry and remote sensing techniques are increasingly utilized in mining, as they can provide rapid, high-resolution mapping of blast sites and measure outcomes like fragmentation and throw with centimeter-level precision. Studies have shown UAV-based monitoring to be efficient, removing the need for personnel in hazardous areas and improving data quality. For example, Bamford et al. [53,54] used UAVs to continuously monitor open-pit blasts, capturing detailed images for fragmentation analysis and flyrock observation, which offered insights for process improvement that traditional manual monitoring could miss. The proliferation of such technologies, as reviewed by Colomina and Molina [55] and Minh and Dung [56], underscores that UAVs have become an excellent tool in mining for tasks ranging from surveying to blast monitoring. In the context of flyrock prediction, better data acquisition means more reliable models and the possibility of near-real-time model updating and validation during blasting operations.

In conclusion, the application of AI and machine learning to flyrock prediction in open-pit blasting has yielded marked improvements in predictive accuracy and practical risk management. Diverse models—ANNs, SVMs, fuzzy systems, tree ensembles, and numerous hybrids—have each demonstrated high accuracy (with $R^2$ often in the high 0.9 range and errors reduced by 50–80% compared to older methods) in various case studies [27,29,33,45]. Hybrid models that combine techniques (ANN + GA, ANFIS + PSO, optimized SVR, etc.) tend to achieve the best performance, reflecting a common finding that no single method is universally best, but combining algorithms can harness their complementary strengths for superior results [37,38,41,42,57]. A recurring theme across studies is the identification of crucial parameters: in almost all models, powder factor (or specific charge) and stemming length emerge as primary controls on flyrock distance, whereas parameters like burden, spacing, or rock quality may have secondary effects. This consistency builds confidence in the physical relevance of the models. On the other hand, researchers consistently warn about limitations: models are usually data-driven and thus valid only within the domain of the training data. For instance, Armaghani et al. [28] explicitly state that their site-specific model should not be directly applied to other conditions without recalibration. To improve generalizability, recent works have embraced cross-validation, larger multi-site datasets, and input reduction to avoid overfitting [37,38]. Another concern is data representativeness—since flyrock events are stochastic and sometimes rare (extreme flyrock incidents might not occur in every dataset), ensuring the model has seen enough "bad cases" is important for reliable hazard prediction. Despite these challenges, the trend in the literature is toward increasingly sophisticated and validated models, often coupled with modern monitoring (e.g., UAV and high-speed cameras) for both model input collection and output verification. Many studies have reported successful field implementations—from adjusting blast designs based on model insights to deploying software that can simulate flyrock before a blast is carried out [23,26,35]. These advancements suggest that AI-driven flyrock prediction is maturing from academic research into practical technology [57].

Although this study does not directly apply artificial intelligence methods for flyrock prediction, the comprehensive literature review included in the introduction aims to highlight the dynamic development of this research area and underscore the need for integrating advanced computational tools with geospatial data acquired through UAV-based surveying. The presented case study represents a step toward such integration, demonstrating the potential of precise geometric analysis as a foundation for the future development of predictive models grounded in real-world spatial data.

## 2. Materials and Methods

### 2.1. Geodetic Technologies Supporting Blast Design

The design of a blasting hole pattern that meets safety requirements for ongoing operations must be based on precise geometric measurements of the burden zone to be excavated from the active deposit. The selection of appropriate geometric parameters for the blasthole grid—in correlation with the type of explosive used and the charge configuration—is a key condition for ensuring operational safety while optimizing the utilization of detonation energy from the explosive charges placed in the holes.

In the blast design process, geodetic tools play a critical role, particularly in the open-pit extraction of hard rock raw materials. Among the most commonly used devices are 3D scanners (e.g., laser total stations), which enable accurate mapping of surface topography and bench face geometry. Global navigation satellite systems (GNSSs) are also of significant importance, providing high-precision positioning of blastholes and other critical points within the mining area. Increasingly widespread is the use of UAVs, which, through rapid photogrammetric data acquisition, support the geometric inventory of the excavation site and enable efficient planning of blasting operations.

Geospatial data obtained from these tools are increasingly being integrated directly into specialized blast design software, such as BlastPlan, JKSimBlast, QuarryX, Strayos, or O-Pitblast. These platforms allow for detailed modeling of the blast layout, forecasting of blasting outcomes (e.g., fragmentation, muckpile displacement, and vibration levels), and verification of compliance with applicable safety standards and environmental constraints.

### 2.2. Laser Total Stations for 3D Terrain Modeling

Laser total stations are currently among the most widely used instruments for conducting the types of measurements described above, offering high precision in determining horizontal and vertical angles as well as distances. Classical tacheometry involves polar coordinate-based surveying to determine the planimetric and altimetric position of detail points, using trigonometric leveling. Based on the collected point cloud data, a 3D model of the bench face or its section can be generated in appropriate processing software. Figure 1 presents a sample 3D model visualized as a point cloud, showing the roof, bench face, and floor of the excavation wall along with the planned layout of the blasthole pattern.

In the case of highly irregular bench faces, there is a high probability that an optimal measurement position for the instrument cannot be found—one that would enable complete geometric reconstruction of the entire face. As seen in Figure 1, the perspective from which the scan was conducted limited the ability to capture the full dimensions of individual surfaces—roof, floor, and face. In such cases, supplementary scans must be carried out from additional vantage points to dimension surfaces that were not visible during the initial measurement.

**Figure 1.** Example visualization of a point cloud acquired using a 3D laser scanner.

Another issue involves the coordinate systems required by the instrument operator. While mapping the model in a standard Cartesian coordinate system (x, y, z) is generally unproblematic for commercially available devices, mapping into custom or local coordinate systems requires the establishment of well-defined reference points. These must precisely define geographic length, width, and elevation. Moreover, these reference points must be visible from all scanning perspectives used during the survey of a given area of the excavation site.

These considerations highlight the fact that measurements performed using stationary laser-based instruments may, under certain conditions, prove insufficiently functional. This can directly affect the quality and completeness of the collected data, which are subsequently used to determine the input parameters for blast design.

### *2.3. UAV-Based Photogrammetry: Capabilities and Limitations*

In recent years, a significant increase has been observed across various industries in the use of measurement techniques based on photogrammetry—a scientific and technical discipline that reconstructs real-world dimensions, shapes, and spatial relationships of objects through the analysis of image sequences captured from multiple viewpoints. Particularly notable is the rapid development of close-range photogrammetry using UAV, equipped with specialized cameras that minimize optical distortions, such as chromatic aberration (Figure 2).

This trend is clearly reflected in the mining sector as well. A comprehensive literature review by Minh Dang Tuyet and Nguyen Ba Dung [56] analyzed 113 scientific publications over the past 12 years, documenting the systematic increase in UAV applications across key mining activities. These include mine surveying and inventory (application in surveying mines), design and monitoring of drilling and blasting operations, mine safety and risk management, ground subsidence monitoring, rock slope stability analysis and monitoring, detection of underground coal fires, and erosion detection.

An optimal—though not exclusive—solution for blasting design is the use of aerial photogrammetry. This technique, by virtue of its mobile measurement platform, inherently eliminates issues related to limited perspective or access typical of ground-based scanning. Figure 3 illustrates an example flight path planned for UAV scanning of a selected production bench.

**Figure 2.** UAV conducting a flight over a designated section of the excavation site.

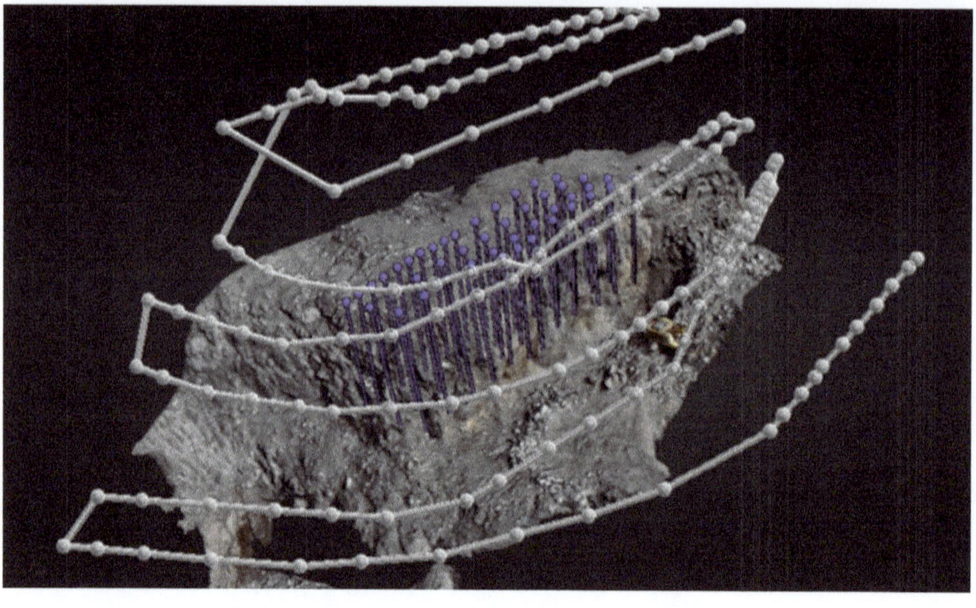

**Figure 3.** Example flight trajectory designed for the UAV.

Key rules to keep in mind before flying a bench [58] are as follows:

- At least two-thirds of the photos should be taken with the camera pointing straight down (nadir images).
- Capture images along flight paths with 70–80% overlap.
- Effective results have been reported at flight altitudes of 75 ft (23 m), 150 ft (46 m), and 250 ft (76 m).
- Always overshoot the mapping area by at least 20 ft (6 m) to avoid stitching or reconstruction issues at the edges.
- Fly the UAV forward with the camera tilted 5° forward; when beginning the next pass, rotate the UAV instead of flying backward.

As previously mentioned, the use of stationary terrestrial laser scanners for single or series-based surveys may present operational limitations due to structural and functional constraints of the device. These limitations often result in extended measurement durations, increasing the risk of incomplete geometric data acquisition, which in turn can compromise the accuracy of blast design.

UAV deployment significantly mitigates problems related to instrument positioning and scanning geometry, thereby improving overall operational efficiency. Photogrammetric techniques based on UAV imagery are typically more functional and offer potentially higher spatial resolution. However, UAV-based measurements also come with certain technical and regulatory limitations. These include high sensitivity to weather conditions (e.g., wind, rain, and fog), limited operational time due to battery capacity, the requirement to maintain stable data transmission with the ground control station, and—under most regulations—the necessity of maintaining visual line of sight (VLOS).

Additionally, photogrammetric data quality can be affected by lighting conditions and surface characteristics—low contrast, homogeneous textures, or reflective surfaces may degrade the final 3D model, requiring supplementary ground-based measurements for validation.

It should be noted that stationary terrestrial laser scanning does not require formal operator licensing. In contrast, UAV operations in public airspace are legally regulated. Operators must obtain UAV pilot certification, which includes theoretical and practical training followed by an official examination. The required qualification level depends on UAV mass and the operational risk category.

Furthermore, effective 13 November 2025, national aviation regulations [59] will mandate liability insurance for all UAV operators. Operating UAVs without valid insurance or in violation of legal regulations may result in substantial financial penalties [59].

## 3. Results and Case Study: UAV-Assisted Blast Design in a Basalt Quarry

### *3.1. Site Description and Context*

The methodology for designing a blasthole pattern using photogrammetric techniques was implemented in one of the basalt quarries located in the Lower Silesian region of Poland. The exploited deposit is characterized by a high degree of fracturing, which significantly complicates rock fragmentation using blasting methods. Accurate determination of the actual and optimal burden values on the working face is critical to ensuring the safety of blasting operations. This must account for the potential risk of excessive flyrock and the necessity to control seismic impacts on the surroundings of the mining site.

An increase in the geometric parameters of the burden directly correlates with a rise in the resistance of the rock mass, which in turn affects both the intensity of seismic vibrations and the degree of rock fragmentation. Precise determination of key parameters, such as blasthole length, angle and direction of inclination, and subdrill depth, is an essential step in blast design. Errors introduced at this stage are difficult if not impossible to correct in subsequent phases of the process, such as selecting the type and mass of explosives, determining the initiation sequence, and setting millisecond delays.

Therefore, it is essential to use measurement and analytical tools that enable continuous verification of the design assumptions throughout all stages of blasting operations from planning to execution.

The presented case study aims to illustrate the scope of challenges involved in designing blasting operations with a focus on controlling the extent of flyrock. Work on implementing and optimizing the use of UAV-based photogrammetry for blasting applications began in late 2023. Since then, each blast series has been designed using actual

3D data obtained from the excavation site. The 3D models obtained from UAV flights were validated using reference measurements conducted at fixed control points within the quarry area, utilizing precise geodetic instruments, such as a total station. This verification confirmed the high geometric accuracy and reliability of the generated models.

### 3.2. UAV Equipment and Photogrammetric Data Acquisition

To define the design assumptions and monitor their implementation, photogrammetric measurement data were collected using a DJI Mavic 3 Enterprise (DJI Sky City, No.55 Xianyuan Road, Nanshan District, Shenzhen, China) unmanned aerial vehicle (Figure 1). This UAV is equipped with a wide-angle camera featuring a 4/3 CMOS sensor. The 20-megapixel sensor includes a mechanical shutter capable of speeds up to 1/2000 s and an image capture interval of just 0.7 s. These features prevent motion blur and enable the operator to conduct photogrammetric missions at flight speeds of up to 15 m/s.

In addition to the wide-angle camera, the DJI Mavic 3 Enterprise is equipped with a 12 MP telephoto lens that enables up to 56× hybrid zoom. This makes it suitable not only for standard photogrammetric missions but also for detailed inspections from considerable distances.

The UAV can remain airborne for up to 45 min, allowing it to cover areas of up to 2 km$^2$ in a single mission. It uses the next-generation DJI O3 Enterprise (DJI Sky City, No.55 Xianyuan Road, Nanshan District, Shenzhen, China) transmission system, ensuring a stable video feed (1080 p/30 fps) even under challenging environmental conditions.

For enhanced safety, the Mavic 3 Enterprise is equipped with the upgraded DJI APS 5.0 system, which uses six wide-angle "fish-eye" optical sensors to provide omnidirectional obstacle avoidance without blind spots. An advanced return-to-home (RTH) system automatically calculates the most efficient return route to the launch point, conserving time, energy, and battery life.

Precise measurements are further supported by an integrated real-time kinematic (RTK) module, which enables centimeter-level positioning accuracy [60].

The acquired data were processed using Strayos software [58], which is dedicated to blast design and analysis.

### 3.3. Blast Design Using 3D Geospatial Data

A total of 30 blastholes were included in the designed blasthole pattern, with individual hole lengths ranging from 15.4 to 15.8 m. Figure 4 presents the parameters of two example blastholes, developed based on a 3D model of the selected quarry section.

Thanks to the integration of an RTK module, each point in the 3D model is precisely georeferenced. This enables accurate determination of blasthole lengths while accounting for terrain irregularities, hole inclination, and orientation—with centimeter-level precision in reaching the target bench elevation (Figure 5). The designed blasthole pattern and the actual positions of the drilled holes are verified by importing drilling data from the HNS (Hole Navigation System) into the Strayos software, where the final burden verification is performed based on the true geometry of the executed pattern.

Analysis of the burden geometry based on the 3D model enables a reliable calculation of actual burden values for each segment of the bench face. These values are obtained through photogrammetric evaluation using a fully three-dimensional visualization of the geometric dataset. As shown in Figure 6, the burden is defined as the shortest distance from the blasthole axis to the nearest point on the bench face, measured continuously along the entire bench height.

In contrast to the conventional 2D burden measured in a single vertical plane perpendicular to the hole axis, the 3D burden provides a more accurate representation of local resistance to detonation by incorporating variations in face geometry along all spatial axes. This enhances the quality of blast assessments, particularly for charge placement optimization and flyrock risk reduction.

**Figure 4.** Summary of blasthole parameters adjustable in the design phase using Strayos software.

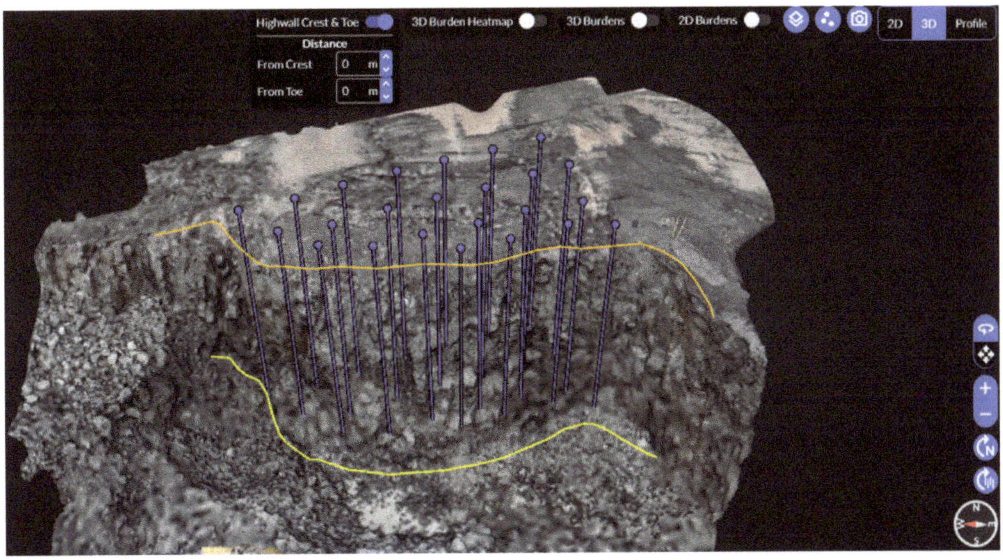


**Figure 5.** 3D model generated from photogrammetric data with a marked blasthole grid.

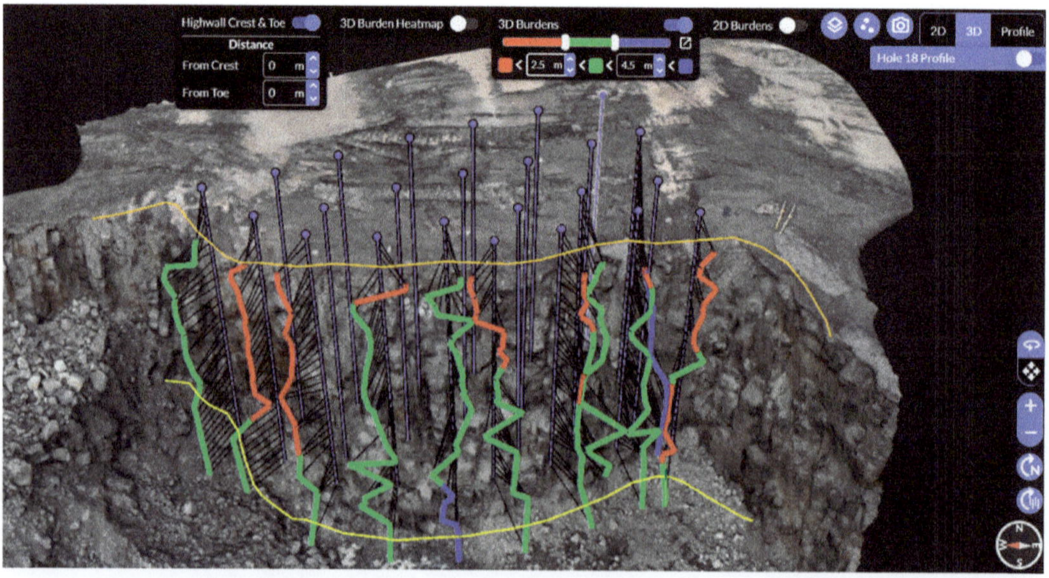


**Figure 6.** 3D model with visualized actual burden values.

Figure 6 presents the results of the spatial burden analysis, indicating that the minimum burden value required under the assumed safety criteria is 2.5 m. The vertical lines represent the positions of the blastholes, while the thin black horizontal lines correspond to burden lines at different elevation levels. Colored lines—red, green, and blue—highlight the locations on the bench face where the minimum burden occurs. In this case study, the 3D model revealed several areas where the burden fell below the safety threshold. Red zones represent areas with insufficient burden (<2.5 m), which significantly increase the risk of flyrock. Green zones denote adequately designed burden values (2.5–4.0 m), considered optimal for the given geological and operational conditions. Blue zones indicate excessive burden (>4.5 m), which may result in poor rock fragmentation and elevated seismic effects, especially near the toe of the bench. Access to detailed geospatial data allows blast designers to proactively revise layouts before drilling begins. In the execution phase, these data support the selection of appropriate explosives, charge masses, and configurations, matched to the actual burden in each blasthole.

Figure 7 compares burden estimates derived from 3D and 2D analyses. In the 3D model, purple indicates areas with undersized burden, while blue denotes properly dimensioned regions. The 2D burden is illustrated as a black line through yellow reference points. Notably, hole No. 8 shows a partially detached rock fragment in the profile, an anomaly unlikely to be detected using traditional 2D laser scanning. Furthermore, the toe zone exhibits a significant discrepancy: the 2D burden value is more than double the actual 3D burden, emphasizing the risk of underestimating flyrock potential in conventional assessments.

The most comprehensive assessment is achieved using a 3D burden heatmap, which visualizes actual burden values not just along individual hole profiles but across the entire blast area (Figure 8). In this representation, green indicates burden values considered optimal based on geological and operational context, red highlights zones with insufficient burden, and blue identifies areas with excessive burden.

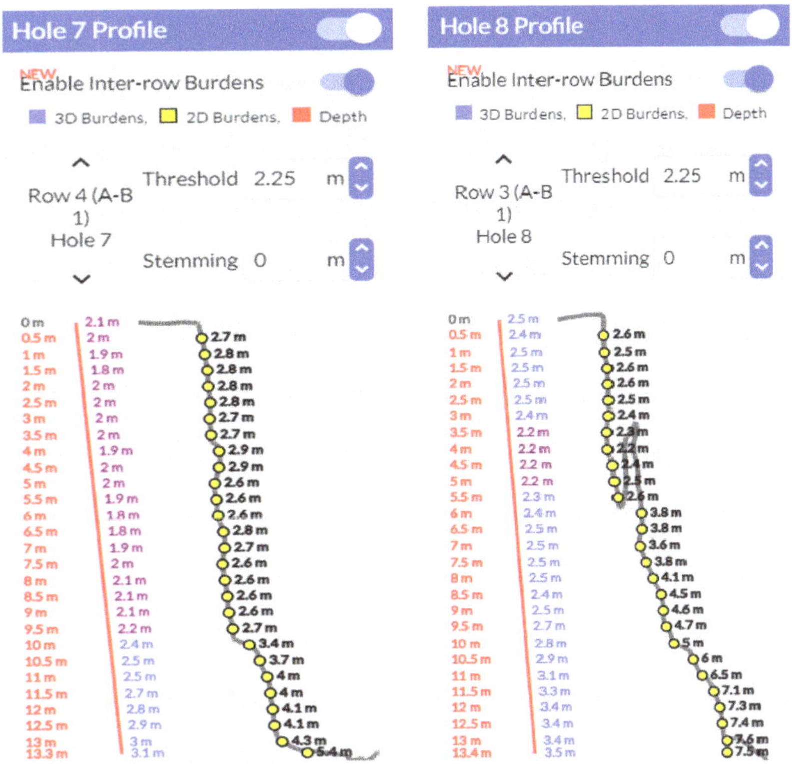


**Figure 7.** Cross-sectional views of two blastholes showing comparison of 2D and 3D burden estimations.

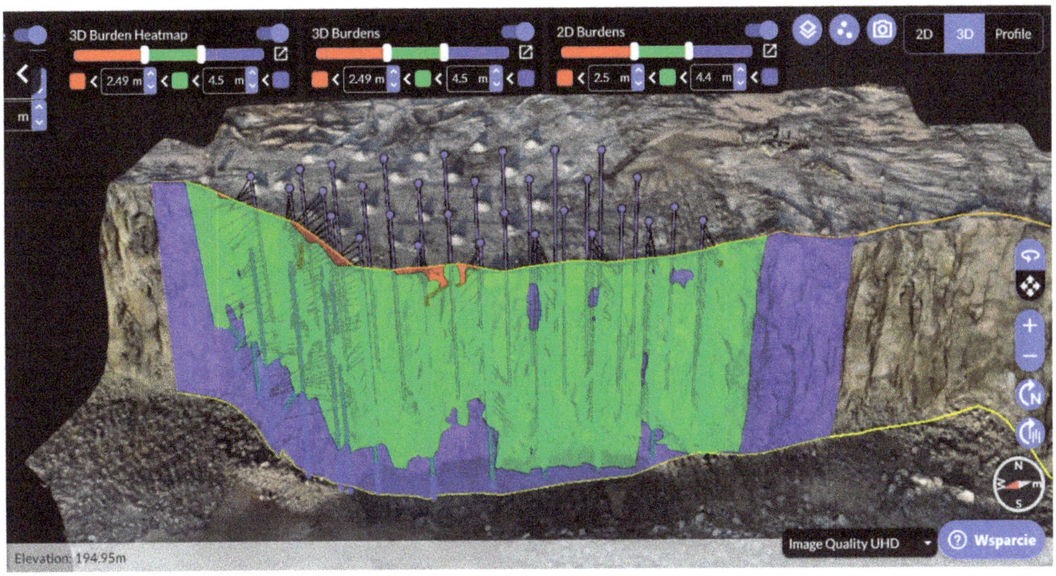


**Figure 8.** 3D burden heatmap of the planned blasthole array.

Overall, the integration of UAV-based 3D data with specialized blast design software allowed for dynamic adjustment of the drilling layout, minimizing risk zones and improving overall blast performance.

## 4. Discussion

The conducted study confirms that the use of photogrammetry supported by unmanned aerial vehicles (UAVs) significantly improves the accuracy of acquiring geometric data for blast design and contributes to enhanced safety by reducing the risk of excessive flyrock. These findings are consistent with the general trend in surface mining, where remote sensing technologies and UAV-based measurement systems are increasingly applied to support operational planning. Minh and Dung [56], in their comprehensive literature review, highlighted the growing importance of UAVs across key mining activities, including drill and blast design.

In the presented case study carried out at a basalt quarry in Lower Silesia (Poland), UAV-supported methods enabled the creation of a highly detailed 3D model of the bench face and the generation of centimeter-precision burden heatmaps. This allowed the identification of zones with inadequate burden (<2.5 m) and excessive burden (>4.5 m), which is critical both in terms of flyrock risk and blasting efficiency. Compared to conventional 2D measurements, the 3D analysis revealed burden discrepancies exceeding 100%, particularly in toe zones, which underscores the insufficiency of flat projections for safe blast pattern design.

The reduction in the number of required images for 3D modeling by nearly a factor of ten, along with an 80% decrease in processing time, confirmed the operational efficiency of this approach. Moreover, the average reduction of flyrock range by an average of 42% (the average value was estimated based on photographic and video documentation) in areas near protected structures demonstrated the practical effectiveness of blast optimization using 3D spatial analysis.

However, certain limitations of the proposed methodology should be acknowledged. UAV-based photogrammetry is sensitive to environmental factors, such as wind, precipitation, fog, lighting conditions, surface reflectivity, and visibility of ground control points, all of which can affect data quality. Additionally, UAV operations require operator certification and are subject to legal regulations, including the obligation to hold valid liability insurance in Poland starting November 2025 [60].

It is important to note that this study did not incorporate predictive modeling methods, such as AdaBoost, random forest, or artificial neural networks, which have been shown to provide highly accurate forecasts of flyrock extent. These techniques allow probabilistic estimation of hazardous areas based on multiple input variables, such as burden, spacing, powder factor, and hole deviation. Combining such models with 3D measurements could further improve prediction accuracy. However, the performance of machine learning models is heavily dependent on the quality and representativeness of the training data, which may be limited in field practice.

A review of the current scientific literature clearly shows a growing trend in the use of advanced mathematical and computational models for predicting flyrock throw in blasting operations. In many studies, the reported coefficients of determination ($R^2$) exceed 0.95, indicating an almost perfect fit between predicted and observed values. Such high accuracy, however, raises valid concerns from a practical standpoint, considering the significant variability and unpredictability of geological conditions in real-world mining operations. Rock formations differ not only in lithology, but also in bedding, weathering degree, presence of joints, and voids, all of which critically affect the propagation of detonation waves and rock fragmentation.

Another methodological concern relates to how the influence of different design parameters is interpreted. Some studies suggest that the powder factor is the most influential parameter in predicting flyrock distance, while burden has a negligible impact. Such conclusions are questionable because these two parameters are directly interrelated—burden represents the resistance of the rock mass to explosive energy, which in turn dictates the amount of energy (and hence explosive material) required for effective fragmentation and displacement. Ignoring this relationship leads to oversimplified and potentially misleading predictive models.

Incorporating accurate burden values into predictive models remains a particularly challenging task. Figure 7 clearly illustrates the substantial variation in actual burden along the profile of a single blasthole. Traditional approaches that assume a constant burden along the entire hole length introduce significant errors in estimating the true rock resistance. Similar issues arise with the powder factor, which is typically calculated as an average across the entire blast pattern and does not account for localized variations in charge distribution, influenced by bench geometry or the presence of detached rock blocks.

For blasting engineers responsible for the planning and execution of blast operations, it is essential to have access to tools that are both simple and effective—tools that allow for the rapid, high-resolution mapping of bench face geometry and the subsequent adaptation of explosive charge mass and design to actual field conditions. Only with such tailored design can flyrock throw be minimized effectively.

It is also important to recognize that even the most sophisticated predictive models cannot replace the experience and local geological knowledge of a skilled blasting engineer. Such an expert, familiar with the specific characteristics of the deposit and its geological variability, can design blasting operations that are both efficient and safe. Therefore, the future of effective flyrock mitigation lies not only in the continued development of predictive tools but also in their intelligent and responsible application by trained professionals.

At the same time, it should be emphasized that no tool currently offers complete assurance against the occurrence of excessive flyrock. One of the main challenges remains the lack of fast and reliable methods for detecting voids between the blasthole pattern and the face geometry. In this context, the precision of geodetic measurements and their integration with specialized blast design software become critical components in ensuring safety in mining operations.

When analyzing only equipment-related costs, the use of UAV technology proves economically favorable compared to traditional surveying methods, such as laser total stations (based on average pricing in the Polish market). The purchase of a high-quality laser total station typically ranges between USD 40,000 and 55,000, with an additional software cost of approximately USD 4000–6000. In contrast, the cost of acquiring a UAV system with RTK functionality ranges from USD 5000 to 8000. Additional expenses include UAV operator training (approximately USD 500, valid for five years), annual insurance (~USD 200), and software licensing fees ranging from USD 200 to 1000 per month, depending on the software's capabilities. Considering average values, the total investment in UAV-based surveying becomes comparable to that of a total station after approximately 5–6 years of use, assuming no major equipment failures occur. Furthermore, operating a total station usually requires two personnel, while UAV data acquisition can be conducted by a single trained operator. The time required to survey the same section of a bench face is also significantly shorter with UAVs, resulting in greater operational efficiency and reduced labor costs.

In summary, the integration of photogrammetric measurements, 3D modeling, and advanced blast design tools significantly enhances the reliability and safety of blasting operations. Future research should focus on hybrid approaches that combine real-time

3D measurements with machine learning algorithms, especially in geologically complex environments or areas close to protected infrastructure. Nevertheless, increasing technical awareness and accountability among blast designers and supervisors remains essential, as human expertise continues to be the most vital link in the decision-making and execution chain.

## 5. Conclusions

The presented work focused on the application of UAV-based photogrammetry in the design of blasthole patterns in a basalt quarry. The use of high-resolution aerial imagery and RTK-supported 3D modeling enabled precise reconstruction of the geometry of the bench face, including critical parameters, such as hole length, inclination, and actual burden distribution. Compared to traditional 2D approaches, the 3D analysis revealed significant spatial variations, especially in toe zones, directly impacting blast efficiency and flyrock risk.

The method led to improved safety conditions, including a documented reduction in flyrock range by an average of 42% near protected structures. In addition, the photogrammetric approach allowed for a ten-fold reduction in the number of images and an 80% decrease in model processing time without compromising mapping accuracy. It should be noted that results may vary depending on specific mining conditions, and it is recommended that similar UAV-supported blast design approaches be further tested in other quarry environments. All UAV flights were conducted in accordance with Strayos software recommendations regarding image overlap, flight path, and camera orientation.

Preliminary cost comparisons indicate that, over time, UAV-based methods offer a cost-effective and efficient alternative to traditional surveying techniques, particularly in operations requiring frequent and rapid terrain documentation.

Despite operational constraints, such as weather sensitivity, licensing requirements, and surface reflectivity issues, UAV photogrammetry proved to be a robust and efficient solution for operational blast planning. Further integration with predictive models and machine learning may offer even greater control over risk assessment and blast outcome optimization.

**Author Contributions:** Conceptualization, J.P. and T.Ż.; methodology, J.P. and T.Ż.; formal analysis, T.Ż.; investigation, J.P.; resources, J.P. and T.Ż.; data curation, J.P. and T.Ż.; writing—original draft preparation, J.P. and T.Ż.; writing—review and editing, J.P. and T.Ż. All authors have read and agreed to the published version of the manuscript.

**Funding:** This research received no external funding.

**Institutional Review Board Statement:** Not applicable.

**Informed Consent Statement:** Not applicable.

**Data Availability Statement:** The data presented in this study are available upon request from the corresponding author.

**Conflicts of Interest:** Author Tomasz Żołądek was employed by Holcim Kruszywa Sp. z o.o. The remaining authors declare that the research was conducted in the absence of any commercial or financial relationships that could be construed as a potential conflict of interest.

## References

1. Lusk, B.; Worsey, P. Explosives and Blasting. In *SME Mining Engineering Handbook*, 3rd ed.; Darling, P., Ed.; Society for Mining, Metallurgy, and Exploration (SME): Englewood, CO, USA, 2011; pp. 443–459.
2. Bhandari, S. *Engineering Rock Blasting Operations*; A. A. Balkema: Rotterdam, The Netherlands, 1997; pp. 135–153.

3. Raina, A.K.; Murthy, V.M.S.R.; Soni, A.K. Flyrock in bench blasting: A comprehensive review. *Bull. Eng. Geol. Environ.* **2014**, *73*, 1199–1209. [CrossRef]
4. Raina, A.K.; Chakraborty, A.K.; Choudhury, P.B.; Sinha, A. Flyrock danger zone demarcation in opencast mines: A risk based approach. *Bull. Eng. Geol. Environ.* **2011**, *70*, 163–172. [CrossRef]
5. Ratnesh, T.; Singh, T.N.; Neel, G. Prediction of Blast-Induced Flyrock in Opencast Mines Using ANN and ANFIS. *Geotech. Geol. Eng.* **2015**, *33*, 875–891. [CrossRef]
6. Mohamad, E.T.; Armaghani, D.J.; Hajihassani, M.; Faizi, K.; Marto, A. A simulation approach to predict blasting-induced flyrock and size of thrown rocks. *Electron. J. Geotech. Eng.* **2013**, *18*, 365–374.
7. Faramarzi, F.; Mansouri, H.; Farsangi, M.A.E. Development of Rock Engineering Systems-Based Models for Flyrock Risk Analysis and Prediction of Flyrock Distance in Surface Blasting. *Rock Mech. Rock Eng.* **2014**, *47*, 1291–1306. [CrossRef]
8. Raina, A.K.; Murthy, V.M.S.R. Prediction of Flyrock Distance in Open Pit Blasting Using Surface Response Analysis. *Geotech. Geol. Eng.* **2016**, *34*, 15–28. [CrossRef]
9. Faradonbeh, R.S.; Jahed, A.D.; Monjezi, M. Development of a new model for predicting flyrock distance in quarry blasting: A genetic programming technique. *Bull. Eng. Geol. Environ.* **2016**, *75*, 993–1006. [CrossRef]
10. Winzer, J.; Sołtys, A.; Pyra, J. *Oddziaływanie na Otoczenie Robót z Użyciem Materiałów Wybuchowych*, 1st ed.; The AGH University of Science and Technology Press: Cracow, Poland, 2016. (In Polish)
11. Pyra, J.; Dworzak, M.; Biessikirski, A.; Twardosz, M.; Kłósko, P. Prognozowanie zasięgu strefy rozrzutu odłamków skalnych dla robót strzałowych w kopalniach odkrywkowych. *Zesz. Nauk. Inst. Gospod. Surowcami Miner. I Energią PAN* **2017**, *101*, 247–264. (In Polish)
12. Wyższy Urząd Górniczy. Available online: https://www.wug.gov.pl/ (accessed on 30 May 2025).
13. Mohamad, E.T.; Yi, C.S.; Murlidhar, B.R.; Murlidhar, B.R.; Saad, R. Effect of Geological Structure on Flyrock Prediction in Construction Blasting. *Geotech. Geol. Eng.* **2018**, *36*, 2217–2235. [CrossRef]
14. Seccatore, J.; Vigna, S.; Marin, T.; Cardu, M. The Role of Delays in the Performance of Blasting. *Appl. Sci.* **2024**, *14*, 4657. [CrossRef]
15. Olofsson, S.O. *Applied Explosives Technology for Construction and Mining*, 2nd ed.; APPLEX AB: Stockholm, Sweden, 1990; pp. 139–144.
16. Jimeno, L.C.; Jimeno, L.E.; Carcedo, F.J. *Drilling and Blasting of Rocks*, 1st ed.; Taylor & Francis: London, UK, 1995; pp. 366–370.
17. Roth, J. *A Model for the Determination of Flyrock Range as a Function of Shot Conditions*; Report no. NTIS PB81-222358; U.S. Department of Commerce: Washington, DC, USA, 1997.
18. Hałat, W.; Morawa, R. Metoda prognozowania zasięgu strefy rozrzutu przy prowadzonych robotach strzałowych. *Bezpieczeństwo Pr. I Ochr. Sr. W Górnictwie* **2007**, *9*, 28–31. (In Polish)
19. Lewicki, J. Prognozowanie wielkości zagrożeń powstałych przy prowadzeniu robót strzałowych w budownictwie. *Górnictwo I Geoinżynieria* **2004**, *28*, 251–267. (In Polish)
20. Barański, K.; Morawa, R. Technologiczne możliwości zmniejszenia zasięgu strefy rozrzutu w górnictwie odkrywkowym. *Górnictwo Odkryw.* **2015**, *56*, 19–26. (In Polish)
21. Grześkowiak, A.; Patla, S. Przyczynek do wyznaczania zasięgów oddziaływań i dopuszczalnych wielkości ładunków materiałów wybuchowych w górnictwie skalnym. *Min. Sci.* **2016**, *23*, 47–58. (In Polish)
22. Trivedi, R.; Singh, T.N.; Raina, A.K. Simultaneous prediction of blast-induced flyrock and fragmentation in opencast limestone mines using back propagation neural network. *Int. J. Min. Miner. Eng.* **2016**, *7*, 237–252. [CrossRef]
23. Monjezi, M.; Bahrami, A.; Yazdian Varjani, A.; Reza Sayadi, A. Prediction and controlling of flyrock in blasting operation using artificial neural network. *Arab. J. Geosci.* **2011**, *4*, 421–425. [CrossRef]
24. Monjezi, M.; Amini Khoshalan, H.; Yazdian Varjani, A. Prediction of flyrock and backbreak in open pit blasting operations: A neuro-genetic approach. *Arab. J. Geosci.* **2012**, *5*, 441–448. [CrossRef]
25. Trivedi, R.; Singh, T.N.; Raina, A.K. Prediction of blast-induced flyrock in Indian limestone mines using neural networks. *J. Rock Mech. Geotech. Eng.* **2014**, *6*, 447–454. [CrossRef]
26. Amini, H.; Gholami, R.; Monjezi, M.; Rahman Torabi, S.; Zadhesh, J. Evaluation of flyrock phenomenon due to blasting operation by suport vector machine. *Neural. Comput. Appl.* **2012**, *21*, 2077–2085. [CrossRef]
27. Khandelwal, M.; Monjezi, M. Prediction of flyrock in open pit blasting operation using machine learning method. *Int. J. Min. Sci. Technol.* **2013**, *23*, 313–316. [CrossRef]
28. Armaghani, D.J.; Mahdiyar, A.; Hasanipanah, M.; Shirani Faradonbeh, R.; Khandelwal, M.; Bakhshandeh Amnieh, H. Risk Assessment and Prediction of Flyrock Distance by Combined Multiple Regression Analysis and Monte Carlo Simulation of Quarry Blasting. *Rock Mech. Rock Eng.* **2016**, *49*, 3631–3641. [CrossRef]
29. Guo, H.; Nguyen, H.; Bui, X.-N.; Armaghani, D.J. A new technique to predict fly-rock in bench blasting based on an ensemble of support vector regression and GLMNET. *Eng. Comput.* **2021**, *37*, 421–435. [CrossRef]
30. Li, C.; Zhou, J.; Du, K.; Armaghani, D.J.; Huang, S. Prediction of flyrock distance in surface mining using a novel hybrid model of harris hawks optimization with multi-strategies-based support vector regression. *Nat. Resour. Res.* **2023**, *32*, 2995–3023. [CrossRef]

31. Koopialipoor, M.; Fallah, A.; Armaghani, D.J.; Azizi, A.; Mohamad, E.T. Three hybrid intelligent models in estimating flyrock distance resulting from blasting. *Eng. Comput.* **2019**, *35*, 243–256. [CrossRef]
32. Rezaei, M.; Monjezi, M.; Yazdian Varjani, A. Development of a fuzzy model to predict flyrock in surface mining. *Saf. Sci.* **2011**, *49*, 298–305. [CrossRef]
33. Ghasemi, E.; Amini, H.; Ataei, M.; Khalokakaei, R. Application of artificial intelligence techniques for predicting the flyrock distance caused by blasting operation. *Arab. J. Geosci.* **2014**, *7*, 193–202. [CrossRef]
34. Hudaverdi, T.; Agan, Y. Application of intelligent models for flyrock prediction considering design parameters and bench face characteristics. *Min. Metall. Explor.* **2023**, *40*, 2331–2347. [CrossRef]
35. Hudaverdi, T.; Akyildiz, O. A new classification approach for prediction of flyrock throw in surface mines. *Bull. Eng. Geol. Env.* **2019**, *78*, 177–187. [CrossRef]
36. Hudaverdi, T. Prediction of flyrock throw distance in quarries by variable selection procedures and ANFIS modelling technique. *Environ. Earth Sci.* **2022**, *81*, 281. [CrossRef]
37. Nguyen, H.; Bao, T.D.; Bui, X.-N.; Pham, V.-V.; Nguyen, D.-A.; Do, N.-H.; Hoa, L.T.T.; Le, Q.-T.; Le, T.-N. Measuring and Predicting Blast-Induced Flyrock Using Unmanned Aerial Vehicles and Lévy Flight Technique-Based Jaya Optimization Algorithm Integrated with Adaptive Neuro-Fuzzy Inference System. *Nat. Resour. Res.* **2025**, *34*, 1773–1806. [CrossRef]
38. Nguyen, H.; Van Thieu, N. Measurement and Prediction of Blast-Induced Flyrock Distance Using Unmanned Aerial Vehicles and Metaheuristic-Optimized ANFIS Neural Networks. *Nat. Resour. Res.* **2025**, *34*, 1169–1198. [CrossRef]
39. Armaghani, D.J.; Hajihassani, M.; Mohamad, E.T.; Marto, A.; Noorani, S.A. Blasting-induced flyrock and ground vibration prediction through an expert artificial neural network based on particle swarm optimization. *Arab. J. Geosci.* **2014**, *7*, 5383–5396. [CrossRef]
40. Marto, A.; Hajihassani, M.; Jahed Armaghani, D.; Tonnizam Mohamad, E.; Mahir Makhtar, A. A novel approach for blast-induced flyrock prediction based on imperialist competitive algorithm and artificial neural network. *Sci. World J.* **2014**, *2014*, 643715. [CrossRef] [PubMed]
41. Murlidhar, B.R.; Kumar, D.; Jahed Armaghani, D.; Tonnizam Mohamad, E.; Roy, B.; Thai Pham, B. A Novel Intelligent ELM-BBO Technique for Predicting Distance of Mine Blasting-Induced Flyrock. *Nat. Resour. Res.* **2020**, *29*, 4103–4120. [CrossRef]
42. Bhatawdekar, R.M.; Kumar, R.; Sabri Sabri, M.M.; Roy, B.; Mohamad, E.T.; Kumar, D.; Kwon, S. Estimating Flyrock Distance Induced Due to Mine Blasting by Extreme Learning Machine Coupled with an Equilibrium Optimizer. *Sustainability* **2023**, *15*, 3265. [CrossRef]
43. Zhou, J.; Koopialipoor, M.; Murlidhar, B.R.; Alireza Fatemi, S.; Tahir, M.M.; Jahed Armaghani, D.; Li, C. Use of Intelligent Methods to Design Effective Pattern Parameters of Mine Blasting to Minimize Flyrock Distance. *Nat. Resour. Res.* **2020**, *29*, 625–639. [CrossRef]
44. Saghatforoush, A.; Monjezi, M.; Shirani Faradonbeh, R.; Armaghani, D. Combination of neural network and ant colony optimization algorithms for prediction and oprimization of flyrock and back-break induced blasting. *Eng. Comput.* **2016**, *32*, 255–266. [CrossRef]
45. Yari, M.; Armaghani, D.J.; Maraveas, C.; Ejlali, A.N.; Mohamad, E.T.; Asteris, P.G. Several Tree-Based Solutions for Predicting Flyrock Distance Due to Mine Blasting. *Appl. Sci.* **2023**, *13*, 1345. [CrossRef]
46. Hasanipanah, M.; Faradonbeh, R.S.; Armaghani, D.J.; Amnieh, H.B.; Khandelwal, M. Development of a precise model for prediction of blast-induced flyrock using regression tree technique. *Environ. Earth Sci.* **2017**, *76*, 27. [CrossRef]
47. Nguyen, H.; Bui, X.-N.; Nguyen-Thoi, T.; Ragam, P.; Moayedi, H. Toward a State-of-the-Art of Fly-Rock Prediction Technology in Open-Pit Mines Using EANNs Model. *Appl. Sci.* **2019**, *9*, 4554. [CrossRef]
48. Zhang, Y.; Qiu, Y.; Du, K.; Nguyen, H.; Jahed Armaghani, D.; Zhou, J. Optimizing Flyrock Forecasting in Open-Pit Blasting Using Hybrid Machine Learning Models. *Rock Mech. Rock Eng.* **2025**. Available online: https://link.springer.com/article/10.1007/s00603-025-04730-2#article-info (accessed on 29 July 2025). [CrossRef]
49. Chen, Y.; Wang, M.; Yin, H.; Zhang, T. Prediction of flyrock distance induced by blasting using particle swarm optimization and multiple regression analysis: An engineering perspective. *Acta Geophys.* **2024**, *72*, 287–301. [CrossRef]
50. Han, H.; Jahed Armaghani, D.; Tarinejad, R.; Zhou, J.; Tahir, M.M. Random forest and bayesian network techniques for probabilistic prediction of flyrock induced by blasting in quarry sites. *Nat. Resour. Res.* **2020**, *29*, 655–667. [CrossRef]
51. Lawal, A.I.; Ojo, O.J.; Kim, M.; Kwon, S. Determination of blast-induced flyrock using a drone technology: A bibliometric overview with practical soft computing implementation. *Arab. J. Geosci.* **2022**, *15*, 1581. [CrossRef]
52. Mishra, R.; Mishra, A.K.; Choudhary, B.S. High-Speed Motion Analysis-Based Machine Learning Models for Prediction and Simulation of Flyrock in Surface Mines. *Appl. Sci.* **2023**, *13*, 9906. [CrossRef]
53. Bamford, T.; Esmaeili, K.; Schoellig, A.P. A real-time analysis of post-blast rock fragmentation using UAV technology. *Int. J. Min. Reclam. Environ.* **2017**, *31*, 439–456. [CrossRef]
54. Bamford, T.; Medinac, F.; Esmaeili, K. Continuous Monitoring and Improvement of the Blasting Process in Open Pit Mines Using Unmanned Aerial Vehicle Techniques. *Remote Sens.* **2020**, *12*, 2801. [CrossRef]

55. Colomina, I.; Molina, P. Unmanned aerial systems for photogrammetry and remote sensing: A review. *ISPRS J. Photogramm. Remote Sens.* **2014**, *92*, 79–97. [CrossRef]
56. Minh, D.T.; Dung, N.B. Applications of UAVs in mine industry: A scoping review. *J. Sustain. Min.* **2023**, *22*, 128–146. [CrossRef]
57. Bhatawdekar, R.M.; Armaghani, D.J.; Azizi, A. Applications of AI and ML Techniques to Predict Backbreak and Flyrock Distance Resulting from Blasting. In *Environmental Issues of Blasting. SpringerBriefs in Applied Sciences and Technology*; Springer: Singapore, 2021. [CrossRef]
58. Available online: https://help.strayos.com/hc/en-us/articles/360037276553-Flying-a-Bench-with-a-Drone (accessed on 2 June 2025).
59. Ustawa z dnia 3 lipca 2002 r. Prawo lotnicze. Dz. U. 2002 Nr 130 poz. 1112. Available online: https://isap.sejm.gov.pl/isap.nsf/download.xsp/WDU20021301112/U/D20021112Lj.pdf (accessed on 4 June 2025).
60. Available online: https://enterprise.dji-ars.pl/product-pol-26054-DJI-Mavic-3-Enterprise.html (accessed on 4 April 2025).

applied sciences

MDPI

Article

# Research on the Impact of Blasting Vibration in Mining Areas on Surrounding Railway Structures

**Shuai Zhang [1], Zhiyu Zhang [1], Kai Wang [2], Defu He [2] and Yonghui Huang [3,*]**

1 Faculty of Land Resources Engineering, University of Science and Technology, Kunming 650093, China; zzs@stu.kust.edu.cn (S.Z.); szw@stu.kust.edu.cn (Z.Z.)
2 Faculty of Public Safety and Emergency Management, Kunming University of Science and Technology, Kunming 650093, China; aka_53@163.com (K.W.); 13330532081@163.com (D.H.)
3 Faculty of Electric Power Engineering, Kunming University of Science and Technology, Kunming 650093, China
* Correspondence: 20130151@kust.edu.cn

**Abstract:** To study the safety impact of open-pit blasting on railway structures, blasting vibration tests were conducted on the railway. Numerical simulation methods were employed to establish a three-dimensional finite element model of the railway structure, encompassing the rail, sleepers, and subgrade. By applying the measured blasting vibration waves from the railway platform and amplifying their velocities to various multiples, the variation patterns of the vibration velocity and displacement of the railway structure were calculated and analyzed. The research findings suggest that the rail in the railway structure is less affected by the vibration, whereas the sleepers and subgrade are more significantly influenced by the blasting vibration. The peak velocities and displacements generated by the structure are well below the requirements of the relevant regulations. Upon amplifying the loading velocities to different multiples and conducting the analysis, it was observed that as the vibration velocity increases, the velocity and displacement of the structure increase to varying extents. It is recommended to control the blasting vibration velocity below 3 cm/s and to adopt corresponding security technology measures.

**Keywords:** blasting vibration; railway; numerical simulation; vibration speed; displacement

## 1. Introduction

Railways are an integral part of the land transportation network, playing a pivotal role in economic growth and national security, and are essential for societal advancement [1]. Blasting is a prevalent technique in open-pit mining, where detonations release substantial energy, some of which travels as vibration waves through the surrounding environment. These vibration waves, upon reaching the railway subgrade, can induce tremors in the railway infrastructure, thus impacting its safety [2–4]. With the expansion of mining operations, the scale of which is growing, mining depths are gradually increasing, and the boundaries of mining zones are extending, and the proximity to nearby railways diminishes. The vibrations from daily blasting activities could affect the surrounding railway structures, potentially causing damage or even jeopardizing public safety [5]. Consequently, it is imperative to analyze the structural safety of railways adjacent to mines when subjected to vibrations from blasting.

The purpose of this research is to investigate the dynamic response of railway structures to blasting vibrations and to establish safe operational limits for blasting activities. By combining on-site vibration testing with numerical simulations, this study aims to provide a comprehensive understanding of how blasting vibrations propagate through railway structures and affect their integrity. The findings of this research are significant for both engineering practice and policy-making, as they contribute to the development of safer blasting protocols and help minimize the risk of structural damage to railway infrastructure.

Currently, scholars have employed various methods to study the impact of vibrations on railway structures, primarily focusing on theoretical research, numerical simulation, and field testing, achieving certain results. Some scholars have studied the deformation effects of train-induced vibrations on railway tracks and subgrades, proposed different analysis models, simulated train loading during passage, established a train vibration source function, simulated vibration waves from high-speed trains, and investigated track vibrations under different load conditions [6–9]. Fattah et al. [10] investigated the effects of railway track vibrations on saturated and unsaturated subgrade layers through laboratory models and cyclic loading experiments, examining the influence of varying saturation levels and load frequencies on subgrade stresses and deformations, and providing recommendations for track design optimization. Pierre et al. [11] introduced a novel method combining a device for measuring deformation at multiple depths with a satellite-based system for precise positioning. This approach enhances the accuracy of track settlement monitoring by integrating absolute displacement data from the satellite system with relative deformation measurements from the multi-depth measurement device. Qu et al. [12] investigated the subgrade vibrations and long-term stability of an embankment-bridge transition zone in a non-ballasted high-speed railway. They conducted two field tests 32 months apart, analyzed FFT, and calculated dynamic shear strains to evaluate stability. Results showed that the abutment's influence diminishes with distance, leading to increased vibration amplitudes and lower dominant frequencies. The subgrade's dynamic stability remained within acceptable limits after 32 months of operation. This study provides insights for designing and maintaining high-speed railway transition zones. Their study provides insights into the impact of blasting vibrations on railway structures.

Monitoring the displacement and velocity responses of railway structures to vibrations in the field is challenging. As an effective research method, numerical simulation has been widely applied. Park et al. [13] analyzed the dynamic characteristics of high-speed trains passing through transition zones using finite element models and on-site measurement data, validating the accuracy of numerical models and exploring the influence of train speed on track stresses and wheel-rail interactions. Khajehdezfuly et al. [14] used a vehicle/track interaction model to obtain force-time history curves for road and railway vehicles, which were then input into a 3D finite element model to analyze the influence of different parameters on vibration caused by vehicles. Jiang et al. [15] proposed a differential evolution algorithm based on 3DEC dynamic analysis, establishing a 3DEC model for the complex ore body of an eastern sulfur iron mine. By analyzing field-monitored vibration displacement and velocity data, waveforms and parameters were simulated, revealing that when the safety vibration velocity of the railway was 4.5 cm/s, the maximum allowable single explosive charge in the area was 44.978 kg.

However, most studies focus only on specific structures or aspects, with limited research on the impact of blasting vibrations on the overall railway structure. This paper focuses on a railway located near a mine site, analyzing vibration response signals from blasting-induced vibrations through monitoring data and field surveys. The signals were pre-processed and analyzed in the frequency domain, then incorporated into a 3D finite element model of the entire railway structure, including the steel rails, rail ties, and subgrade. This study discusses key parameters such as displacement and vibration velocity caused by blasting-induced vibrations, comparing and analyzing the vibration deformation and safety impact on the railway structure.

## 2. Methods

To investigate the impact of blasting vibrations on railway structures, this study employed a systematic approach combining on-site vibration testing and numerical simulation. The methodology is outlined as follows:

(1) Install vibration testing instruments during normal railway operations to record the vibration data of the railway tracks as trains pass, as well as the vibration response of the surrounding environment.
(2) Building a 3-D model of the railway structure based on collected parameters, meshing the model, applying boundary constraints, and considering actual material properties.
(3) Conduct modal analysis on the 3-D model to determine the resonance frequency range.
(4) Integrate the actual vibration data from on-site testing into the 3-D finite element model. Subsequently, perform additional analysis and calculations, considering material and geometric nonlinearity, to simulate the stress, strain, and displacement responses of the railway structure. Using the results and relevant safety standards, assess the safety of the railway structure under various vibration conditions and establish the safe vibration speed threshold.

## 3. Project Overview and Blasting Vibration Testing

### *3.1. Overview of the Mine and Surrounding Environment*

The mine is situated in the southwest of Haikou Town, Xishan District, Kunming City, on the western side of Dianchi Lake. It is part of a low-to-mid-mountain terrain. The highest point is at Shansongyuan in the western section of the site, reaching an elevation of 2482 m, whereas the lowest point is at the northwest edge of the surveyed area, in the Mingyi River Valley, with an elevation of about 1860 m. The maximum elevation variation is 622 m, with elevations generally ranging from 2100 to 2250 m and a typical elevation difference between 240 and 390 m. The rock in the blasting area is predominantly dolomitic sandstone, featuring relatively stable structural conditions and no observed fault zones. Its geographical location is distinctive, being adjacent to Dianchi Lake, and it possesses abundant groundwater resources.

The primary protection targets in the vicinity of the mine are the residential buildings in Haifeng Village and Ercai Village, as well as the adjacent railway. The railway is a standard Chinese railway with a track gauge of 1.435 m. The ties are constructed from reinforced concrete, measuring 220 mm in width, 160 mm in thickness, and 2500 mm in length. The total depth of the subgrade is 2.3 m, utilizing layered fill materials. Blasting operations near the railway employ No. 1 rock emulsion explosives, with a total of 110 blast holes arranged in five rows. The holes are spaced 6 m apart, and the rows are spaced 5 m apart. The total quantity of explosives used amounts to 9792 kg, with a maximum charge per section not exceeding 90 kg. Digital electronic detonators are used for delayed blasting, with a delay interval of 45 ms between holes, and the explosive consumption rate is 0.32 kg per cubic meter.

As the mining area continues to expand and mining depths gradually increase, the boundaries of the mining area also gradually extend, bringing it closer to the surrounding Kunyang branch line railway. The blasting vibrations from the mine's daily production may affect the structural safety of the surrounding railway and residential buildings. Should the railway structure experience displacement or deformation, it could result in significant losses for the enterprise and endanger the safety of people's lives and property. Consequently, it is essential to analyze the structural safety of the Kunyang branch line railway and residential buildings near the mine, considering the effects of daily production blasting vibrations.

### 3.2. *Blasting Vibration Testing*

The precision of the testing instruments is closely related to the outcomes of the blasting vibration tests. The on-site blasting vibration monitoring experiment utilizes the TC-4850 blasting vibration meter in conjunction with three-direction vibration velocity sensors that are compatible with the meter. The blasting vibration testing system is depicted in Figure 1. The system comprises several key components as follows: a velocity sensor positioned at the measurement point with its x-axis aligned toward the blast area, a vibration meter connected to the sensor for signal acquisition, a data acquisition unit for converting the signal into a digital format, and a computer for data processing and analysis. The sensor's y-axis is oriented perpendicular to the x-axis in the horizontal plane, and the z-axis is vertical, ensuring comprehensive measurement of vibrations in three orthogonal directions. The data collected on-site were imported into a computer for processing, enabling the acquisition of the raw blasting vibration waveform data.

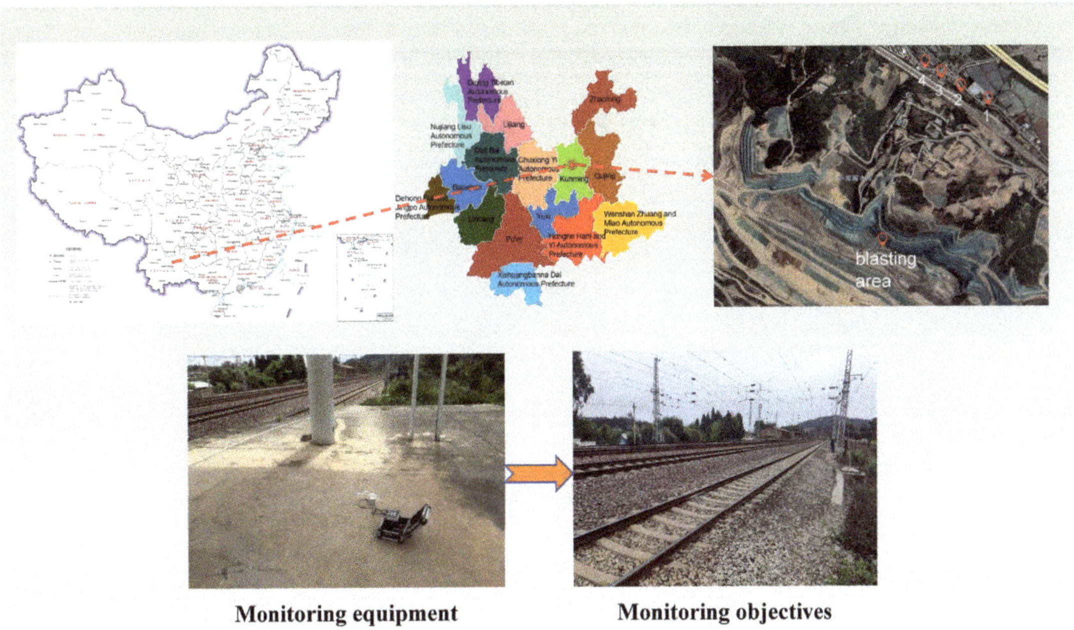


**Figure 1.** Study area and on-site blasting vibration testing.

The surrounding area of the mining site consists of water-rich rock layers, and the propagation of blasting vibrations is influenced by various factors, including the maximum charge per delay, distance from the blast center, site conditions, and the propagation medium. To more effectively monitor railway vibrations around the mining area, multiple measurement points were arranged in a straight line as much as possible between the railway platform measurement points. Each measurement point is labeled as 1, 2, 3, and 4, as shown in Figure 1. At each measurement point, a three-direction velocity sensor was fixed using gypsum at the subgrade or bedrock to monitor the vibration velocity in three directions. The equipment trigger value was set at 0.015 cm/s, with a 10-s recording duration and a 16 kHz sampling rate. The sensor's x-axis was aligned toward the blast area, with the positive direction pointing from the measurement point toward the epicenter of the explosion. The y-axis was oriented perpendicular to the x-axis in the horizontal plane, and the z-axis was oriented vertically, perpendicular to both the x-axis and y-axis. The velocity sensor converted the blasting vibration signal into an electrical signal, which was transmitted to the vibration meter for real-time processing. To establish the time correlation between the measurement and the explosion, the vibration duration was defined as the time from the arrival of the blast wave to the decay of the signal amplitude to 1/e of its maximum value. The trigger value ensured immediate recording upon detecting the initial blast wave, capturing the full time history of the vibration. For multi-hole blasting, delay intervals were optimized to minimize overlapping vibration effects, and the time delay between blast initiation and signal arrival at the sensor was accounted for based on the distance from the blast source to the sensor and wave propagation speed. After processing the on-site test data, the raw data and waveforms of the blasting vibration were obtained, providing a basis for further analysis and evaluation of the vibration impact on nearby structures.

The peak vibration velocity in each direction for different measurement points is shown in Table 1. A comparative analysis of the data collected from four railway measurement points revealed that Measurement Point 1, being closer to the blast area, had a larger peak vibration velocity compared to the other measurement points. Therefore, the data from Measurement Point 1 was used for further analysis.

**Table 1.** Monitoring results of measurement points.

| Measurement Point | Direction | Maximum/(cm·s$^{-1}$) | Main Frequency/Hz | Distance/m |
|---|---|---|---|---|
| 1 | X | 0.169 | 12.32 | 688 |
| | Y | 0.167 | 14.49 | |
| | Z | 0.098 | 9.61 | |
| 2 | X | 0.119 | 17.55 | 725 |
| | Y | 0.136 | 19.95 | |
| | Z | 0.043 | 6.38 | |
| 3 | X | 0.035 | 18.99 | 787 |
| | Y | 0.036 | 6.83 | |
| | Z | 0.033 | 15.70 | |
| 4 | X | 0.050 | 17.21 | 877 |
| | Y | 0.051 | 6.62 | |
| | Z | 0.073 | 36.20 | |

### 3.3. Vibration Data Processing

The impact of blast vibrations on structures is closely related to the peak velocity and frequency. The FFT filtering method is used to filter the raw waveform data [16]. By performing a Fourier transform, the time-domain data are converted into the frequency domain. The low-pass filter eliminates unnecessary data and high-frequency anomalies, reducing the influence of noise and interference on the blast vibrations. This results in waveform data in the x, y, and z directions, as shown in Figure 2. The processed data are then imported into the railway finite element model for further analysis and computation.

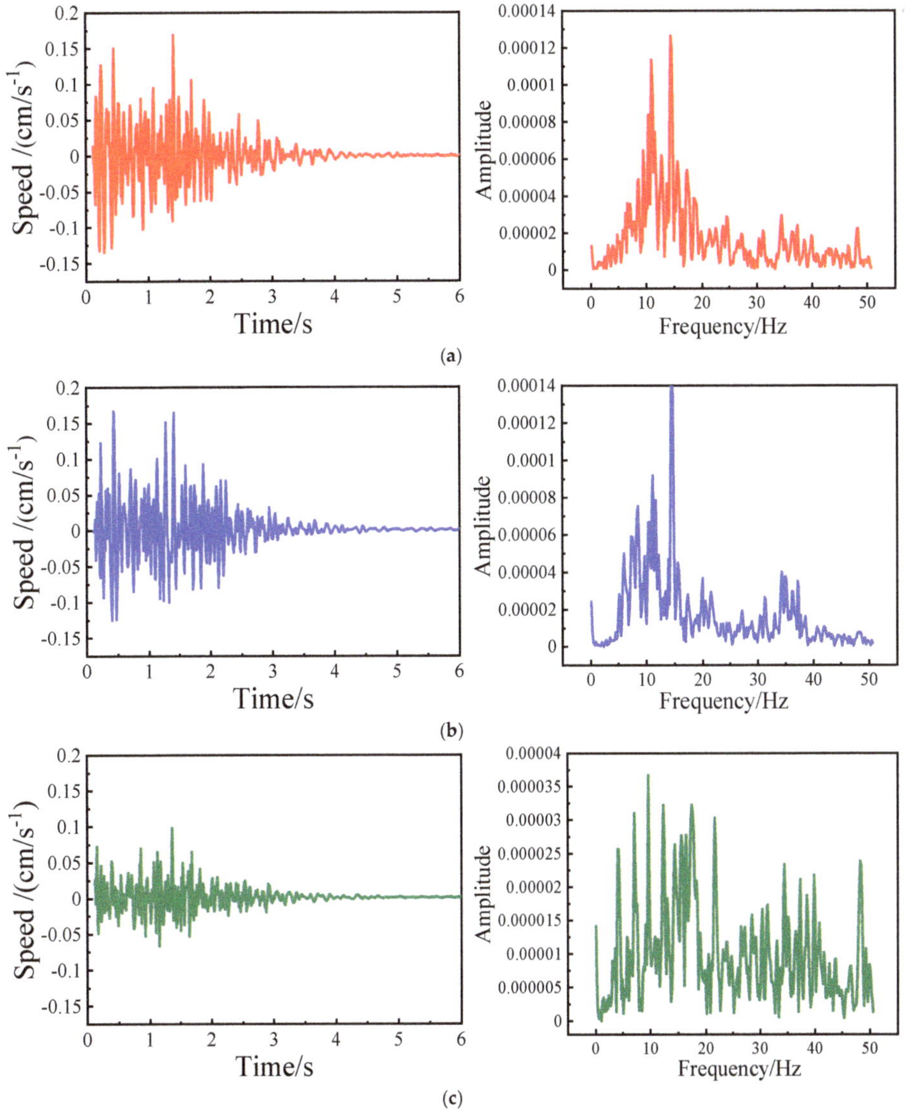


**Figure 2.** Time domain and frequency domain curves of three directional velocities. (**a**) Velocity time-domain and amplitude frequency-domain curve in the X direction; (**b**) Velocity time-domain and amplitude frequency-domain curve in the Y direction; (**c**) Velocity time-domain and amplitude frequency-domain curve in the Z direction.

## 4. Simulation Analysis

### *4.1. Dynamic Response Time History Analysis Method*

The vibration differential equation for a multi-degree-of-freedom elastic system under vibration [17] (see Equation (1)):

$$[\mathrm{M}]\{\ddot{\mathrm{x}}\} + [\mathrm{c}]\{\dot{\mathrm{x}}\} + [\mathrm{K}]\{\dot{\mathrm{x}}\} = [\mathrm{M}]\{\ddot{\mathrm{x}}_{\mathrm{g}}\} \quad (1)$$

where [M] represents the mass matrix, in kilograms (kg); [c] is the damping matrix, in kilograms per second (kg/s); [K] denotes the damping matrix, in Newtons per meter (N/m); $\{\ddot{x}\}$ is the system's relative horizontal acceleration vector, in meters per second squared (m/s$^2$); $\{\dot{x}\}$ represents the system's relative horizontal velocity vector, in meters per second (m/s); $\{\dot{x}\}$ is the system's relative horizontal displacement vector, in meters (m); and $\{\ddot{x}_g\}$ denotes the ground horizontal acceleration vector, in meters per second squared (m/s$^2$).

The time history analysis method involves solving equations through numerical integration over several continuous time intervals, based on the structural restoring force characteristic curve. By inputting vibration waveform data and employing numerical computation methods from structural dynamics and finite element analysis, the system performs incremental integration from the initial state to calculate the velocity, acceleration, and displacement dynamic responses of each mass point as they evolve over time.

The dynamic response of the structure is significantly affected by damping, with the effect of the damping ratio being inversely proportional to the displacement response. Rayleigh damping [18] is selected as follows (see Equation (2)):

$$C = \alpha M + \beta K \tag{2}$$

where C represents the damping matrix in the dynamic equation, measured in kg/s; α is the mass-proportional damping constant, measured in 1/s; β is the stiffness-proportional damping; M is the mass matrix, measured in kg; K is the stiffness matrix, measured in N/m; and αM denotes the mass component; and βK signifies the stiffness component.

### *4.2. Railway Model Establishment*

Using SolidWorks for 3D modeling and HyperMesh for meshing, the steel rails, railroad ties, and subgrade in the vicinity of the mining area have been modeled and meticulously meshed. The dimensions of the model are consistent with the actual site conditions, as depicted in Figure 3. The rail spacing measures 1435 m, the railroad ties have dimensions of 220 mm in width, 160 mm in thickness, and 2500 mm in length, and the total thickness of the subgrade is 2.3 m. The meshes of various structures are connected by shared nodes, and the overall model consists of 333,292 elements and 385,637 nodes.

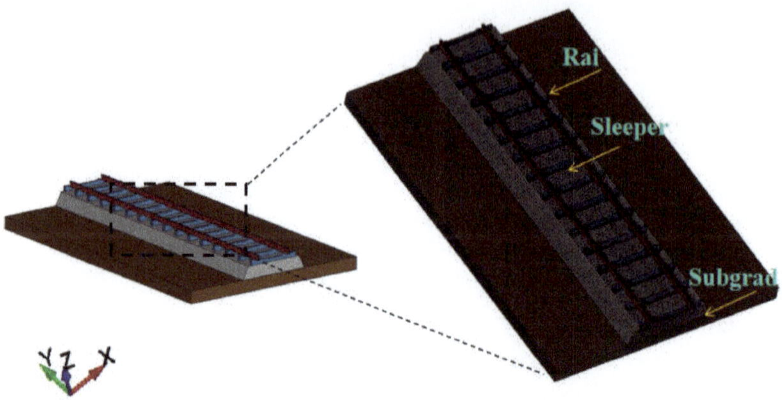


**Figure 3.** Overall railway model.

To simplify the analysis, the steel rails, railroad ties, and subgrade are each calculated using three different materials. The specific parameters are shown in Table 2.

**Table 2.** Parameters of railway structural materials.

| Material | Density/(kg·m$^{-3}$) | Elastic Modulus/Pa | Poisson's Ratio | Damping Ratio |
|---|---|---|---|---|
| Rail | 7830 | $2.06 \times 10^{11}$ | 0.3 | 0.05 |
| Sleeper | 2500 | $2.82 \times 10^{10}$ | 0.2 | 0.08 |
| Subgrade | 2130 | $2.53 \times 10^{7}$ | 0.28 | 1 |

Linear elastic behavior was assumed with considerations for nonlinearity. Fixed boundary conditions were applied at the bottom nodes to prevent displacement/rotation, along with non-reflective boundaries to minimize wave reflections.

After adding the corresponding keywords, the ANSYS/LSDYNA software version R11 is used for analysis and computation.

### 4.3. Analysis and Calculation

Considering the actual situation, when analyzing and calculating the railway model, the impact of the self-weight of the railway structure itself must be taken into account. Initially, the self-weight stress calculation for the railway should be performed, as depicted in Figure 4. The simulation of the railway structure should adhere to the self-weight distribution law, which can provide initial conditions for the subsequent vibration calculations.

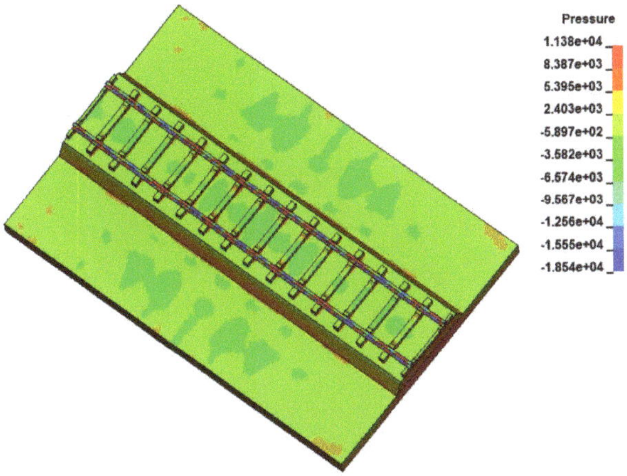


**Figure 4.** Distribution of model weight.

The vibration waves generated by blasting are transmitted to the railway structure through the ground. The bottom nodes of the model are selected, and after performing the self-weight calculation, the time history curves of the horizontal radial, horizontal tangential, and vertical velocities are imported and applied to the calculation, respectively.

#### 4.3.1. Modal Analysis

When studying the dynamic response of railway structures to blasting vibration waves, it is essential to consider the influence of their inherent characteristics, such as mass and the distribution of elastic properties. These characteristics are manifested through natural frequencies and mode shapes (Yang et al. [19]).

ANSYS is utilized to conduct a modal analysis on the railway structure, calculating its natural frequencies and mode shapes. The results for the initial six modes are presented in Table 3. Each mode represents a unique vibration pattern at a specific frequency. Here, "6 orders" refers to the first six vibration patterns, ranked from the lowest to the sixth frequency.

**Table 3.** Results of modal analysis of the first 6 orders of railway structure.

| Mode | 1st | 2nd | 3rd | 4th | 5th | 6th |
|---|---|---|---|---|---|---|
| Frequency/Hz | 3.106 | 6.317 | 6.796 | 7.437 | 7.557 | 8.809 |

For a given structure, the frequency at which resonance occurs is commonly known as the first-order natural frequency [20]. In this instance, the resonance frequency of the railway structure is 3.106 Hz. The closer the frequency of the vibration waves produced by an explosion is to the resonance frequency, the greater the impact on the structure.

#### 4.3.2. Structural Velocity Response Analysis

Based on the simulation results, the velocity response of the railway structure is analyzed at the moments corresponding to the maximum velocity in the three directions from the blasting vibration test data. By comparing the velocity distribution contour maps at different moments, as shown in Figure 5, it can be observed that the steel rails of the railway structure are least affected by the vibration, while the rail ties and subgrade experience a greater impact from the vibration. When affected by the horizontal radial direction, the vibration response of the structure is more pronounced.

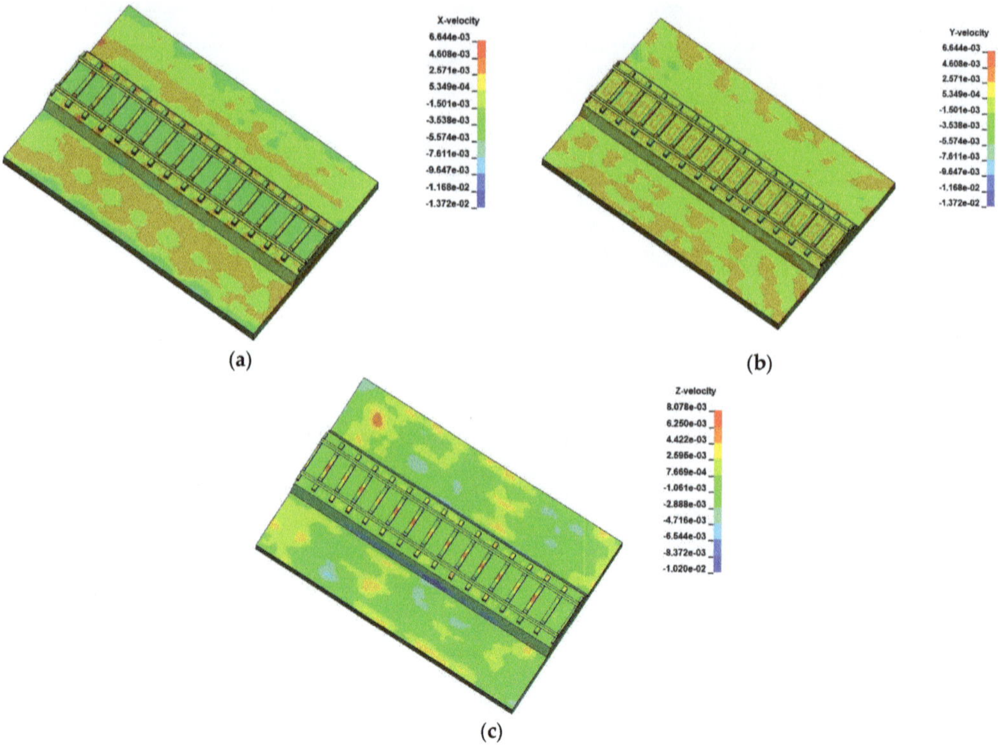


**Figure 5.** Three directional velocity distribution. (**a**) Structural velocity distribution in the X-direction. (**b**) Structural velocity distribution in the Y-direction. (**c**) Structural velocity distribution in the Z-direction.

To accurately analyze the specific velocities and variation patterns of the structure affected by blasting vibrations, velocity–time history curves under the influence of different directional vibrations are extracted for the subgrade, rail ties, and steel rails, respectively.

Figure 6a–c represent the velocity–time history curves in three directions under the influence of blasting vibrations. It is observed that the velocity–time history curves of the railway structure in all three directions generally align with the waveforms of the vibration curves produced by the blasting. In the x-direction velocity response curve of the structure, the velocity response value is highest at the rail ties, followed by the subgrade, and lowest at the steel rails. The peak value occurs at 1.65 s at the rail ties, with a value of 0.54 cm/s. In the y-direction and z-direction velocity response curves of the structure, the velocity response value is highest at the subgrade, followed by the rail ties, and lowest at the steel rails. The peak values in the y-direction and z-direction occur at 2.75 s (0.35 cm/s) and 1.95 s (0.51 cm/s), respectively. This indicates a directional variation in the subgrade response, with the y-direction exhibiting a later peak compared to the z-direction. The peak vibration velocities in each direction are all significantly lower than the blasting vibration safety limits specified in TB10313-2019 "Technical Regulations for Railway Engineering Blasting Vibration Safety [21]" (see Table 4).

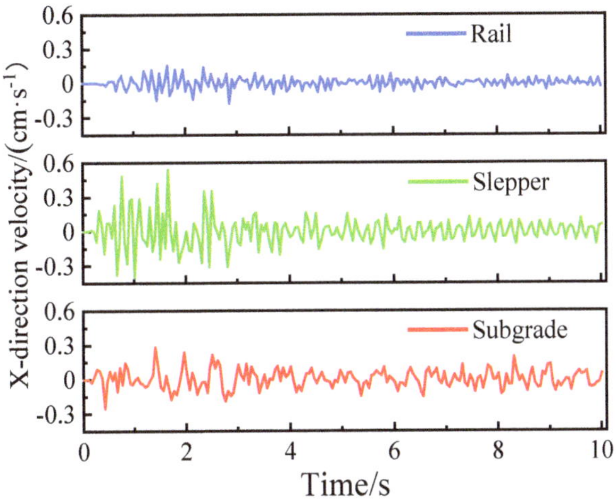


(a)

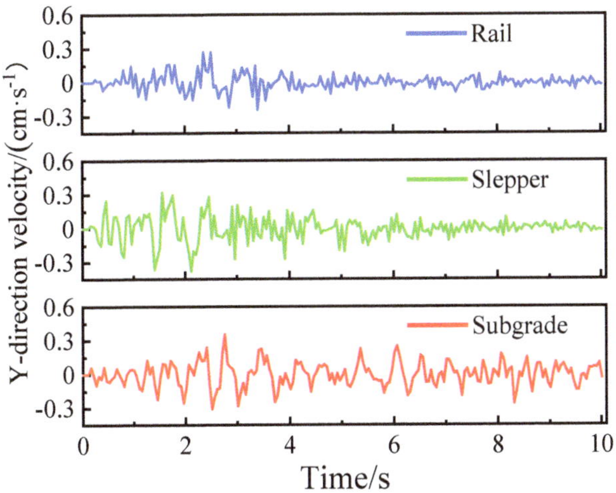


(b)

**Figure 6.** *Cont.*

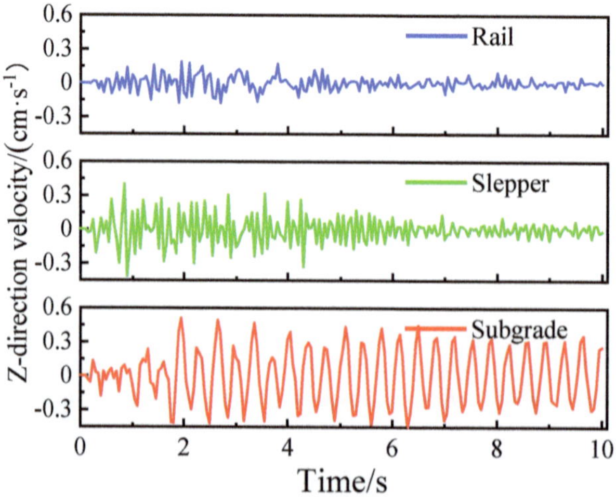


(c)

**Figure 6.** Time history curves of response speed of structures in different directions. (**a**) Time history curve of velocity in the X-direction of the structure; (**b**) Time history curve of velocity in the Y-direction of the structure; (**c**) Time history curve of velocity in the Z-direction of the structure.

**Table 4.** Safety allowable values for blasting vibration of railway subgrade.

| Category | Particle Vibration Velocity Safety Limit V/(cm·s$^{-1}$) | | |
|---|---|---|---|
| | f ≤ 10 Hz | 10 Hz < f ≤ 50 Hz | f > 50 Hz |
| Ballasted track subgrade | 5~6 | 6~7 | 7~8 |
| Unballasted track subgrade | 3~4 | 4~5 | 5~6 |

Upon comparing the three directional velocity response curves, it becomes evident that the x-direction velocity response value exceeds those in the other directions. This is attributed to the larger peak horizontal radial velocity produced by the blasting, coupled with a smaller main frequency in comparison to the y-direction. Although the peak velocity in the vertical direction is lower, its main frequency is closer to the structure's natural frequency, leading to a more significant velocity response in the z-direction than in the y-direction.

#### 4.3.3. Structural Displacement Response Analysis

The impact of blasting vibrations on the displacement deformation of the railway structure cannot be overlooked. Figure 7 depicts the displacement distribution contour map at the moment of maximum velocity in the three directions.

The displacement response observed at the rail location within the railway structure is notably diminished compared to that at the rail tie and subgrade. Displacement time history curves for the subgrade, rail tie, and rail were extracted under the influence of vibrations in different directions, as shown in Figure 8a–c. The peak displacement values in the x, y, and z directions are 0.185 cm, 0.155 cm, and 0.118 cm, respectively. The peak displacement values in all three directions are much smaller than the limits specified in the National Railway Administration's standard TB10001-2016 "Code for Design of Railway Subgrades", which states that "the dynamic deformation of the surface layer of the subgrade should meet $\omega \leq C\omega$, where $\omega$ is the calculated deformation value (mm) and $C\omega$ is the deformation limit (mm), with 1 mm for ballasted tracks [22]".

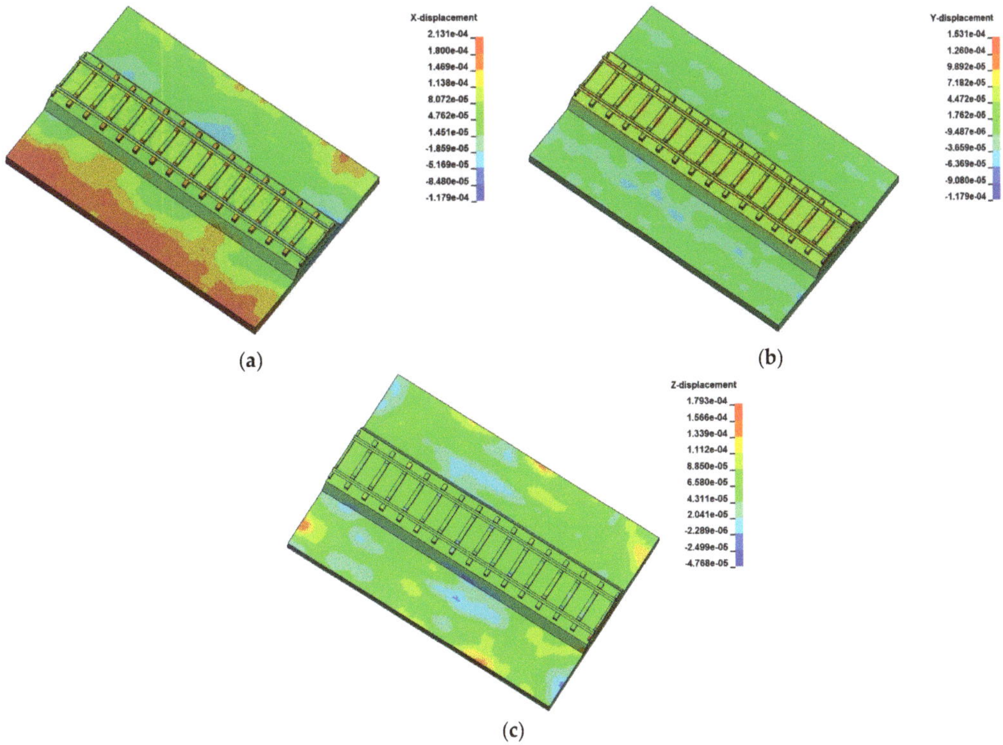


**Figure 7.** Three directional displacement distribution. (**a**) Structural displacement distribution in the X-direction; (**b**) Structural displacement distribution in the Y-direction; (**c**) Structural displacement distribution in the Z-direction.

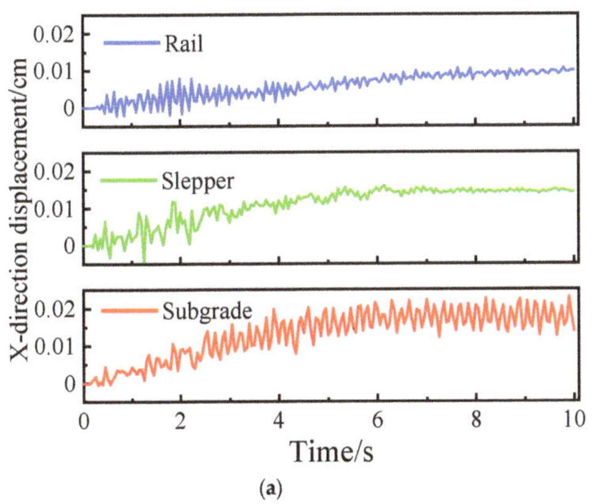


**Figure 8.** *Cont.*

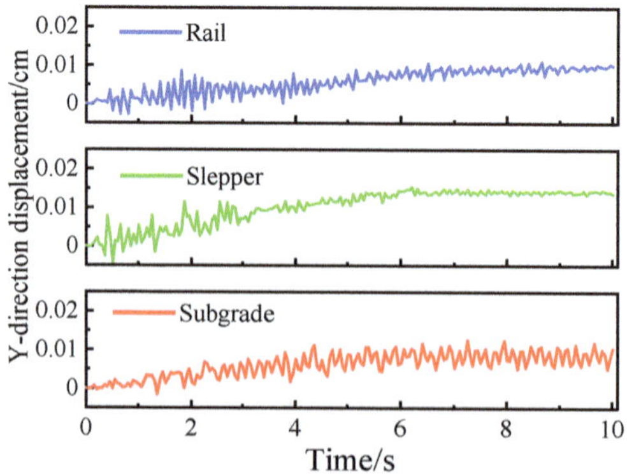


(**b**)

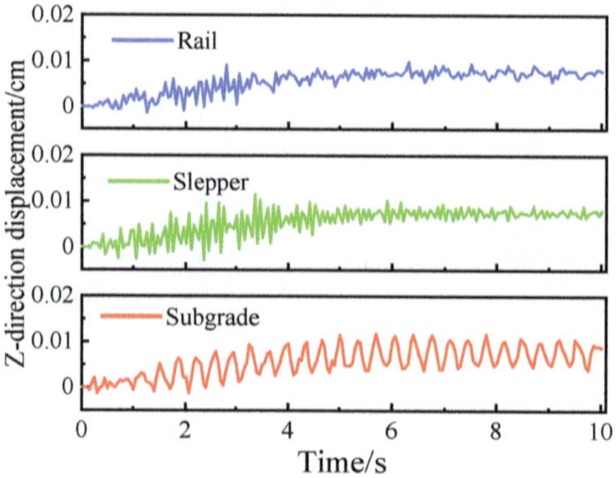


(**c**)

**Figure 8.** Time history curves of displacement in different directions of the structure. (**a**) Structural displacement time history in the X-direction. (**b**) Structural displacement time history in the Y-direction. (**c**) Structural displacement time history in the Z-direction.

By comparing the displacement time history curves in all three directions, it can be observed that the displacement curves induced by blasting vibrations show an initial increase followed by stabilization. The displacement impact in the z direction is relatively small. In the displacement response curves for the x and z directions, the displacement response is greatest at the subgrade, followed by the rail tie, and smallest at the rail. However, in the displacement response curve for the y direction, the rail tie shows the largest displacement response, followed by the subgrade, and the smallest at the rail. This indicates that the structural displacement deformation is not only related to the characteristics of the vibration waves but is also significantly influenced by the structural properties of the railway.

4.3.4. Analysis of the Impact of Different Earthquake Wave Amplitudes on Structures

A multitude of studies have indicated that the peak velocity and primary frequency of blasting vibration waves are two critical parameters influencing the safety of structures [23]. In this study, the overall effect of the blasting vibration wave on the railway structure is relatively minor. Nevertheless, when the maximum single-section charge increases or the distance to the blasting area diminishes, the impact on the railway structure becomes more significant. To investigate the impact pattern of vibration waves on railway structures, while maintaining the main frequency and duration constant, the peak vibration velocity of the selected three-direction vibration waves is amplified by various multiples as detailed in Table 5 and subsequently applied to the railway structure model for analysis.

**Table 5.** Speed amplification factor.

| Magnification Factor | 1 | 2 | 4 | 8 | 16 | 32 | 64 |
|---|---|---|---|---|---|---|---|
| X-direction/(cm·s$^{-1}$) | 0.169 | 0.338 | 0.676 | 1.352 | 2.704 | 5.408 | 10.816 |
| Y-direction/(cm·s$^{-1}$) | 0.167 | 0.334 | 0.668 | 1.336 | 2.762 | 5.344 | 10.688 |
| Z-direction/(cm·s$^{-1}$) | 0.098 | 0.196 | 0.392 | 0.784 | 1.568 | 3.136 | 6.272 |

Figure 9 illustrates the relationship between the peak vibration velocity of various structures and the loading speed across three axes. Analysis of Figure 9a–c indicates that as the loading speed in different directions rises, the overall peak vibration velocity of the structures increases to varying extents. The peak vibration velocity in the z-axis is relatively low. The peak vibration velocity at the rail location is also low, and its rate of increase is gradual as the loading speed escalates, suggesting that blasting vibrations have a minimal impact on the rail's vibration. Conversely, the peak vibration velocity at the subgrade is considerably higher than at other structures. Should the loading speed escalate further, the subgrade is likely to be the first to sustain damage. When the loading speed in the x-axis reaches or exceeds 5.408 cm/s, the peak vibration velocity at the subgrade surpasses the permissible standard set by relevant regulations, posing a potential safety risk to the structure.

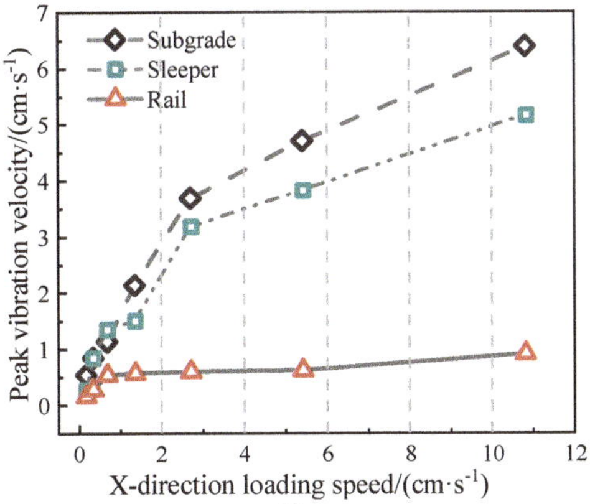


(a)

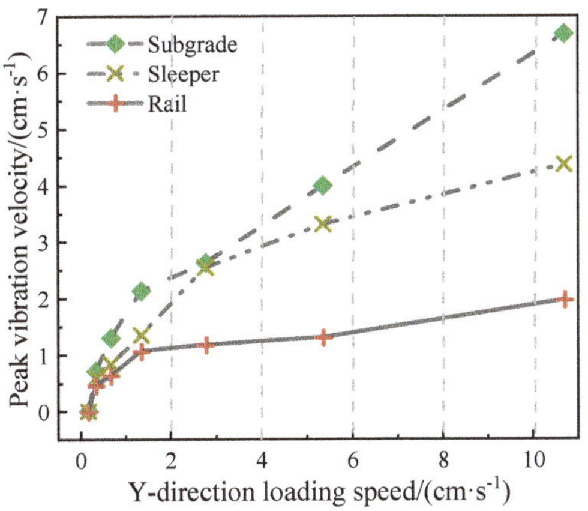


(b)

**Figure 9.** *Cont.*

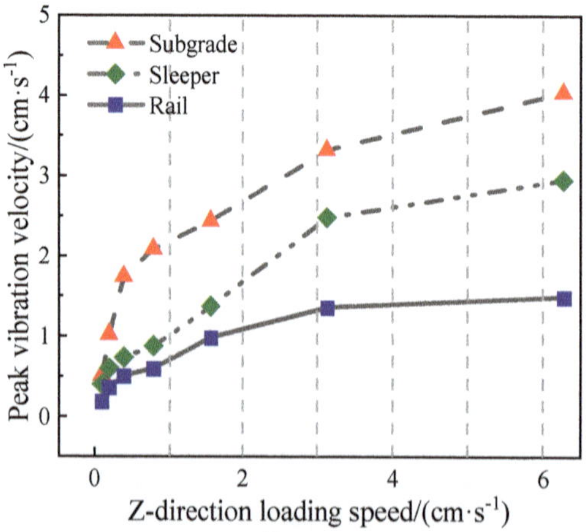


(c)

**Figure 9.** Relationship between peak vibration velocity of different structures and loading speed. (**a**) Variation in structural peak vibration velocity with X-direction loading speed; (**b**) Variation in structural peak vibration velocity with Y-direction loading speed; (**c**) Variation in structural peak vibration velocity with Z-direction loading speed.

The relationship between the maximum displacement of various structures and the loading speed in three directions is depicted in Figure 10. Analysis of Figure 10a–c indicates that the maximum displacement in all three directions increases to varying extents with the rise in loading speed. When the loading speed in all three directions is low, there is no significant difference in the maximum displacement among different structures. However, as the loading speed gradually increases in the x and z directions, the maximum displacement at the subgrade is notably higher than at the other two locations, and the displacement changes of the track slab and rail are similar. In the y direction, the difference in the maximum displacement change between structures is more pronounced, and the overall displacement response of the structure in the z direction is smaller. When the loading speed in the x and y directions exceeds 3 cm/s, the maximum displacement at the subgrade surpasses 1 mm, exceeding the allowable standard as per relevant regulations, rendering it susceptible to damage.

Under the influence of peak velocity and dominant frequency of blast vibrations, the railway structure exhibits various displacement and velocity responses due to its structural characteristics. Numerical simulations indicate that the inherent frequency of the railway structure is 3.016 Hz. As the peak velocity of blast vibrations rises and the frequency approaches, the structure's velocity and displacement increase at varying rates, contingent on the structure's inherent characteristics. The increase becomes more significant with higher loading speeds. Considering the peak velocity and maximum displacement of the vibrating structure, it is advisable to control the blast vibration velocity below 3 cm/s. Additionally, based on the maximum single charge used in mining area blasting, the minimum safe distance can be calculated using the Sadovski formula.

The Sadovski formula [24] can be expressed as follows:

$$V = K\frac{\sqrt[3]{Q}^{\alpha}}{R} \tag{3}$$

In this formula, V represents the vibration velocity of the medium particles, measured in cm/s; Q is the maximum amount of explosive charge in a single segment, measured in kg; R is the distance from the blast center, measured in meters; and K and $\alpha$ are coefficients and attenuation indices related to the terrain and geological conditions between the blast area and the measurement point, respectively.

The maximum single charge amount for the mine is 95 kg. Based on the blasting vibration velocity detection results, it is known that K and α are 208 and 1.69, respectively. After calculation, the minimum safety distance is found to be 138 m.

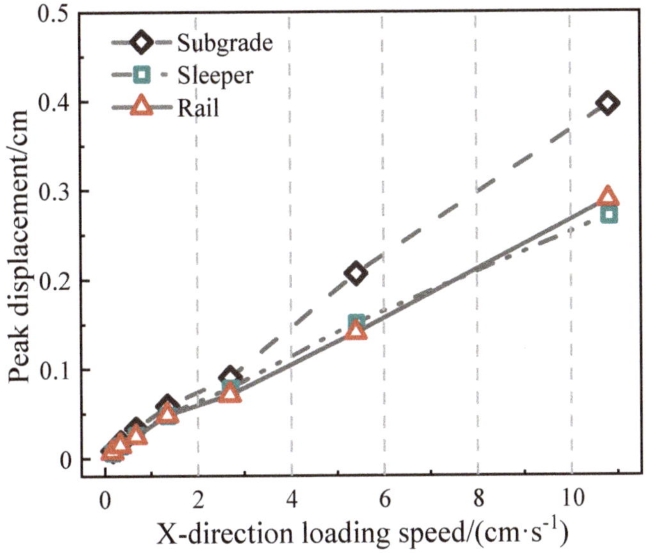


(**a**)

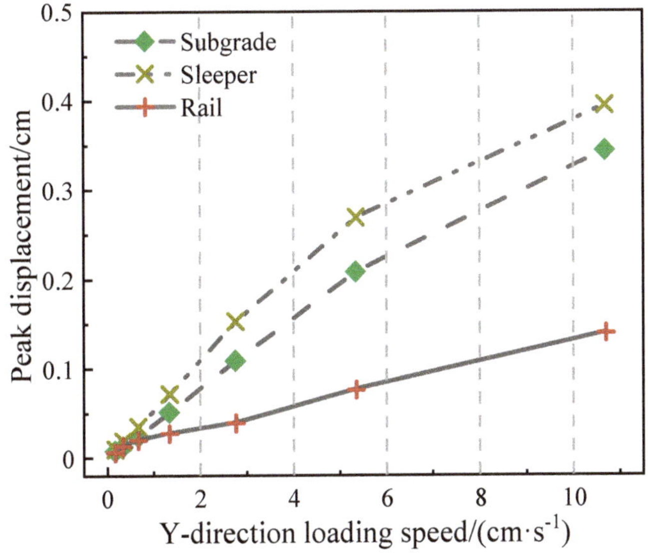


(**b**)

**Figure 10.** *Cont.*

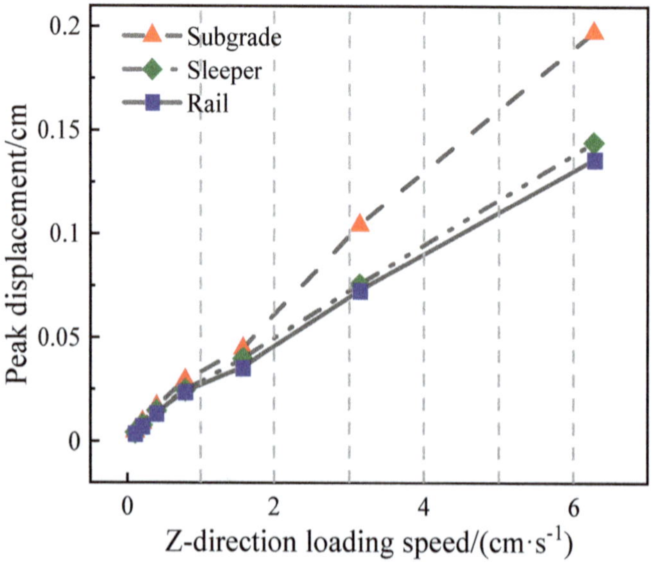


(c)

**Figure 10.** The relationship between the maximum displacement of different structures and the loading speed. (**a**) Variation in structural maximum displacement with X-direction loading speed; (**b**) Variation in structural maximum displacement with Y-direction loading speed; (**c**) Variation in structural maximum displacement with Z-direction loading speed.

## 5. Conclusions

(1) The analysis of vibration test data and simulation results indicates that the peak velocities and displacements generated in the railway structure are significantly below the regulatory limits specified in relevant standards. Specifically, the maximum recorded vibration velocity of 0.54 cm/s at the rail-sleeper interface in the x-direction and the maximum displacement of 0.185 cm at the subgrade are well within safe operational thresholds. These findings confirm that the current blasting activities have a minimal impact on the structural integrity of the railway.

(2) Among the three vibration directions, the x-direction has the most significant impact on the railway structure. This is attributed to the larger peak horizontal radial velocity generated by blasting and the relatively lower dominant frequency compared to other directions. The z-direction also exhibits notable responses due to its closer proximity to the natural frequency of the railway structure.

(3) A comparative analysis of the displacement and velocity responses of the structure under different amplitudes of vibration waves reveals that the structure exhibits varying increasing trends, which are significantly correlated with the characteristics of the vibration waves and the structure itself. Based on relevant codes and standards, and considering the peak velocity and displacement generated by the structure, it is recommended to control the blast vibration velocity to below 3 cm/s and maintain a minimum distance of 138 m between the blast zone and the railway.

(4) Although the impact of blasting in the mining area on the safety of nearby railways is currently relatively small, the cumulative effects of multiple blasts on the railway structure should be taken into account. To minimize the impact of blast vibrations on nearby railway structures, it is recommended to optimize drilling and explosive charge arrangements by controlling charge height and regulating explosive quantities per blast and per segment. Additionally, implementing pre-split blasting can create buffer cracks

to mitigate vibration propagation. Furthermore, the blasting workface should be oriented perpendicular to the railway alignment, and a safe distance must be maintained between the blasting zone and railway structure to ensure structural integrity.

**Author Contributions:** This manuscript was authored through the collaborative efforts of all contributors. Conceptualization, Y.H.; methodology, K.W. and D.H.; software, K.W.; validation, K.W., D.H., S.Z. and Z.Z.; formal analysis, S.Z. and Z.Z.; investigation, S.Z. and Z.Z.; resources, Y.H.; data curation, S.Z. and Z.Z.; writing—original draft preparation, S.Z. and Z.Z.; writing—review and editing, Y.H., K.W., D.H., S.Z. and Z.Z.; visualization, S.Z. and Z.Z.; supervision, Y.H.; project administration, Y.H.; funding acquisition, Y.H. All authors have read and agreed to the published version of the manuscript.

**Funding:** The research received financial support from the National Natural Science Foundation of China (Grant No. 52064025).

**Institutional Review Board Statement:** Not applicable.

**Informed Consent Statement:** Not applicable.

**Data Availability Statement:** The raw data supporting the conclusions of this article are available from the authors on request.

**Acknowledgments:** Thanks are extended to the editors and reviewers. Thanks to Shuai Zhang and Yonghui Huang for contributing to the article.

**Conflicts of Interest:** The authors declare no conflicts of interest.

## Abbreviations

The following abbreviations are used in this manuscript:

| | |
|---|---|
| FFT | fast Fourier transform |

## References


1. Fang, R.; Zheng, L.; Yao, H.; Luo, X.; Yang, M. Study on Dynamic Responses of Unsaturated Railway Subgrade Subjected to Moving Train Load. *Soil Dyn. Earthq. Eng.* **2018**, *115*, 319–323. [CrossRef]
2. Komadja, G.C.; Rana, A.; Glodji, L.A.; Anye, V.; Jadaun, G.; Onwualu, P.A.; Sawmliana, C. Assessing Ground Vibration Caused by Rock Blasting in Surface Mines Using Machine-Learning Approaches: A Comparison of CART, SVR and MARS. *Sustainability* **2022**, *14*, 11060. [CrossRef]
3. Zhang, X.; Huang, T.; Ge, Z.; Man, T.; Huppert, H.E. Infiltration Characteristics of Slurries in Porous Media Based on the Coupled Lattice-Boltzmann Discrete Element Method. *Comput. Geotech.* **2024**, *177*, 106865. [CrossRef]
4. Fissha, Y.; Ikeda, H.; Toriya, H.; Owada, N.; Adachi, T.; Kawamura, Y. Evaluation and Prediction of Blast-Induced Ground Vibrations: A Gaussian Process Regression (GPR) Approach. *Mining* **2023**, *3*, 659–682. [CrossRef]
5. Zhou, Z.; Jia, B.; Yuan, W.; Zhang, J. Vibration Response Prediction and Vibration Safety Assessment Method for Building Structures around Viaducts under the Action of High-Speed Railway Trains. *Constr. Build. Mater.* **2024**, *422*, 135877. [CrossRef]
6. Auersch, L. Different Types of Continuous Track Irregularities as Sources of Train-Induced Ground Vibration and the Importance of the Random Variation of the Track Support. *Appl. Sci.* **2022**, *12*, 1463. [CrossRef]
7. Wu, Y.; Zhang, X.; Li, Z.; Gao, W.; Xu, Z.; Zhang, Y.; Zhou, J. Effects of In-Situ Stress on Damage and Fractal during Cutting Blasting Excavation. *Fractal Fract.* **2024**, *8*, 450. [CrossRef]
8. Yang, J.; Zhu, S.; Li, Q.; Cai, C.; Zhai, W. A Spatial Train-Track-Tunnel Coupled Dynamics Model for Assessing Influences of Curved Railway Line on Circular Tunnel Vibrations. *Appl. Math. Model.* **2024**, *140*, 115892. [CrossRef]
9. Eades, R.Q.; Perry, K. Understanding the Connection between Blasting and Highwall Stability. *Int. J. Min. Sci. Technol.* **2019**, *29*, 99–103. [CrossRef]
10. Fattah, M.Y.; Majeed, Q.G.; Joni, H.H. Stresses in Saturated and Unsaturated Subgrade Layer Induced by Railway Track Vibration. *Infrastructures* **2024**, *9*, 64. [CrossRef]
11. Sołtys, A.; Pyra, J. The Influence of Vibrations Induced by Blasting Works in an Open-Pit Mine and Seismic Events in an Underground Mine on Building Structures—A Case Study. *Appl. Sci.* **2024**, *14*, 4414. [CrossRef]
12. Qu, C.; Tan, X.; Xiao, Y.; Wang, Z.; Wei, L. Subgrade Vibrations and Long-Term Stability of an Embankment-Bridge Transition Zone in Non-Ballasted High-Speed Railway. *Transp. Geotech.* **2024**, *45*, 101199. [CrossRef]

13. Park, S.; Kim, J.Y.; Kim, J.; Lee, S.; Cho, K.-H. Analysis of Dynamic Characteristics of Deformed Concrete Slab Track on Transition Zone in High-Speed Train Line according to Train Speeds. *Appl. Sci.* **2020**, *10*, 7174. [CrossRef]
14. Khajehdezfuly, A.; Shiraz, A.; Sadeghi, J. Assessment of Vibrations Caused by Simultaneous Passage of Road and Railway Vehicles. *Appl. Acoust.* **2023**, *211*, 109510. [CrossRef]
15. Lichun, J.; Liuqing, H.; Xiuying, L. Investigation on the Threshold Control of Safety Blasting Vibration Velocity for the Extraction of Complicated Orebody under Railway. *Min. Sci. Technol. (China)* **2011**, *21*, 169–174. [CrossRef]
16. Li, Z.; Hu, Y.; Wang, G.; Zhou, M.; Hu, W.; Zhang, X.; Gao, W. Study on Cyclic Blasting Failure Characteristics and Cumulative Damage Evolution Law of Tunnel Rock Mass under Initial In-Situ Stress. *Eng. Fail. Anal.* **2023**, *150*, 107310. [CrossRef]
17. Karnovsky, I.A.; Lebed, O. Dynamics of Elastic Systems: Forced Vibration. In *Advanced Methods of Structural Analysis*; Springer: Cham, Switzerland, 2021; pp. 615–690. [CrossRef]
18. Cruz, C.; Miranda, E. Evaluation of the Rayleigh Damping Model for Buildings. *Eng. Struct.* **2017**, *138*, 324–336. [CrossRef]
19. Yang, M.; Wu, J.; Zhang, Q. Inclination and Acceleration Data Fusion for Two-Dimensional Dynamic Displacements and Mode Shapes Identification of Super High-Rise Buildings Considering Time Delay. *Mech. Syst. Signal Process.* **2024**, *223*, 111938. [CrossRef]
20. Jin, H.; Hao, H.; Hao, Y.; Chen, W. Predicting the Response of Locally Resonant Concrete Structure under Blast Load. *Constr. Build. Mater.* **2020**, *252*, 118920. [CrossRef]
21. *TB10313-2019*; The National Standards Compilation Group of People's Republic of China, Safety Technical Specification for Blasting Vibration of Railway Engineering. Chinese Standard Press: Beijing, China, 2019. (In Chinese)
22. *TB10001-2016*; The National Standards Compilation Group of People's Republic of China, Code for Design of Railway Earth Structure. Chinese Standard Press: Beijing, China, 2016. (In Chinese)
23. Liu, X.; Zeng, Y.; Li, H.; Xia, X.; Li, J.; Li, Z. Research on Blasting Vibration Effect and Time-Frequency Characteristics of Vibration Signals in a Road Corridor at Xianning Nuclear Power Station. *J. Vibroeng.* **2021**, *23*, 823–846. [CrossRef]
24. Singh, P.K.; Roy, M.P. Damage to Surface Structures due to Blast Vibration. *Int. J. Rock Mech. Min. Sci.* **2010**, *47*, 949–961. [CrossRef]

applied sciences

MDPI

*Article*

# The Investigation of Porcelain Plates Roughness Influence on Determination of Explosives Friction Sensitivity

**Sonia Nachlik * and Mateusz Pytlik**

Conformity Assessment Body, Central Mining Institute-National Research Institute, 1 Gwarków Square, 40-166 Katowice, Poland; mpytlik@gig.eu
* Correspondence: snachlik@gig.eu

**Abstract:** The influence of the roughness of porcelain plates on the results of the friction sensitivity test of explosives was investigated. For this purpose, the roughness of selected batches of plates from several manufacturers (Julius Peters, OZM Research s.r.o., Deltima Precision s.r.o.) was determined. Subsequently, according to the standards EN 13631-3:2005 and STANAG 4487 JAIS (edition 2), friction sensitivity tests of PETN (Pentaerythritol Tetranitrate, penthrite) and RDX (Royal Demolition Explosive, hexogen) were carried out. No statistically significant correlation was established under the experimental conditions—although sanded plates had the lowest roughness (mean value 5.07 μm) and simultaneously gave the lowest sensitivity results (168 N for RDX and 80 N for PETN according to the EN 13631-3 while 216 N for RDX and 121 N for PETN according to the STANAG 4487), Julius Peters plates with a similar level of roughness (mean value 6.07 μm) did not reflect the pattern and results of the sensitivity tests that were surprisingly high (120 N for RDX and 64 N for PETN according to the EN 13631-3 while 182 N for RDX and 67 N for PETN according to the STANAG 4487). Due to these results, the human factor is indicated as a key factor in the obtained discrepancies; however, further research in this matter may be needed.

**Keywords:** roughness; porcelain plates; friction sensitivity; explosives

## 1. Introduction

Due to the development of the explosives industry, there is a need to define the requirements for their safe manufacturing, storage, transport, and use. The discrepancies between the effects of different mechanical, thermal, or electrical stimuli on the tendency of materials to undergo explosive decomposition have been noticed quite quickly. Consequently, different methods have been developed to determine the sensitivity of explosives depending on the chosen stimulus.

The most common and impossible to completely eliminate stimulus is friction, which is present during the handling, mixing, and packaging of a substance. Moreover, the ability to convert a part of the frictional energy into thermal energy must be considered when manipulating a given material. Consequently, several methods have been developed to determine the frictional sensitivity of an explosive, while minimizing the risk of other stimuli accompanying the frictional force, such as heat release and additional impact on the sample. Avrami and Hutchinson, in agreement with Brown [1,2], listed three categories of friction sensitivity testing methods:

1. In which a thin layer of material is placed between two rigid elements, which then act on the material in a reciprocating or rotary-continuous motion;

2. In which a block of material is rubbed violently against an abrasive surface;
3. In which the explosive sample is deformed significantly during impact (a combination of frictional and impact stimuli suitable only for secondary explosives).

Most tests used worldwide belong to the first category and differ in the design of the apparatus in which the explosive is placed, the type of frictional surfaces, and the nature of the movement acting on the sample.

Among these is the Bundesanstalt für Materialforschung und -prüfung (BAM) friction test developed by Germans and standardized in Europe. It is carried out using a Julius Peters apparatus. The apparatus has a table for placing grooved porcelain plates and a notched arm on which weights of a chosen mass and at a chosen distance from the end of the arm are hung so as to obtain a target value of the friction force acting on the sample. Approximately 10 $mm^3$ of the test substance is placed on a porcelain plate attached to the apparatus carriage. A porcelain peg attached to the arm acts on the sample from above [3]. Depending on the type of tested material (primary or secondary explosive), a smaller- or larger-scale Julius Peters apparatus is used. Loads for primary materials range from 0.1 to 10 N, while for secondary materials the range is 5 to 353 N [4,5]. Each time, the test result for a given load is assessed in terms of the occurrence of flame, sparkle, crackling, odor of decomposition products, and/or left trace on the plate. The final sensitivity of the material is determined based on the chosen standard.

There is a known issue with a certain degree of subjectivity in the evaluation of test results. The BAM friction test provides less clear results than the analogous impact test, particularly in the case of an inexperienced researcher. The method of preparing the samples for testing and the size of the crystals also have a significant impact on the obtained sensitivity parameters. For azides with different crystal sizes, Bowden and Singh noticed the divergent nature of decomposition dynamics: larger crystals displayed a tendency to decompose rapidly during heating and were also more sensitive to impact [6]. The influence of the method of sample preparation and crystal size on friction sensitivity was analyzed by Matyas et al. [7]. Radacsi et al. [8] observed that the BAM friction test yields unreliable results for sub-micrometer-sized RDX and HMX (High Molecular weight Explosive, octogen) particles, as the crystals diffuse across the surfaces of the plates and fill the grooves.

The results of the 2014 edition of the Round Robin Test on Explosives (RRTE) indicate significant differences in the results of sensitivity determinations (shown in Table 1) depending on the laboratories carrying out tests and the porcelain components used (sensitivity was determined for identical samples of double-base propellant). Limiting the number of manufacturers of plates and pegs during these studies significantly reduced the discrepancies from the range of 96–288 N to the range of 96–160 N, leaving only room for differences caused by the human factor (Table 2) [9].

Furthermore, another discrepancy in the results of friction sensitivity determinations for penthrite and hexogen (according to the statistical Bruceton's method) is mentioned in STANAG 4487 (edition 2) [10]. The sensitivity of PETN ranges between 56 and 102 N, whereas the range for RDX is 110–185 N. Such differences may be due to possible wetting of the samples (resulting in the phlegmatisation of the material [11]); nevertheless, the influence of the applied porcelain components may also be the reason. The results of comparative tests between Polish laboratories (carried out in 2008) also show differences in the obtained values, with results ranging between 107 and 240 N for crystalline RDX samples tested according to the European standard [12].

Le Roux highlighted the variation in the roughness of the available porcelain plates. He suggested using an emery paper while testing primary explosives owing to its normalized surface (in comparison with the lack of uniformity of roughness for plates) [13].

**Table 1.** Friction sensitivity of double-base propellant determined according to the European standard during the RRTE 2014 edition [9].

| No. of Laboratory | Porcelain Components Manufacturer | Determined Sensitivity [N] |
|---|---|---|
| 1 | No data | 192 |
| 2 | OZM Research s.r.o. | 168 |
| 3 | Reichel and Partner GmbH | 240 and 252 [1] |
| 4 | OZM Research s.r.o. | 240 |
| 5 | No data | 168 and 216 |
| 6 | OZM Research s.r.o. | 96, 108 and 128 [1] |
| 7 | OZM Research s.r.o. | 288 |
| 8 | Julius Peters | 96 |
| 9 | OZM Research s.r.o. | 168 |

[1] From various technicians.

**Table 2.** Friction sensitivity of double-base propellant determined after choosing one manufacturer of porcelain components [9].

| No. of Laboratory | Porcelain Components Manufacturer | Determined Sensitivity [N] |
|---|---|---|
| 1 | OZM Research s.r.o. | 128 |
| 3 | OZM Research s.r.o. | 108 |
| 5 | OZM Research s.r.o. | 160 |
| 6 | OZM Research s.r.o. | 96 |
| 10 | OZM Research s.r.o. | 112 |

Both the plates and pegs for the BAM friction test have standardized dimensions. The grooves are made with a special sponge imprinted before firing the ceramics. The European standard [14] also specifies requirements for the surface roughness of the plates: the arithmetic mean of the plate profile deviations (*Ra* parameter) should be between 6 and 15 µm, with individual readings between 5 and 19 µm. STANAG 4487 (edition 2) does not specify the roughness of plates.

Currently, there are no requirements for checking the roughness of plates prior to friction sensitivity testing. However, the discrepancies in the results raise questions as to whether such a procedure should be introduced, especially when the examined explosives are on the edge of admission for civil use. According to European standards secondary explosives should not be sensitive to loads lower than 80 N (EN 13631-1:2005 [15]) whereas in the case of explosives used for detonating cords, the limit is 30 N (EN 13630-1:2003 [16]).

This paper presents an attempt to determine the effect of porcelain plate roughness on friction test results. To this end, the profiles of selected types of plates available on the European market were measured and the number of scratches on the plates was visually assessed. The results of the measurements were compared with the results of subsequent sensitivity tests for penthrite and hexogen, according to Bruceton's statistical method and the European standard EN 13631-3:2005.

## 2. Materials and Methods

Porcelain plates produced by the following manufacturers were selected for testing:

- Julius Peters;
- OZM Research s.r.o. (Hrochův Týnec, Czech Republic);
- Deltima Precision s.r.o. (Praha, Czech Republic).

For each batch of plates, OZM pegs were used. Julius Peters plates were noticeably thicker (approx. 5.5 mm, while it was approx. 5 mm for the others). In addition, only one side of the plates was designed for measurements; significant scratches were carved on the other side (Figure 1).

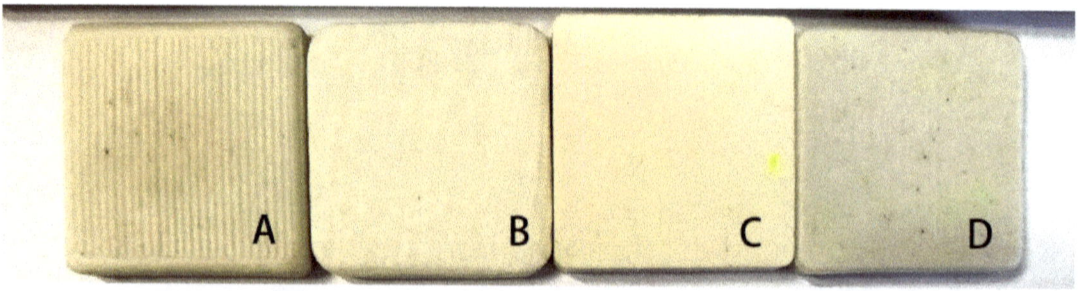


**Figure 1.** Selected plates: (**A**) Julius Peters; (**B**) old OZM; (**C**) new OZM; (**D**) Deltima Precision.

Due to the large differences between the two batches of the owned OZM plates, it was decided to carry out the experiments for both of them. Older batch plates had an uneven shape, and their thickness, distribution, and depth of grooves varied within a single plate. The newer batch did not have such flaws. Moreover, a part of the new batch was sanded with emery paper (60 grit) to purposely reduce its roughness and compare the obtained results. Fifty plates from each plate type, including sanded ones, were tested (with the exception of the single-sided Julius Peters plates, for which one hundred units were chosen). Each time, each side of the plate was tested in triplicate.

To determine the profile of a given surface, a number of contact or non-contact methods may be applied, each of which differs in the simplicity of the application under given laboratory conditions, the range of roughness to be measured, and the ease of handling the output data [17]. In addition, more precise methods for the measurement and processing of surface profiles are still being developed [18,19]. In this study, the contact method for roughness measurement was chosen. For this purpose, a SURFTEST SJ-410 surface roughness tester (software Ver. 2.00) from Mitutoyo (Kanagawa, Japan) was used, and the following three parameters were measured:

- *Ra*—arithmetical mean of absolute profile deviations;
- *Rt*—total height of profile;
- *Rz1max*—maximum roughness depth [20].

A visual assessment of the roughness was also carried out. First, magnified photographs of the plates were taken to apply a filter. The number of grooves per measurement section of the plate was then assessed.

In the second part of the study, material for the indication of friction sensitivity, which included crystalline RDX (provided by NITRO-CHEM S.A., Bydgoszcz, Poland) and crystalline PETN (provided by SSE Group, Brig, Switzerland), was chosen. The results of the granulometric analysis of the substances used are shown in Figures 2 and 3. The distributions of RDX and PETN fractions were obtained using a laser analyzer ANALYSETTE 22 MicroTec plus (Fritsch GmbH, Idar-Oberstein, Germany).

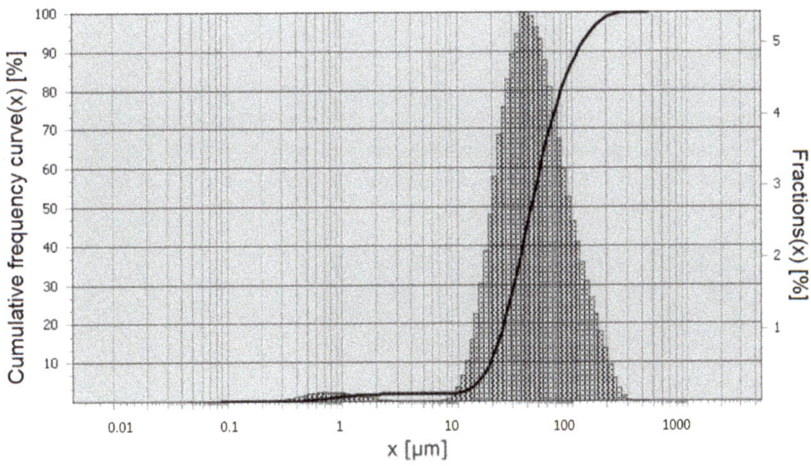


**Figure 2.** Fractions distribution of used RDX.

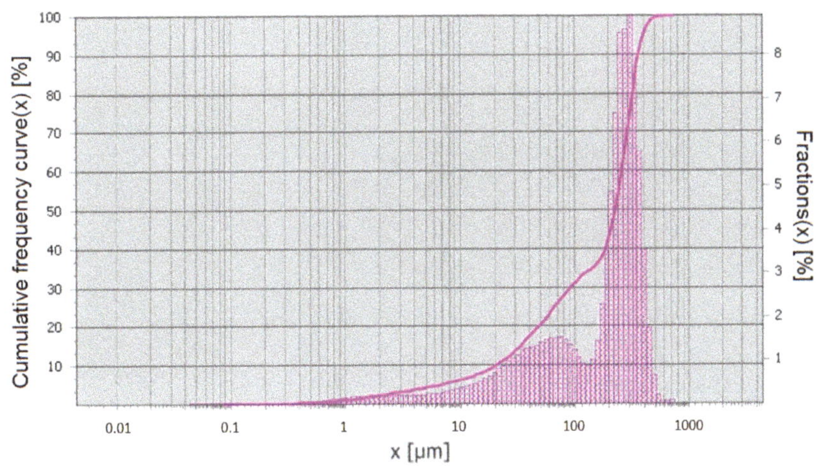


**Figure 3.** Fractions distribution of used PETN.

The assessment of friction sensitivity was carried out by a single technician (in order to avoid the variable of different operators) according to the standards EN 13631-3:2005 and STANAG 4487 JAIS (Editon 2), using Julius Peters apparatus. In the European standard friction sensitivity is defined as the lowest peg load at which a reaction occurs in at least one of the six tests [14]. However, in STANAG the sensitivity of an explosive is described with a 50% point, i.e., a load calculated using the Bruceton statistical method (also known as the up-and-down method), for which the probability of initiation is 50%. To obtain the 50% point, at least 25 tests are required [10]. Noteworthy, the up-and-down method, established in the 1940s, has many limitations (e.g., the condition of choosing an appropriate interval for loading the arm during the test, so that the value of the interval divided by the obtained standard deviation is between 0.5 and 2), which has led to the development of new methods for calculating the 50% point, like the Neyer D-optimal method or the probit analysis [21–23]. Nevertheless, Bruceton's method is still commonly used to determine explosives' sensitivity, e.g., in the United States [24].

# 3. Results

## *3.1. Roughness Evaluation*

### 3.1.1. Plates Visual Comparison

During the visual assessment of the plates surfaces, magnified photographs were taken. Photographs before and after the filter was applied are shown in Figures 4–7. The red sections visible in the photos are the measurement sections of 1 cm, for which the number of grooves was counted (Table 3).

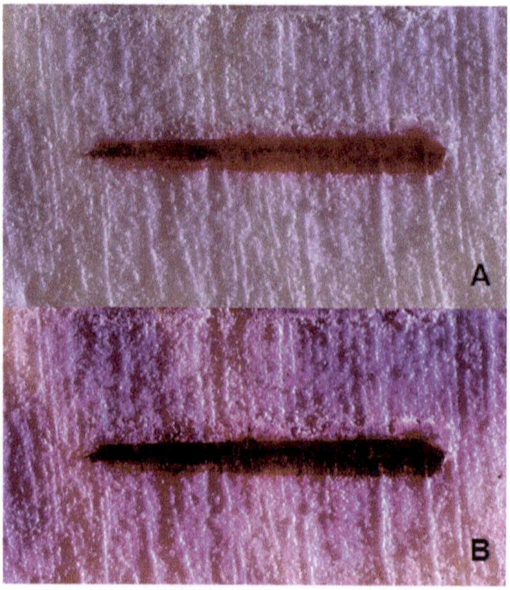


**Figure 4.** Julius Peters plate: (**A**) without filter; (**B**) with filter.

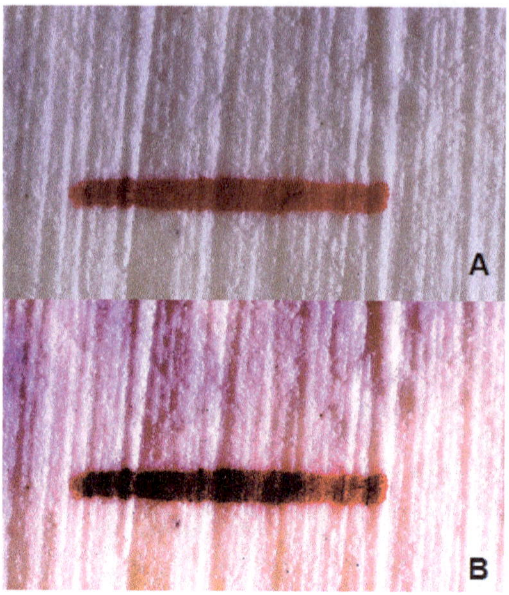


**Figure 5.** Old OZM plate: (**A**) without filter; (**B**) with filter.

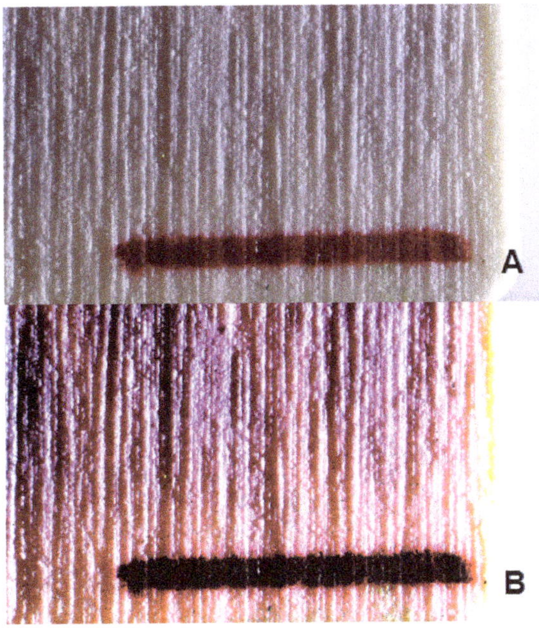


**Figure 6.** New OZM plate: (**A**) without filter; (**B**) with filter.

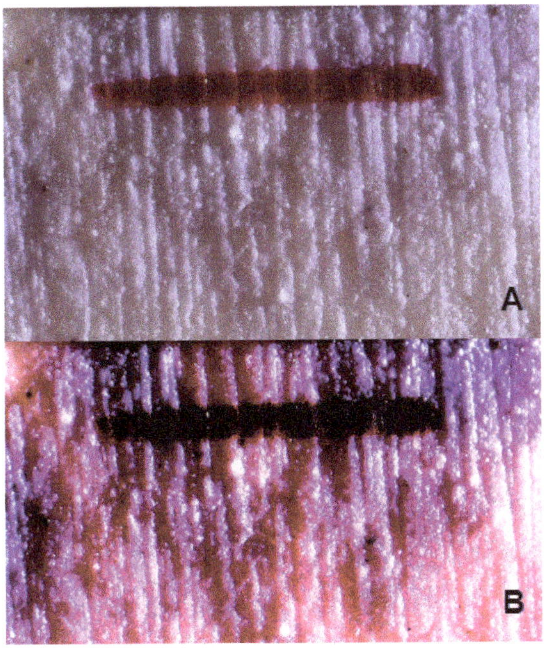


**Figure 7.** Deltima Precision plate: (**A**) without filter; (**B**) with filter.

**Table 3.** Results of plates visual comparison.

| Plate Type | Grooves per 1 cm (Mean Value) | Easy Identification of Single Groove | Surface Homogeneity |
|---|---|---|---|
| Julius Peters | 16 ± 2 | No | No |
| Old OZM | 16 ± 3 | Yes | No |
| New OZM | 18 ± 2 | Yes | Yes |
| Deltima Precision | 15 ± 2 | Yes | Yes |

3.1.2. Determining Roughness Parameters

Roughness measurements were carried out under the following parameters: cutoff wavelength $\lambda$ = 8 µm, a stylus feed rate of 0.5 mm/s, measuring distance of five single measured lengths (5 × 2.5 mm), a measuring reference $Ra$ = 3 µm, filter type: Gauss. Sanded OZM plates were also analyzed. Table 4 contains the aggregate results for all plate types. The plate surface profiles generated by the Mitutoyo software from the measurements (one example for each batch of plates tested) are shown in Figures 8–11.

**Table 4.** Roughness parameters for each plate type.

| Plate Type | Parameter | Mean Value [µm] | Minimum Value [µm] | Maximum Value [µm] | Most Popular Range [µm] |
|---|---|---|---|---|---|
| Julius Peters | *Ra* | 6.07 ± 0.79 | 4.76 | 7.80 | 6 to 7 |
| | *Rt* | 54.43 ± 8.36 | 42.57 | 76.61 | 50 to 55 |
| | *Rz1max* | 51.62 ± 8.41 | 40.40 | 72.89 | 50 to 55 |
| Old OZM | *Ra* | 10.89 ± 1.60 | 7.99 | 14.44 | 10 to 11 |
| | *Rt* | 84.99 ± 13.46 | 62.93 | 131.56 | 75 to 80 |
| | *Rz1max* | 80.98 ± 13.53 | 58.11 | 125.82 | 75 to 80 |
| New OZM | *Ra* | 9.87 ± 1.01 | 7.69 | 12.83 | 9 to 10 |
| | *Rt* | 83.01 ± 12.19 | 60.07 | 119.56 | 80 to 85 |
| | *Rz1max* | 78.40 ± 10.38 | 60.07 | 103.59 | 80 to 85 |
| Deltima Precision | *Ra* | 9.67 ± 1.07 | 6.67 | 12.47 | 9 to 10 |
| | *Rt* | 81.00 ± 10.26 | 64.35 | 114.03 | 80 to 85 |
| | *Rz1max* | 76.43 ± 9.55 | 58.75 | 108.79 | 70 to 75 |
| Sanded OZM | *Ra* | 5.07 ± 1.22 | 2.99 | 7.14 | 4 to 5 |
| | *Rt* | 51.69 ± 6.81 | 42.14 | 62.69 | 45 to 50 |
| | *Rz1max* | 51.15 ± 6.72 | 40.13 | 62.69 | 45 to 50 |

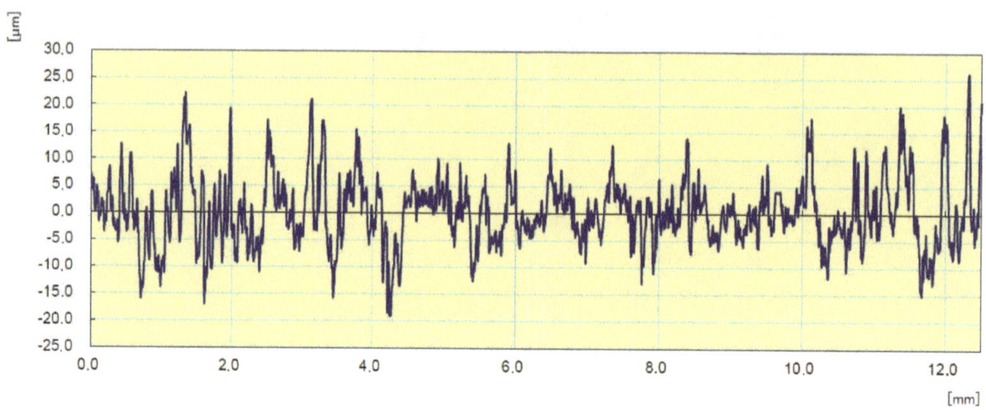


**Figure 8.** Exemplary surface profile of Julius Peters plate.

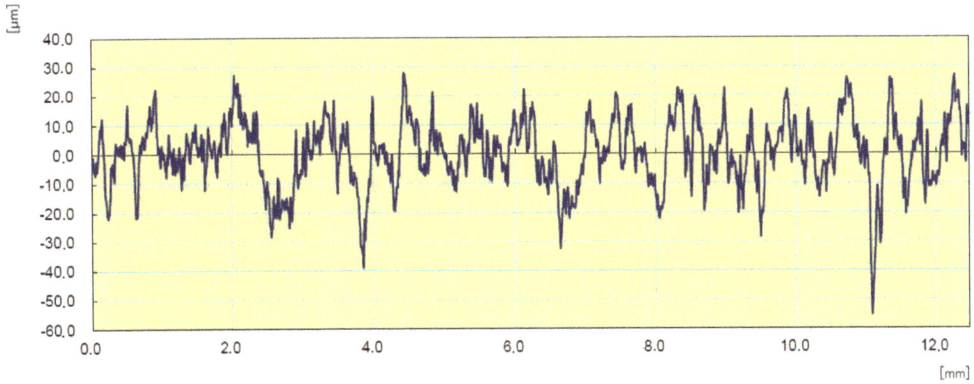


**Figure 9.** Exemplary surface profile of old OZM plate.

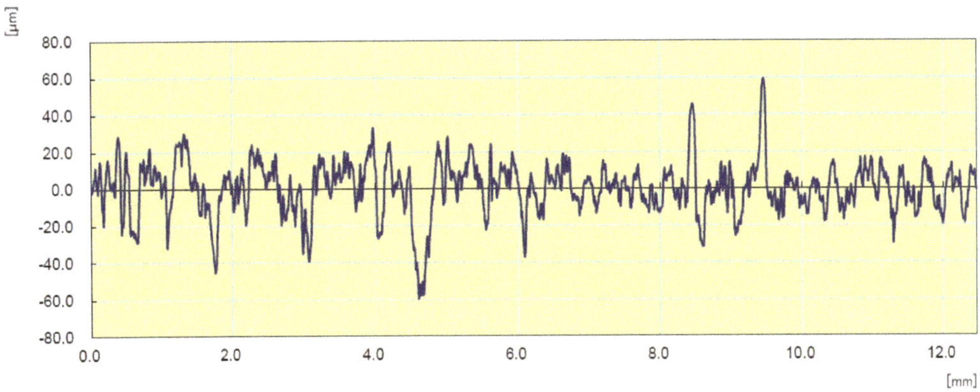


**Figure 10.** Exemplary surface profile of new OZM plate.

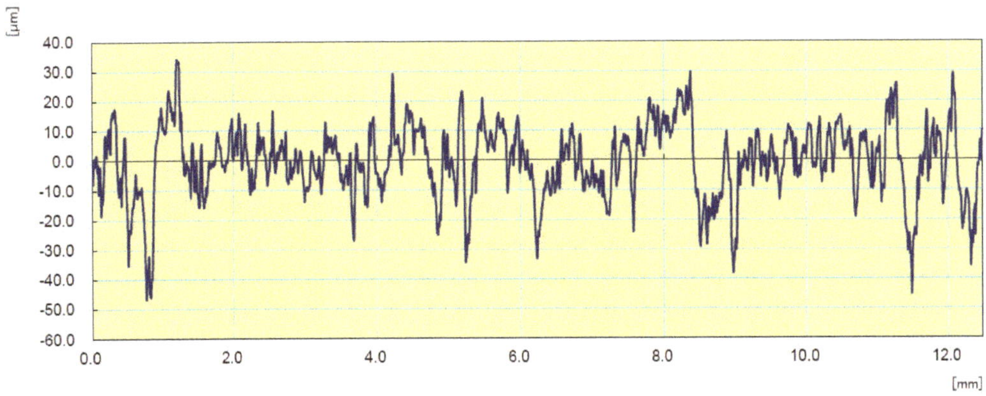


**Figure 11.** Exemplary surface profile of Deltima Precision plate.

### *3.2. Friction Sensitivity*

A summary of the friction test results for RDX and PETN is presented in Tables 5 and 6. To compare it with the plates surface differences, the parameter *Ra* and the number of grooves per 1 cm were chosen. Parameters *Rt* and *Rz1max* turned out to be relatable to *Ra*, e.g., plates with the lowest *Ra* value had also the lowest *Rt* and *Rz1max* values.

The obtained friction sensitivities are comparable to the hexogen and penthrite sensitivities available in the literature and STANAG 4487 (Table 7).

**Table 5.** Friction sensitivity of hexogen for all plate types.

| Plate Type | Sensitivity (EN 13631-3) [N] | 50% Point (STANAG 4487) [N] | *Ra* (Mean Value) [μm] | *Ra* Range [μm] | Grooves per 1 cm (Mean Value) |
|---|---|---|---|---|---|
| Julius Peters | 120 | 182 ± 9 | 6.07 ± 0.79 | 4.76 to 7.80 | 16 ± 2 |
| Old OZM | 144 | 206 ± 89 | 10.89 ± 1.60 | 7.99 to 14.44 | 16 ± 3 |
| New OZM | 144 | 196 ± 49 | 9.87 ± 1.01 | 7.69 to 12.83 | 18 ± 2 |
| Deltima Precision | 128 | 215 ± 74 | 9.67 ± 1.07 | 6.67 to 12.47 | 15 ± 2 |
| Sanded OZM | 168 | 216 ± 55 | 5.07 ± 1.22 | 2.99 to 7.14 | Not appli[illegible]ble |

**Table 6.** Friction sensitivity of penthrite for all plate types.

| Plate Type | Sensitivity (EN 13631-3) [N] | 50% Point (STANAG 4487) [N] | *Ra* (Mean Value) [μm] | *Ra* Range [μm] | Grooves per 1 cm (Mean Value) |
|---|---|---|---|---|---|
| Julius Peters | 64 | 67 ± 9 | 6.07 ± 0.79 | 4.76 to 7.80 | 16 ± 2 |
| Old OZM | 60 | 73 ± 21 | 10.89 ± 1.60 | 7.99 to 14.44 | 16 ± 3 |
| New OZM | 64 | 75 ± 19 | 9.87 ± 1.01 | 7.69 to 12.83 | 18 ± 2 |
| Deltima Precision | 56 | 69 ± 9 | 9.67 ± 1.07 | 6.67 to 12.47 | 15 ± 2 |
| Sanded OZM | 80 | 121 ± 29 | 5.07 ± 1.22 | 2.99 to 7.14 | Not applicable |

**Table 7.** Friction sensitivity—RDX and PETN. All figures given in N.

| Explosive | Meyer et al. [5] | 50% Point-Musil et al. * [25] | 50% Point-Künzel et al. * [26] | 50% Point Range-STANAG 4487 [10] |
|---|---|---|---|---|
| Hexogen | 120 | 127 | 127 | 110 to 185 |
| Penthrite | 60 | 75.5 | 75.1 | 56 to 102 |

* Obtained using the probit analysis.

## 4. Discussion

The visual assessment of the plates surfaces and the resulting number of grooves per measurement section are subject to a high degree of uncertainty due to the difficulty in identifying individual grooves (especially for Julius Peters plates). The obtained number of grooves for each plate type is similar due to the standard deviation value. Nonetheless, the results of this assessment highlight the unevenness of the tested surfaces. Therefore, from all tested plate types new OZM plates are mostly recommended.

During the roughness measurements, it was found that the Julius Peters plates do not meet the requirements of the standard EN 13671-3:2005. The individual deviation values from the plate profile are not within the range of 5 to 19 μm. The thickness of the plates is also inappropriate. Therefore, the Julius Peters plates may not be suitable for measurements according to this standard. Plates from other manufacturers meet the standard requirements.

No major differences in sensitivity measurements between the OZM plates from the new and the old batches were noticed, despite the discrepancies in the regularity of surfaces and ease of groove identification. Taking into consideration standard deviation, there is no huge divergence in measured *Ra* (10.89 ± 1.60 μm for the old batch and 9.87 ± 1.01 μm for the new one).

Analysis of the generated by the Mitutoyo software profiles indicates, that on average, the Julius Peters plate deviations have the lowest values (up to 20 μm with several values not exceeding 25 μm). The new batch of OZM plates is characterized by single deviations with large values (up to 40 ÷ 60 μm), while the average deviations reach the level from 20 to 30 μm; for the old OZM batch, the average deviation level is similar, while the single deviations reach 60 μm. For Deltima plates, the largest deviations do not exceed 50 μm with an average level of 10 to 20 μm. These observations are convergent to the measured roughness parameters.

The plate profiles are not completely uniform, with noticeably higher profile deviations or various deviation widths observed in certain areas of the plates. The presence of such irregularities can lead to the accumulation of excess samples, which may distort the results of the sensitivity test. For Deltima and old OZM plates, profile deviations are noticeably wider, while for Julius Peters plates deviations are mostly narrow and, along with their low intensity, make it mostly unlikely to gather crystals in the grooves. In the case of old OZM and Julius Peters plates, the analysis of the obtained profiles (along with a visual assessment of the plates) shows that the surface exhibits an uneven distribution of profile deviations for the same side of one plate, and depending on the selected position of the sample on the plate, a risk of different results of the sensitivity determination occurs.

The sanded OZM plates have the lowest roughness and each time they were used, the determined sensitivities of RDX and PETN were the lowest. However, the use of Julius Peters plates with a roughness at a similarly low level (5.07 ± 1.22 μm for sanded plates while 6.07 ±0.79 μm for Julius Peters) does not reflect the expected pattern. For penthrite, the friction sensitivity ranks at a similar level (European standard: 64 N for Peters and new OZM, 50% Point: 67 ± 9 N for Peters and 75 ± 19 N for new OZM), while for hexogen the sensitivity is even higher than when using plates from other manufacturers.

Deltima plates with slightly lower roughness than new OZM plates give higher friction sensitivities while determining according to the European standard, but this pattern does not coincide with the Bruceton test results for hexogen.

Neither RDX nor PETN was at the edge of admission for civil use. The tested RDX sample passed all the requirements for each type of plates, whereas crystalline PETN each time had the approvable sensibility for use as an explosive core in detonating cords.

## 5. Conclusions

On the basis of the obtained results, no obvious correlation between the plate roughness and the determined value of the friction sensitivity can be established. Therefore, currently, it is not justified to introduce a requirement for a roughness measurement preceding the Peters test into the standard. However, it is recommended to include information on the components used in the test reports.

Nevertheless, discrepancies between various types of plates are still being observed. This may be caused by another valid feature of used components that have not been examined in this study, e.g., the hardness of ceramics or variations in thickness. However, RRTE results have shown that the limitation of the types of plates may help lower the differences in the obtained sensitivity. Despite this, the total unanimity of the results is practically impossible to achieve, where the role of the human factor and subjectivity in the assessment of the test results seems to be currently unavoidable. Therefore, interlaboratory comparison tests are crucial for determining the overall level of sensitivity of the tested explosives.

Applying sound and light detectors during sensitivity to friction tests may partially eliminate the above-mentioned human factor, resulting in more compliant results being obtained. Due to the development of 3D printing, another possibility for minimizing the discrepancies is to use printed plates—3D printers enable obtaining more regular grooves in technical ceramics used as filament. In each case, further tests are required.

Although the explosives tested in this study met the requirements of the standards for their approval in each case, it must be taken into account that, for some of the explosives, the results will place them on the borderline of admissibility, which may be problematic for explosives manufacturers (identical product may exclusively be approved by some certifying laboratories). Hence, further investigation of this issue is necessary to improve the comparability of the obtained results and indirectly increase the safety of the explosives used.

**Author Contributions:** Conceptualization, M.P.; validation, S.N. and M.P.; formal analysis, S.N. and M.P.; investigation, S.N. and M.P.; resources, M.P.; writing—original draft preparation, S.N.; writing—review and editing, S.N.; visualization, S.N. and M.P.; supervision, M.P.; project administration, M.P. All authors have read and agreed to the published version of the manuscript.

**Funding:** This research received no external funding.

**Institutional Review Board Statement:** Not applicable.

**Informed Consent Statement:** Not applicable.

**Data Availability Statement:** The paper contains all original work and findings from the conducted study. Further questions can be addressed to the corresponding author.

**Conflicts of Interest:** The authors declare no conflicts of interest.

## Abbreviations

The following abbreviations are used in this manuscript:

| | |
|---|---|
| BAM | Bundesanstalt für Materialforschung und -prüfung |
| HMX | High Molecular weight Explosive (octogen) |
| PETN | Pentaerythritol Tetranitrate (penthrite) |
| RRTE | Round-Robin Test on Explosives |
| RDX | Royal Demolition Explosive (hexogen) |

## References

1. Avrami, L.; Hutcherson, R. The Sensitivity to Impact and Friction. In *Energetic Materials 2. Technology of the Inorganic Azides*; Fair, H.D., Walker, R.F., Eds.; Plenum Press: New York, NY, USA, 1977; pp. 111–162.
2. Brown, J. *A Study of Friction Fundamentals in Explosives*; Final Technical Report Contract DAM21-69-C-0558; Picatinny Arsenal: Berkeley Heights, NJ, USA, 1970.
3. Sućeska, M. *Test Methods for Explosives*, 1st ed.; Springer: New York, NY, USA, 1995; pp. 27–29.
4. Simpson, R.L.; Foltz, F.M. *LLNL Small-Scale Friction Sensitivity (BAM) Test, Report UCRL-ID-124563*; Lawrence Livermore National Laboratory: Livermore, CA, USA, 1996.
5. Meyer, R.; Köhler, J. *Homburg, A. Explosives*, 5th ed.; Wiley-VCH: Weinham, Germany, 2002; pp. 144–145.
6. Bowden, F.P.; Singh, K. Size effects in the initiation and growth of explosion. *Nature* **1953**, *172*, 378–380. [CrossRef]
7. Matyáš, R.; Šelešovský, J.; Musil, T. Sensitivity to friction for primary explosives. *J. Hazard. Mater.* **2012**, *213–214*, 236–241. [CrossRef] [PubMed]
8. Radacsi, N.; Bouma, R.H.B.; Krabbendam-la Haye, E.L.M.; ter Horst, J.H.; Stankiewicz, A.I.; van der Heijden, A.E.D.M. On the Reliability of Sensitivity Test Methods for Submicrometer-Sized RDX and HMX Particles. *Propellants Explos. Pyrotech.* **2013**, *38*, 761–769. [CrossRef]
9. Van de Velde, C.; Lefebvre, M.H. *Test Report CECOC-RRTE-2014-B*; CECOC European Explosives and Pyrotechnics Notified Body: Brussels, Belgium, 2017.
10. *STANAG 4487*; JAIS (EDITION 2)-EXPLOSIVE, FRICTION SENSITIVITY TESTS. NATO Standardization Agency: Brussels, Belgium, 2009.
11. Klapötke, T.M.; Lemarchand, G.; Lenz, T.; Mühlemann, M.; Stierstorfer, J.; Weber, R. Impact and Friction Sensitivities of PETN: I. Sensitivities of the Pure and Wetted Material. *Propellants Explos. Pyrotech.* **2022**, *47*, e202200150. [CrossRef]

12. Buczkowski, D. Międzylaboratoryjne badania materiałów wybuchowych-ocena wyników. In Proceedings of the Blasting Safety in the Mining Industry, Ustroń, Poland, 13–15 October 2010; Volume 1, pp. 9–14.
13. Le Roux, J. The Dependence of Friction Sensitivity of Primary Explosives upon Rubbing Surface Roughness. *Propellants Explos. Pyrotech.* **1990**, *15*, 243–247. [CrossRef]
14. *EN 13631-3:2005*; Explosives for Civil Uses-High Explosives-Part 3: Determination of Sensitiveness to Friction of Explosives. European Committee for Standardization: Brussels, Belgium, 2004.
15. *EN 13631-1:2005*; Explosives for Civil Uses-High Explosives-Part 1: Requirements. European Committee for Standardization: Brussels, Belgium, 2005.
16. *EN 13630-1:2003*; Explosives for Civil Uses-Detonating Cords and Safety Fuses-Part 1: Requirements. European Committee for Standardization: Brussels, Belgium, 2003.
17. Lyukshin, V.; Shatko, D.; Strelnikov, P. Methods and approaches to the surface roughness assessment. *Mater. Today Proc.* **2021**, *38*, 1441–1444. [CrossRef]
18. Tonietto, L.; Gonzaga, L., Jr.; Veronez, M.R.; Kazmierczak, C.S.; Arnold, D.C.M.; da Costa, C.A. New Method for Evaluating Surface Roughness Parameters Acquired by Laser Scanning. *Sci. Rep.* **2019**, *9*, 15038. [CrossRef] [PubMed]
19. Rifai, A. P.; Aoyama, H.; Huu Tho, N.; Zawiah Md Dawal, S.; Aini Masruroh, N. Evaluation of turned and milled surfaces roughness using convolutional neural network. *Measurement* **2020**, *161*, 107860. [CrossRef]
20. Quick Guide to Surface Roughness Measurement. Reference Guide for Laboratory and Workshop. Available online: https://www.mitutoyo.com/webfoo/wp-content/uploads/1984_Surf_Roughness_PG.pdf (accessed on 17 April 2025).
21. Christensen, D.; Novik, G.P.; Unneberg, E. Estimating sensitivity with the Bruceton method: Setting the record straight. *Propellants Explos. Pyrotech.* **2024**, *49*, e202400022. [CrossRef]
22. Šelešovský J.; Pachmáň J. Probit Analysis—A Promising Tool for Evaluation of Explosive's Sensitivity. *Cent. Eur. J. Energetic Mater.* **2010**, *7*, 269–278.
23. Vesterinen, A. Sensitivity Testing for Quantile Estimatiom under Constraints. Master's Thesis, Tampere University, Tampere, Finland, 2024.
24. Preston, D.; Brown, G.; Skidmore, C.B.; Reardon, B.L.; Parkinson, D.A. Small-scale explosives sensitivity safety testing: A departure from Bruceton. *AIP Conf. Proc.* **2012**, *1426*, 713–716.
25. Musil, T.; Matyás, R. Matyás; Lycka, A.; Růžička, A. Characterization of 4,6-Diazido-N-nitro-1,3,5-triazine-2-amine. *Propellants Explos. Pyrotech.* **2012**, *37*, 275–281. [CrossRef]
26. Künzel, M.; Matyáš, R.; Vodochodský, O.; Pachman, J. Explosive Properties of Melt Cast Erythritol Tetranitrate (ETN). *Cent. Eur. J. Energetic Mater.* **2017**, *14*, 418–429. [CrossRef] [PubMed]

applied sciences

MDPI

Article

# Blasting of Unstable Rock Elements on Steep Slopes

**Marco Casale [1], Giovanna Antonella Dino [2] and Claudio Oggeri [3,*]**

[1] Department of Management, Università di Torino, 10100 Torino, Italy; ma.casale@unito.it
[2] Earth Sciences Department, Università di Torino, 10100 Torino, Italy; giovanna.dino@unito.it
[3] Dipartimento di Ingegneria dell'Ambiente, del Territorio e delle Infrastrutture (DIATI), Politecnico di Torino, 10129 Torino, Italy
* Correspondence: claudio.oggeri@polito.it

**Abstract:** The improvement of safety conditions on hazardous rock slopes in civil work, mining and quarrying, and urban environments can be achieved through the use of explosives for the removal of unstable rock elements and final profiling. This technique is often applied because, in most cases, drill and blast operations, where they can be used, are cheaper and faster than other techniques and require fewer subsequent maintenance interventions. Blasting represents a suitable and effective solution in terms of different geometries, rock formation types, access to site, safety, and the long-term durability of results. The primary purpose of this approach is the improvement of the safety conditions of sites, depending on their local features, as well as the safety of workers, so that the blasting scheme, geometry, and firing can be carefully adapted, thus imposing relevant limitations on the operating techniques. All these constraints associated with complex logistics make it difficult to standardize the demolition technique, due to different situations in terms of extension, location, fracturing state, and associated traffic risk. Considering the significant number of influencing factors for both the rock mass features and for the topography, the present research has been necessarily validated through the analysis of several case histories, thus on an experiential basis focusing on some simple control parameters to help engineers and practitioners regarding the first design and control of blasting schemes.

**Keywords:** blasting; slope protection; natural hazards; powder factor; rockfall; explosive

## 1. Introduction

Unstable rock elements along a slope pose significant hazards linked to the possibility of rockfalls. Managing these situations becomes crucial to ensure safety for nearby infrastructure, communities, and the environment, reducing the risk of rockfalls by eliminating potential sources of instability. The simplest method for addressing unstable rock elements on slopes is "rock scaling", or simply "scaling". To address these challenges effectively, this research focuses on using blasting techniques to improve slope safety conditions by removing unstable rock blocks. Unlike traditional scaling, which is carried out by climbers using portable devices, this method uses carefully designed drill and blast operations tailored to complex geometries and conditions where conventional methods may be less effective. The main limitation of the common scaling along slopes is that it is time-consuming for the workers operating in difficult conditions and does not have the ability to cover large volumes, while it is decisive for the final cleaning of residual small fragments after blasting or demolition. A common feature and requirement among operative methods is that of claiming skilled personnel, for safety and also for care in preparation of the site in terms of access, equipment, and control.

By analyzing multiple blasting cases, this study develops a simplified approach to determine the Powder Factor (PF), which can serve as both a design validation tool and a cost estimation parameter for interventions on unstable slopes. Using explosives to remove unstable rock elements can be an effective technique for improving safety conditions on unstable rock slopes. The use of explosives can quickly break up large volumes of rock, making the removal process faster compared to manual or mechanical methods. This efficiency is especially beneficial for addressing urgent safety concerns on unstable slopes.

Further, drill and blast operations, which utilize explosives, are often more cost-effective than other techniques. While upfront costs are associated with explosives and drilling equipment, the overall project costs may be lower due to reduced time, labor, and equipment usage. In comparison, other techniques can be adopted, such as hydraulic splitting, chemical rock breaking, expansive grout, and quick reactive powder; each solution can exhibit its own advantages in terms of safety, environmental suitability, or vibration reduction, but geometrical precision of cutting and rock fragmentation control are the key factors for conventional blasting agents (both as cartridges and also as detonating cords).

Blasting can target specific areas of instability, minimizing disruption to surrounding areas and infrastructure. This technique can be tailored to suit the site's specific geological conditions and requirements. Depending on the situation, different types of explosives and blasting methods can be adopted to achieve desired outcomes, such as controlled fragmentation or controlled collapse.

Compared to some alternative methods, such as rock bolting or slope reinforcement, drill and blast operations may require less subsequent maintenance interventions. Once unstable rock elements are removed, ongoing monitoring and maintenance efforts may be reduced, leading to long-term cost savings.

However, it is essential to note that the use of explosives for slope stabilization should be carefully planned and executed to ensure safety and minimize environmental impact. This includes conducting thorough site assessments, risk analyses, and blast designs, and implementing strict safety measures throughout the drilling and blasting process [1]. Additionally, environmental regulations and community concerns must be taken into account to mitigate any potential adverse effects.

While explosives can offer significant advantages in certain situations, they are not always the most suitable solution for every slope stabilization project. Alternative methods, such as slope reinforcement, may be more suitable depending on the site's specific circumstances. It is essential to consider all available options and consult experienced professionals to determine the most effective approach for improving safety conditions on unstable rock slopes.

When addressing unstable rock elements on slopes, it is essential to conduct thorough site assessments, consider the underlying geology and hydrology, and prioritize safety at all times. Implementing a combination of stabilization measures tailored to the specific site conditions can help mitigate risks and ensure the long-term stability of the slope.

In a simplified approach, it should be possible to design ordinary blasting and controlled blasting. Ordinary blasting works for tunnel excavation, quarry production, etc., and is associated with well-coded parameters as result of many experiences over the years, depending on the available techniques and results obtained. These parameters are mainly geometrics (blasthole diameter, blasthole depth, spacing, burden, etc.), but there is a synthetic parameter that reflects the characteristics of the blasting, the rock features, and the fragmentation to be obtained, which is the powder factor (PF), which can be defined as the weight of explosives used (in kilograms) per broken rock volume (in cubic meters) [2]. Generally, the PF is a crucial parameter in blasting design, as it helps to determine the amount of explosive needed to efficiently break the rock while minimizing waste and controlling the fragmentation size. A higher PF generally results in more energy being applied to the rock, which can lead to larger fragmentation and better productivity in some cases. However, excessive powder factors can lead to overbreak, increased vibration, and environmental concerns. The optimal PF depends on various factors such as the rock type, geology, desired fragmentation, environmental considerations, and safety requirements. The PF can be estimated and adjusted based on these factors to achieve the desired results with minimal adverse effects. Controlled blasting techniques aim to take great care of the resulting profile, to avoid damage to rock surfaces left in place, and to minimize induced vibrations; they are used to efficiently distribute explosive charge in the rock mass volume, thereby minimizing the fracturing of rock beyond the crestline of the highwall or designed boundary of main excavation zones.

The value of the PF hinges significantly on the presence or absence of free surfaces that facilitate the blasting process. Rock blasting leads to volume expansion. When a free surface is available, this volume increase can dissipate outward. Otherwise, in scenarios lacking a free surface, additional energy is required to displace the broken material and create space for the material blasted sequentially by a series of micro-delayed mines.

This fundamental principle underscores why, as a rule, a blind tunnel excavation typically yields shows higher PF values compared to a stepped quarry configuration [1]. In the latter, where a free surface extends for the entire height of each step, the possibility of material displacement is notably enhanced. In standard geometric situations, the PF has become a synthetic control parameter indicative of the accuracy of the blasting design.

In more complex cases, where the geometry is irregular and the presence of free surfaces varies from case to case, it is much more difficult to standardize the PF and then use it as a control parameter. Typically, the demolition of unstable blocks on a slope reflects this last situation, with shapes, positions, geological and structural characteristics, and free surfaces being extremely variable from case to case. Blasting for isolated rock elements differs significantly from other projects, such as quarry or tunnel blasting, due to its focus on minimizing collateral damage and ensuring post-blast stability. These operations prioritize precise detachment over fragmentation, requiring tailored designs that account for irregular geometries, limited free surfaces, and environmental constraints. Unlike bulk fragmentation, the goal here is often safe displacement rather than maximizing yield or productivity.

This research, which is focused on novel features, proposes a first procedure to determine in advance the PF related to the demolition of unstable rock blocks in complex situations by proposing a simplified approach. In this way, the PF can be used to quickly check the correct blasting design and to estimate the intervention costs in advance.

The study has been carried out by analyzing several blasting cases across different contexts and scenarios. These cases serve as the basis for determining how PF varies in response to changes in blasting conditions. Through this analysis, the authors aim to elucidate the relationship between PF and the various factors influencing the blasting design.

## 2. Materials and Methods

### 2.1. Experimental Approach to Powder Factor (PF)

Several approaches have been developed for estimating the PF in tunnel blasting and in open-air bench blasting, but not for the blasting of a single element on a slope. The PF permits summarizing information in a single parameter, even if in a very simplified way; it captures several conditions of the blasting pattern and provides an estimation regarding the blasting efficiency. Many contributions are available in the technical literature seeking correlations between PF and induced vibration, fragment size distribution, residual contour [3], and stability; as an example, key objectives in the drilling and blasting design are the determination of the powder factor, as well as the explosive weight and distribution in rock mass [4]; other studies examine the improvement of the blasting design for a more satisfying fragmentation, a more tolerable oversized fragments rate in relation to the overall blasted rock volume, and better bench stability [5–7], as well as the adoption of numerical methods for the quantitative assessment of the effectiveness of vibration reduction in line-drilling as a screening approach [8]. Interdependent factors should be studied by careful monitoring during field tests; fragmentation and environmental effects are influenced by the amount of powder factor; a higher amount of powder factor improves fragmentation and, thus, productivity—the blast-induced vibration and air-blast also rise with an increase in the powder factor. This impasse calls for an optimum powder factor for sustainable blasting operations [9]. Finally, effective parameters that influence the powder factor can be divided into three contributions, namely, rock mass, geometric, and explosive parameters [10].

Referring to the specific technical literature, [11] states regarding tunnel blasting that the PF depends on the blasting section, rock type, explosive type, and blasting pattern. The blasting section is the most relevant parameter and the PF can be estimated as follows:

$$PF = A * B * C * [(10/S) + 0.6] \quad (1)$$

where

- A is a coefficient that varies according to the rock type;
- B depends on the employed explosive type;
- C is related to the blasting pattern adopted;
- S is the area of the excavation section.

A, B, and C were tabulated based on the analysis of several real cases.

More generally, referring to the existing formulas for the dimensioning of explosive charges, according to empirical study, these formulas include the following [12]:

- A coefficient characteristic of the rock to be blasted;
- A coefficient characteristic of the employed explosive;
- A coefficient of "effect" referring to the result to be obtained with the blasting (size distribution of the blasted material, width of the zone of influence of the charge);
- A characteristic dimension of the blasting pattern geometry (normally, the line of least resistance), eventually raised to a power, or a polynomial combination of several characteristic dimensions.

The more general expression of such formulas may be written as follows:

$$C = k' * k'' * k''' * f(I) \quad (2)$$

where

- C is the explosive charge;
- $k'$ represents a coefficient referring to the rock type;
- $k''$ depends on the employed explosive type;
- $k'''$ is the "effect" coefficient for fragmentation;
- I is the characteristic dimension of the rock block to be blasted;
- f(I) is a function of "I" raised to a power of a number varying from 2 to 3, with lower values associated with blasting along a defined failure surface and higher values referring to blasting volume.

When the exponent of f(I) is 2, Equation (2) refers to a simple detachment effect, mainly based on the tensile strength of the rock. When, instead, the exponent is 3, (2) refers to a volume to be blasted.

As was to be expected, in the first case, C is much smaller than in the second case.

The study was carried out through an in-depth analysis of numerous real-world blasting cases, aiming to discern and extrapolate the key factors governing the powder factor. By meticulously examining these cases, we seek to unveil the intricate interplay of various elements influencing the efficiency and efficacy of blasting operations.

Through careful observation and analysis, patterns began to emerge, shedding light on the nuanced factors that dictate the PF in blasting. Factors such as rock type, geological structure, blast design parameters, explosive properties, and environmental conditions were precisely scrutinized and evaluated.

This comprehensive approach allowed for the identification of critical variables and their relative impact on the PF. By extrapolating from these real-world scenarios, this study aimed to provide valuable insights and guidelines for optimizing blasting operations on steep rock formations, thereby enhancing safety and efficiency in such challenging environments.

### 2.2. General Constraints of Blasting Operations on Rock Slopes

Following an extensive geostructural survey, the development of a drill hole pattern necessitates a delicate balance among numerous constraints:

- Assessment of the rock quality and lithological and strength properties of the rock formations to ensure effective drilling and blasting outcomes;
- Understanding the joints' orientation (dip, dip direction) and characteristics of joints within the rock mass to optimize drilling angles and minimize potential instability;
- Implementation of measures to mitigate vibrations generated during drilling and blasting operations, reducing the risk of structural damage and instability of additional rock blocks or discomfort to nearby structures and communities;
- Determination of the required fragmentation size of the blasted rocks, in order to enable the required removal of the subsequent muck or material after blasting;
- Assessment of the accessibility of the drilling site to ensure efficient deployment of drilling equipment and personnel;
- Incorporation of time-sensitive requirements, such as the need to swiftly reopen roads or access routes, into the scheduling and execution of drilling and blasting operations.

By carefully addressing these constraints, the design of the drill hole pattern can be optimized to ensure safe, efficient, and effective drilling and blasting operations while meeting project objectives and minimizing potential impacts on the surrounding environment and infrastructure.

### 2.3. Risks and Undesired Effects

To address potential risks and unintended consequences in blasting activities, it is imperative to implement robust safety measures.

By implementing both preventive and protective measures, one can mitigate risks and unwanted effects associated with blasting activities, ensuring the safety of workers, surrounding communities, and nearby infrastructures while minimizing environmental impact; such measures include detailed inspection, localization of weak zones, and respect areas around the zone to be treated.

To achieve these goals, prioritizing prevention strategies, including risk assessments, is necessary.

A comprehensive training program for expert climbers, focusing on advanced safety protocols tailored for blasting operations on steep rock formations, must be included in the design requirements.

Stringent measures must be implemented to limit the occurrence of fly rocks, which pose a significant hazard to both workers and nearby communities.

Blasting operations should take care of residual surfaces to preserve the rock mass quality by adopting appropriate techniques such as presplitting, smooth blasting, etc. [13,14].

Mitigation with bolting and wire mesh or drapery systems, contouring the residual surface remaining after the demolition, can represent the final step of the blasting operations; therefore, local conditions can claim further access to expert rock climbers to remove fragments and to install bolts and drapery, starting from the top of the area.

### 2.4. Design Criteria

Even if, in the case of blasting an unstable rock element on a slope, the design criterion is most of the time more similar to ornamental stone cutting, the most straightforward parameter to check a blasting scheme remains the powder factor (PF).

It is important to conduct possible surveys with direct access to the slope with the help of climbers (specifically for joints conditions) and also by full recognition by means of unmanned vehicles (drones) to fully build a digital optical twin to describe distances and volumes; the use of drones is now complimentary for quick and complete volume descriptions and for supporting blasthole pattern configurations.

Geostructural, visual, and topographical surveys are also essential after blasting in order to verify the residual condition of joint surfaces and remove, by means of scaling techniques, unstable small blocks in order to finally proceed with the eventual protective/reinforcing works (typically drapery systems) and monitoring of major joints (crackmeters, topographic targets). Moreover, residual risk assessment after scaling and reinforcing remains of great concern for civil infrastructure, such as roads and tunnels [15] or for quarry yards.

The blasting criterion aims in this case to simply detach from the mountain the rock to be demolished, exploiting gravity. Compared to a situation where fragmentation is sought (e.g., a blasting operation for aggregates or industrial minerals), the operating philosophy is completely different: here, the effects sought are only those of detachment and, possibly, of rock displacement, which must detach from the original rock mass, falling downstream with the help of gravity; this operation is correctly referred to as cutting and not as demolition.

As proof of this, the specific consumption found in these cases is much lower than normal for industrial materials. The cut is normally obtained by a series of side-by-side holes, often of small diameter (normally ranging from 48 mm to 51 mm). The blasting design essentially consists of determining the distance between the holes and the charge per hole.

A very simple design method based on a quasi-static approach to the explosion phenomenon is now described. It is assumed that the force generated by the pressure of the explosion gases in the holes must overcome the tensile strength of the rock between two adjacent holes, as exemplified in Figure 1.

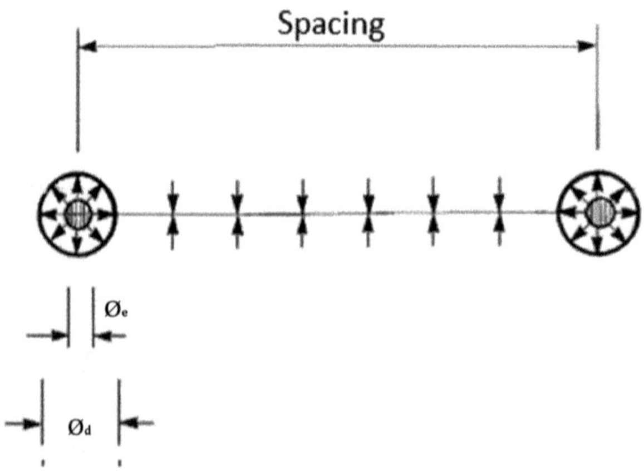


**Figure 1.** Scheme of mechanism of explosive cut-action according to quasi-static approach for new fracture formation between adjacent blastholes—modified from [16].

The pressure Ph inside the hole is a function of the specific pressure of the explosive (Pe). The input parameters are as follows:

- Type of explosive.
- Drilling ($Ø_d$) and explosive ($Ø_e$) diameter (i.e., cartridge diameter).
- Length of hole ($L_h$).
- Explosive length ($L_e$).
- Tensile strength of rock (T).

By imposing the static equilibrium of the system, the following formula for calculating the Spacing (S) between the holes in the same row can be obtained [16]:

$$S = Ø_d + \frac{Ø_d}{kT} P_h \rho_e \left(\frac{L_e}{L_h}\right)\left(\frac{Ø_e}{Ø_d}\right)^2 \quad (3)$$

where

- $\rho_e$ is explosive volumetric mass;
- k is a coefficient that takes into account the effect of the alignment of the holes, which ranges from 0.5 (if the spacing between the holes is less than 10 * $Ø_d$) to 1 when the spacing is higher (more than 30–40 * $Ø_d$).

The tensile strength of rocks is not always easy to obtain, especially when considering factors like schistosity. Schistose rocks indeed pose challenges due to their varying strengths in different directions. In the first hypothesis, we can refer to the uniaxial compressive strength, assuming that the tensile strength is on average 1/20 of the uniaxial compressive strength. However, this value is not very reliable: this underscores the importance of comprehensive testing and understanding of the specific properties of the rock in question. Specific phenomena affect the behavior involving tensile strength: for example, it was found that the stress–strain relation is quasi-linear up to failure, which occurs in a brittle way. The stiffness modulus values and apparent tensile strength depend on the inclination of the schistosity concerning the loading direction [17]. For rock materials, the compressive strength is generally about 8–15 times higher than tensile strength, apart from some exceptions (serpentinite rocks, for example). The scattering is always relevant, even for the same rock type. Moreover, in site conditions, weathering acts on the rock matrix and along joints and microcracks. In order to avoid a final excess in the blasting charge, and in thus rock fragment projections (flying rocks), it is preferable to reduce the ratio and to keep the blastholes spacing (well defined) as it is. It should be suggested, as a rigorous approach, to test a specimen from the site in a lab or by means of a point load testing apparatus in order to properly adjust the blasting scheme parameters. In the authors' experience, obtaining permissions for preliminary blasting trials on-site remains quite hard.

*2.5. Analysis of Case Histories for Validation*

A summary of the specific case histories analyzed in the research is proposed below. Some of them have been designed, applied, and controlled directly by the authors.

1. The first one was presented in [18] and concerns the blasting of a large dangerous block (about 60,000 m$^3$) overhanging the village of Meiringen, Switzerland (Figure 2). The initial intervention stemmed from the looming threat of a significant rockfall poised to engulf the town. Minor detachments from the limestone cliff, situated approximately 340 m above the town, had already been documented. The rock mass was Dogger limestone, exhibiting relatively robust resistance properties with a tensile strength of 5 MPa and a shear strength of 11 MPa. To ensure optimal safety measures, the blasting parameters were meticulously calibrated, employing a dense drill hole pattern to facilitate the fine fragmentation of the material. This is, on the other hand, an ever-present concern in this kind of blasting, as the uncontrolled fall down of large intact blocks can lead to substantial damage to infrastructure and does not allow forecasts of falling trajectories. Hence, the aim was to limit the size of blasted blocks to no more than 1 cubic meter. To protect the town, the construction of a sturdy stone embankment was performed along the forecast fall down path, capable of containing an estimated volume of around 60,000 m$^3$. The blasting was carried out in four distinct rounds.

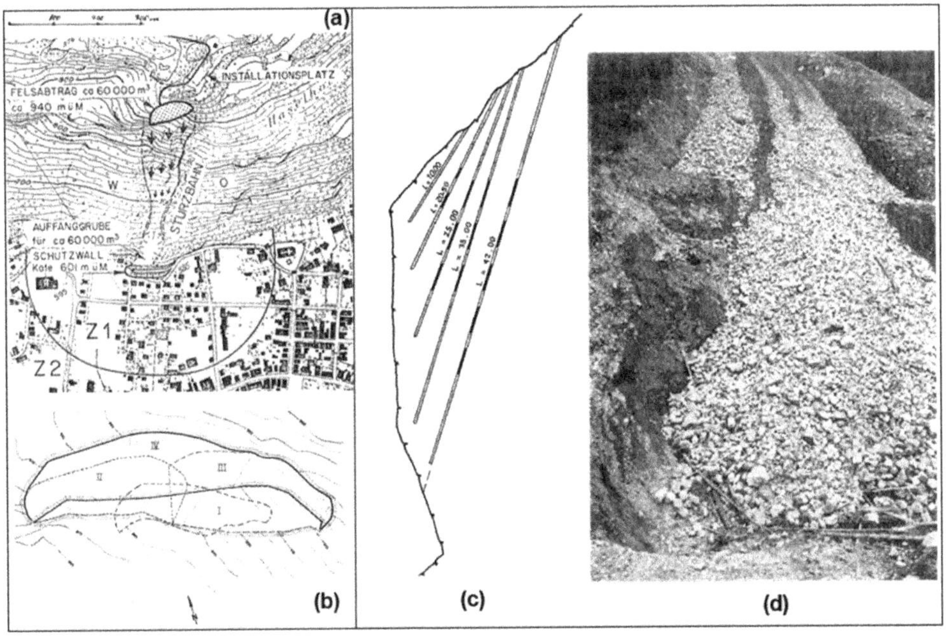

**Figure 2.** The case of Meiringen in Switzerland. (**a**) Plan view of the blasting project. Z1 area to be evacuated; (**b**) detailed plan with the subdivision of the block to be blasted in 4 rounds; (**c**) lateral section of the drill hole pattern from the firing plan of the second round; (**d**) view of the muck fan after the first round (modified after [18]).

2. A second case on Quebec highway 155 (Figure 3) is described in [19], motivated by the danger represented by a potentially unstable rocky dihedral of a volume of about 800 $m^3$ looming over the road, passing at an altitude of 30 m lower. The rock was andesitic gneiss of the Precambrian age. The analysis of stability indicated a very low safety factor, which was variable from 1.16 to 1.06 depending on the assumed friction angle.
The PF adopted, decidedly enormous for European standards, was about 0.77 $kg/m^3$, probably motivated by the need to achieve very fine fragmentation. Blasting has been implemented with micro-retarded electrical detonators. Before the blasting, the road below was protected with a sand bed 1 m thick.

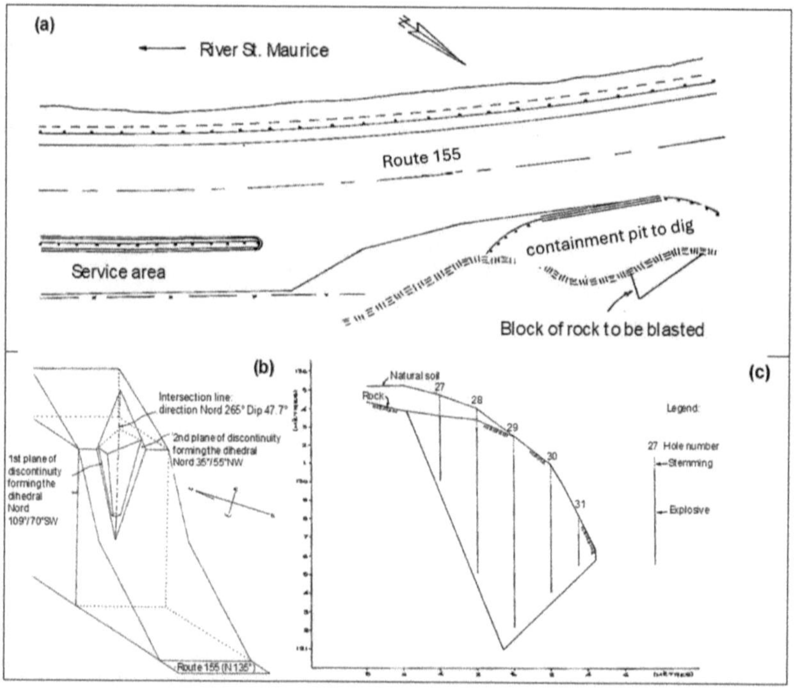


**Figure 3.** The case of Quebec highway 155. (**a**) Plan view of the blasting project; (**b**) perspective view of the dihedral to be blasted; (**c**) lateral section of the drill hole pattern from the firing plan (modified after [19]).

3. A further example related to the blasting of an overhanging block is shown in Figure 4 [20]: a local mountainside road in the northwest of Italy (the provincial road of Val Mastallone in Cravagliana, Vercelli province), which was periodically closed due to minor rockfalls after rainfall events and thawing periods.
The object of the blasting is a spur in unstable conditions with a volume of about 2500 $m^3$, as made clear by a previous collapse. The slope is made of weathered gabbric rock with a uniaxial compressive strength of 90 MPa and tensile strength of 8.5 MPa. This stretch of road was subjected to a systematic collapse of rock blocks of various sizes. Particularly critical was the earlier collapse of 1000 $m^3$, from just below the blasted rock induced by a planar slide. The blasting design, provided by the authors, requested a preliminary careful geostructural survey to clearly define the persistence and location of main joints at the rear of the potential unstable volume. Both drilling and charging have been prepared by using both a hydraulic long-arm

platform and the experience of skilled climbers along the slope. Non-electric ignition has been adopted.

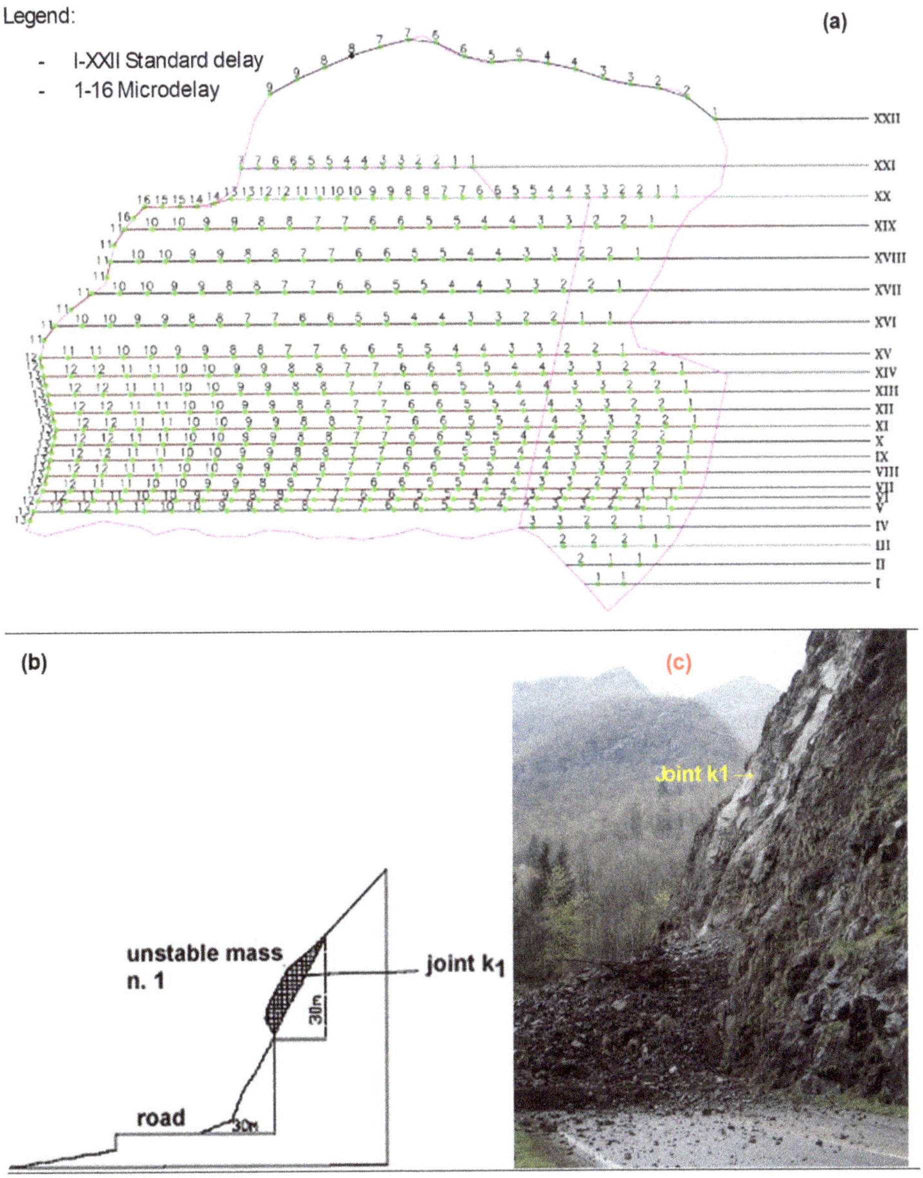


**Figure 4.** The case of the provincial road of Val Mastallone in the northwest of the Alpine range in Italy. (**a**) Firing plan (non-electric ignition has been adopted); (**b**) schematic cross section; (**c**) result of the blasting, with rock fragments accumulated in a regular shape on the road at the base of the slope (paving was protected from impacts with a granular debris cover) (modified after [20]).

4. The case of the demolition of an unstable monolithic slab (Figure 5) [20] of porphyry rock about 10 m wide, 33 m high, and 2–6 m thick, with a global size of about 1300 $m^3$, that threatened a road with a high traffic density of more than 1 car/min: the

provincial road Gattinara-Borgosesia in Serravalle Sesia (Vercelli province—Italy). The monolith was totally isolated from the rock mass of the slope by two open joints and its base was an irregularly and highly fractured rock portion. The authors carried out a local geostructural survey to help with the design of the blast round. The probability of occurrence of a sudden collapse of the slab was very high.

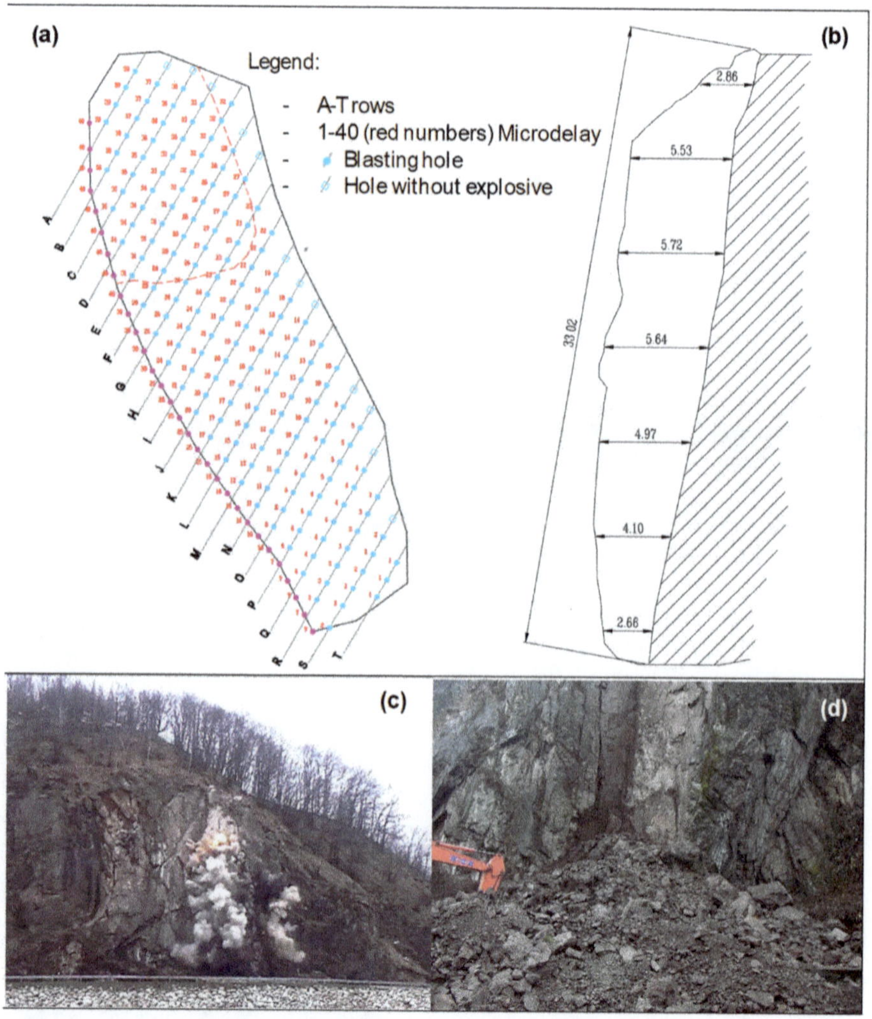


**Figure 5.** The case of the provincial road Gattinara-Borgosesia in the northwest of the Alpine range in Italy. (**a**) Firing plan designed by the authors; (**b**) cross section with measures in m; (**c**) the ignition of the round (non-electric ignition has been adopted); (**d**) result of the blasting in terms of fragments size on the road (paving was protected from impacts with a granular debris cover) at the base of the subvertical slab (modified after [20]).

Other relevant original cases, presented here for the first time, are the following:

5. The demolition of a set of mutually bound blocks above provincial road 169 of Val Germanasca (Turin province) in Northern Italy (Figure 6). The rock was in this case a minute gneiss of mediocre quality and the total volume of the blocks was about 640 $m^3$. In this case, which was known for decades for the inherent hazard it posed to

road traffic, the greatest risk during the blasting was represented by the possibility of shearing the explosion line, following the beginning of the round, for the induced movement of the individual blocks. This phase did not stand alone, but after an arranged detailed local geostructural survey, a blasting design was carried out by the authors; after blasting, manual scaling and a strong and wide reinforcing of the rock slope was carried out (bolting, draperies, cable securing, net fences to intercept fragments fall along the slope). Before and after demolition, scaling operated by climbers was carried out to achieve better safety for work along the slope.

**Figure 6.** The case of provincial road 169 in Val Germanasca in the northwest of the Alpine range in Italy. (**a**) Firing plan; (**b**) overall view of the block set; (**c**) the ignition phase of round; (**d**) result of the blasting resulting in large blocks. Consider that the difference in elevation of the claimed volume and the base road was about 130 m.

6. This case considers the demolition of a highly unstable monolithic block (Figure 7) of about 6000 $m^3$ within a quarry area in Northern Italy, overlooking the quarry square. The rock was represented, in this case, by a gneiss of excellent quality (orthogneiss), with a compressive strength of 185 MPa, mainly used for ornamental purposes. The firing plan was studied by the authors to obtain a fragmentation between 0.5 and 2.0 $m^3$ in order to allow the subsequent reuse of the blasted rock as by-products.
   Obviously, the main purpose of the blasting demolition was to improve the safety conditions of the site; however, when it is possible to recover and recycle blasted rocks, it is a good practice to do so in order to obtain new products/by-products for civil, building, road, and environmental applications.
   In a similar vein, when a rock slide occurs along a mountainside where road or infrastructures are present, it becomes urgent and necessary to remove the collapsed

material, and its possible reuse (as aggregate, for backfilling, for embankments) depends also on the blocks' fragment size distribution, especially to fulfill technical requirements concerning compaction and consistency.

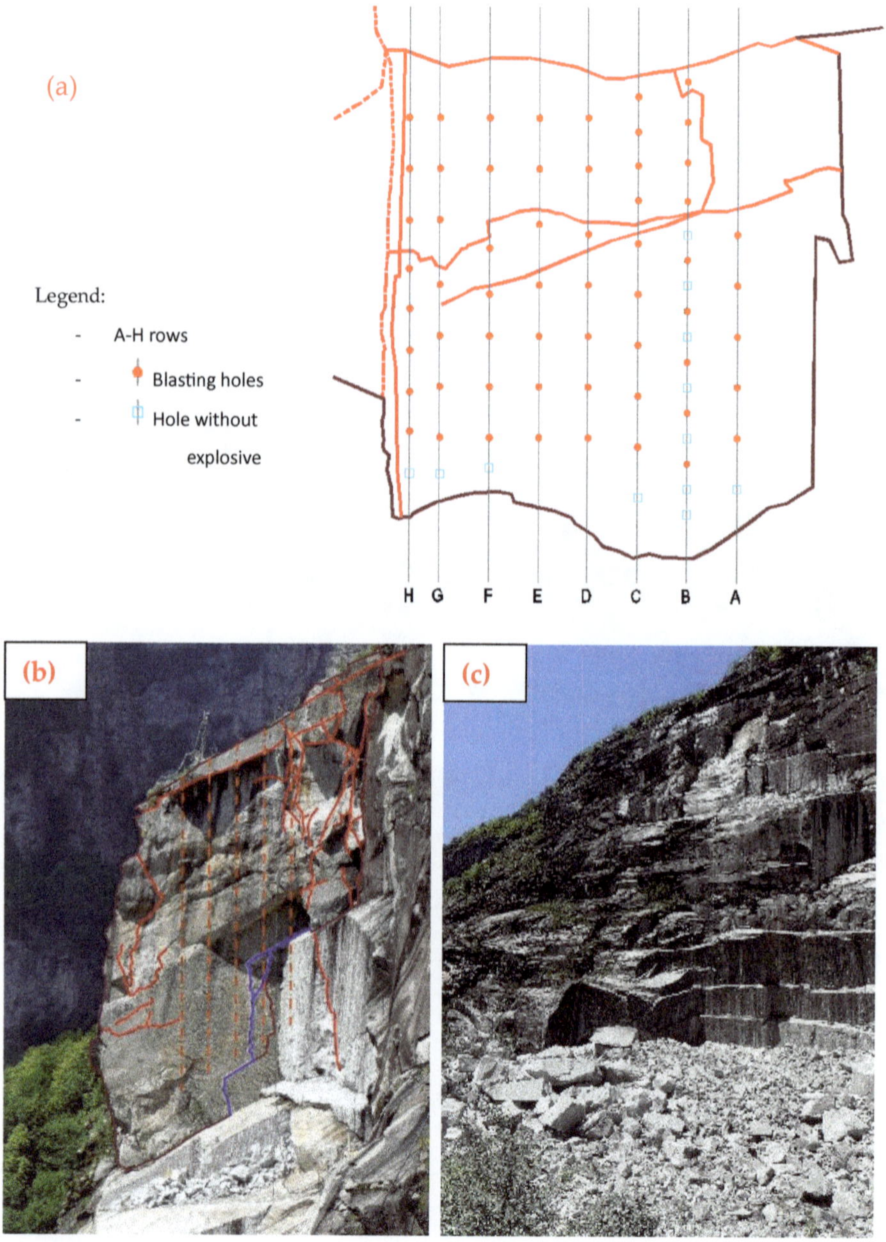


**Figure 7.** The gneiss quarry in Northern Italy. (**a**) Firing plan (top view) with the locations of blastholes drilled vertically, as designed by the authors. (**b**) Cross section with drilled holes; in red, joints and fractures, in orange blastholes, in blue rear fracture; the bench is about 25 m high. (**c**) Result of the blasting: in the upper part of the quarry face, the 'clean' and regular residual surfaces are visible, as is the blasted material in the quarry yard at the base.

In the following, Table 1, a summary of the main parameters for each considered case for our analysis is given.

**Table 1.** Overview of the main parameters.

| Case Studies | Rock Type | Total Volume (m$^3$) | Fragmentation | Prevalent Explosive Type | Drilling Diameter (mm) | Spacing in Row (m) |
|---|---|---|---|---|---|---|
| Meiringen (Switzerland) | Limestone | 60,000 | high | ANFO | 85 | 3.0 |
| Québec motorway (Canada) | Precambrian andesitic gneiss | 800 | very high | ANFO | 63 | 1.5 |
| Cravagliana (VC), Italy | Weathered gabbro rock | 2500 | very high | slurry | 41 | 1.25 |
| Le Cave (VC), Italy—Slab | Porphyry rock | 1300 | high | slurry | 41 | 1.1 |
| Perrero (TO), Italy—Catasta | Minute gneiss | 640 | normal | dynamite | 34 | 1.0 |
| Balmoreglio quarry (VB), Italy | Serizzo (gneiss) | 6000 | normal | watergel | 51 | 2.0 |

## 3. Results

Starting from the antecedents cited in Section 2.4, the formula presented here for determining and dimensioning the blasting of an unstable rock element on a rock slope assumes the following expression:

$$PF = K * (S/Ø_d) * R * E * (F * D) \quad (4)$$

where

- K is a numerical coefficient depending on the rock behavior at failure;
- S is the spacing between two holes of the same row;
- $Ø_d$ is the drilling hole diameter;
- R is a coefficient referring to rock type;
- E is a coefficient referring to explosive features;
- F is the desired fragmentation effect;
- D is the desired displacement of the blasted material.

It can be observed that $(S/Ø_d)$ represents the ratio of the rock's resistance surface to the surface (represented by the diameter of the hole) on which the pressure of the explosion gases is applied.

K is taken as 0.0059. According to [11], the rock coefficient R takes the values shown in Table 2.

**Table 2.** Division into classes for rock types.

| Class | Rock Type | R |
|---|---|---|
| 1 | quartzites and compact porphyries | 1.30 |
| 2 | sound granitoid rocks, gneisses, basalts, gabbro rock | 1.00 |
| 3 | compact limestone and dolomite, grés, highly cemented sandstone | 0.90 |
| 4 | phyllites, hard shale clay, serpentine | 0.80 |
| 5 | marl and soft limestone, gypsum, poorly cemented sandstone | 0.50 |

Table 3, related to explosives, applies the same reference but reproduces the types of explosives currently used.

**Table 3.** Division into classes for explosive types.

| Class | Explosive Type | E |
|---|---|---|
| 1 | nitroglycerine, dynamite | 0.95 |
| 2 | slurries, watergel | 1.00 |
| 3 | ammonium nitrate (ANFO) | 1.10 |

To increase the degree of fragmentation of the blasted rock, as well as to move the heap of the blasted material, additional energy is needed, which is provided with an increase in the employed explosive. The coefficients F and D in Tables 4 and 5 increase the estimated value of PF according to these concepts.

**Table 4.** Increase in FP to achieve greater fragmentation.

| Class | Fragmentation | F |
|---|---|---|
| 1 | normal (several blocks bigger than 1 $m^3$) | 1.00 |
| 2 | high (maximum size up to 1 $m^3$) | 1.20 |
| 3 | very high (maximum size up to 0.5 $m^3$) | 1.40 |

**Table 5.** Increase in FP to obtain displacement of blasted rock.

| Class | Displacement | D |
|---|---|---|
| 1 | no additional displacement in addition to the action of gravity | 1.00 |
| 2 | additional displacement required (few meters) | 1.20 |
| 3 | high additional displacement required (10 m or more) | 1.60 |

By applying Equation (4) to the real cases described in Section 2.4, the following results are obtained:

## 4. Discussion

Generally, a good match between the real (PFr) and calculated (PFe) powder factors can be observed. The only exception is case study n. 2 from Quebéc. The PFr used in this case is much higher than in other cases and has already been reported as anomalous [21]. In order to make the various PF values comparable, they are normalized to the empirical coefficients R, E, F, and D.

Figure 8 shows the normalized values of the real powder factor PFr and the calculated power factor PFe as a function of the ratio between the spacing of the drilled holes along the same row and the selected borehole diameter. The trend of the normalized PF values is represented by the line of equation PF = 0.0059 * (spacing/drilling diameter) + 0.0009. The PF obviously increases with the increase in the ratio between the spacing and diameter of the hole. The latter varies in the range 24–40, which is in good agreement with field experiments in the literature [22,23]. It should be outlined that demolition for securing a slope can follow different partial goals: cases n.3 and n.4 were finalized with a driven fragmentation in order to protect the road pavement, while in case n.5, the aim was the removal of huge unstable and distressed rock elements, with little concern for the size of fragments—something that should have be flagged regarding hazards during drilling by the rock climbers.

Comparing the proposed solution for the blasting of impervious and critical portions of rock slopes with current blasting schemes or other solutions (reinforcing or protection), it appears that this study can help improve the efficiency of results, optimize the consumption of explosives, reduce excess vibration and fly rocks, and enable such operations to occur in safer working conditions: more care put into investigating the rock mass structure and blasting scheme turns into improved final results. For example, in terms of local geostructural characterization, seismic surveys, specifically where the installation of a geophone array becomes possible, should help to detect possible hidden persistent joints at the rear, even if P-wave velocity detection alone does not appear sufficient to evaluate and classify rock masses, because characteristic impedance may be taken as a comprehensive index to evaluate and classify rock masses [24], as it is evident that the properties of rock formations constitute a decisive factor in determining the parameters of drilling and blasting operations [25]. Valuable data from the multichannel analysis of surface waves (MASWs) could be used for the selection of explosives with the desired velocity of detonation and density, so as to match the impedance of rock mass [26]; these data could also be used with the development of algorithms to match the relevant impedance for selecting the right type of explosive [27].

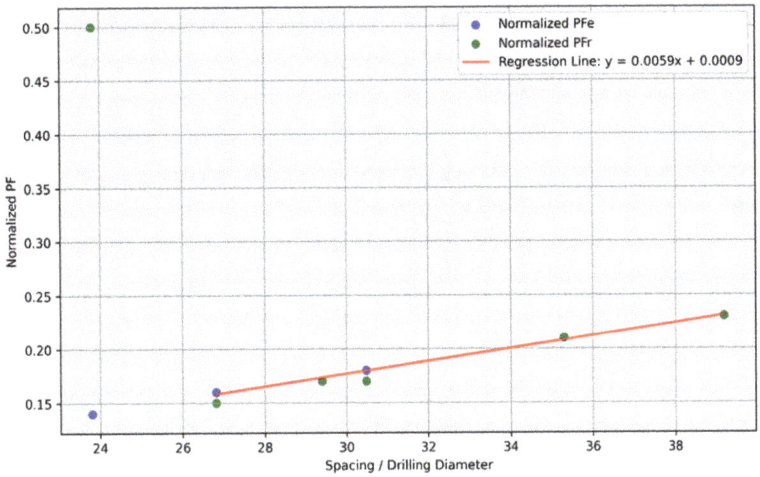


**Figure 8.** Normalized powder factor (PF) vs. ratio spacing/borehole drilling diameter.

Using explosives to remove unstable rock elements on rocky slopes presents a viable and often efficient solution for enhancing safety conditions. This method offers advantages in terms of speed, cost-effectiveness, and minimal subsequent maintenance interventions compared to alternative techniques.

The proposed innovative PF formula provides a simple and robust tool for optimizing blasting designs, allowing engineers to tailor interventions based on site-specific conditions and operational goals. By integrating real-world validations, the method offers actionable guidelines that can streamline blasting designs while ensuring safety and cost-effectiveness.

Through an in-depth analysis of various case studies, this study aimed to establish control parameters, particularly the powder factor (PF), for optimizing blasting schemes on rocky slopes. The PF serves as a synthetic parameter indicative of blasting efficiency, influenced by factors such as rock type, explosive properties, blast design, and desired fragmentation, eventually taking into account the stress or load level, as rock breakage by blasting is directly proportional to the exposed rock's stress level and pre-blasting conditions [28]; on this topic, comparison with other approaches could be applied for cross-validation, namely, after fragmentation determination by image analysis and after constraints are selected, when they are used in optimization control of the performance [29].

While traditional blasting formulas exist for tunnel and open-air bench blasting, specific criteria for blasting single elements on slopes are lacking. The proposed formula for PF calculation incorporates parameters such as hole spacing, rock and explosive characteristics, desired fragmentation, and material displacement. Real-world case analyses revealed a correlation between calculated and real PF values, with exceptions attributed to anomalous conditions. Normalized PF values showed an increasing trend with the ratio of spacing to drilling diameter.

Despite the effectiveness of explosives in slope stabilization, challenges persist, including the variability of rock properties and the need for precise blasting designs to prevent adverse effects such as fly rocks and overbreaking at the rear residual surfaces. Safety measures, such as risk assessments, comprehensive staff training, and protection of in place surfaces, are essential to mitigate risks associated with blasting operations at civil and mining sites, also combined with protective countermeasures [30,31].

The considered experiments and engineering judgments made on available data can allow one to observe that even if explosives offer a valuable tool for improving safety conditions on rocky slopes, their use requires meticulous planning, adherence to safety protocols, and consideration of site-specific factors to ensure optimal outcomes while minimizing risks and environmental impacts; additional features such as the distribution of rock fragments can be considered for comprehensive results [32,33].

## 5. Conclusions

This study's concluding remarks are based on its meaningful field experiments, which demonstrated that while explosives offer a valuable tool for improving safety conditions on rock slopes, their use requires meticulous planning, adherence to safety protocols, and consideration of site-specific factors to ensure optimal outcomes while minimizing risks and environmental impacts.

In particular, the following issues can be pointed out:

(a) The relationship between row spacing, spacing, and detonation sequence is embedded within the PF formula, where these parameters influence fragmentation and displacement outcomes. Empirical correlations derived from case studies (Table 6) highlight how adjustments in spacing and sequencing can optimize energy distribution, reducing overbreak and enhancing safety. Future research will aim to quantify these interdependencies further, potentially integrating numerical simulations for greater predictive accuracy.

(b) Developments and data collection should prioritize the investigation of additional real-world cases to bolster the robustness of the proposed formula for calculating the powder factor. Researchers can refine and validate the formula by analyzing a wider range of scenarios, ensuring its applicability across diverse geological and operational contexts. This expanded dataset will provide a more comprehensive understanding of the factors influencing blasting efficiency and allow for the identification of any additional variables that may impact the accuracy of the formula. Additionally, incorporating data from a variety of case studies will enhance the reliability and generalizability of the findings, ultimately contributing to more effective and precise blasting practices along rock slopes.

(c) The prospects for possible applications in excavation works or in geomechanical cases are related to the refinement of blasting scheme parameters according to a more reliable approach balancing and prioritizing the various involved factors: lab or site rapid testing, care regarding spacing and aperture estimation for joint structures, and the assessment of physical connections between rock volumes (rock bridges) for determining a progressive and complete kinematic evolution. Also, remote surface acquisition (photogrammetry or laser scanning) helps to build up the geometrical model and direct inspection with climbers and with unmanned vehicles (drones) is fundamental to detect possible persistent and hidden joints at the rear of slopes. Civil work (excavation or reclamation of impervious slopes), mining and quarrying bench profiling, and reinforcing work in mountain areas represent possible fields of application.

**Table 6.** Comparison between real FP and those obtained by applying Formula (4) to case studies.

| Case Study | K Coefficient | Rock Coefficient "R" | Explosive Coefficient "E" | Fragmentation Coefficient "F" | Additional Displacement Coefficient "D" | PF Real (kg/m$^3$) PFr | Estimated PF (kg/m$^3$) PFe |
|---|---|---|---|---|---|---|---|
| Meiringen (Switzerland) | 0.0059 | 0.50 | 1.10 | 1.20 | 1.20 | 0.30 | 0.30 |
| Québec motorway (Canada) | 0.0059 | 1.00 | 1.10 | 1.40 | 1.00 | 0.77 | 0.22 |
| Cravagliana (VC), Italy | 0.0059 | 1.00 | 1.00 | 1.40 | 1.00 | 0.24 | 0.25 |
| Le Cave (VC), Italy—Slab | 0.0059 | 1.30 | 1.00 | 1.20 | 1.00 | 0.24 | 0.25 |
| Perrero (TO), Italy—Catasta | 0.0059 | 1.00 | 0.95 | 1.00 | 1.00 | 0.16 | 0.16 |
| Balmoreglio quarry (VB), Italy | 0.0059 | 1.00 | 1.00 | 1.00 | 1.00 | 0.23 | 0.23 |

**Author Contributions:** Conceptualization, M.C. and C.O.; methodology, M.C.; validation, M.C. and C.O.; investigation, M.C. and C.O.; data curation, M.C.; writing—original draft preparation, M.C. and C.O.; writing—review and editing, M.C. and C.O.; visualization, G.A.D.; project administration, G.A.D. All authors have read and agreed to the published version of the manuscript.

**Funding:** This research received no external funding.

**Institutional Review Board Statement:** Not applicable.

**Informed Consent Statement:** Not applicable.

**Data Availability Statement:** The original contributions presented in this study are included in the article. Further inquiries can be directed to the corresponding authors. All citations of other authors have been taken into account.

**Conflicts of Interest:** The authors declare no conflicts of interest.

## References

1. Costamagna, E.; Oggeri, C.; Vinai, R. Damage and contour quality in rock excavations for quarrying and tunnelling: Assessment for properties and solutions for stability. *IOP Conf. Ser. Earth Environ. Sci.* **2021**, *833*, 012137. [CrossRef]
2. Jong, Y.H.; Lee, C.I. Influence of geological conditions on the powder factor for tunnel blasting. *Int. J. Rock Mech. Min. Sci.* **2004**, *41*, 533–538. [CrossRef]
3. Costamagna, E.; Oggeri, C.; Segarra, P.; Castedo, R.; Navarro, J. Assessment of contour profile quality in D&B tunnelling. *Tunn. Undergr. Space Technol.* **2018**, *75*, 67–80. [CrossRef]
4. Gavrilkovich, E.G.; Onika, S.G.; Gets, A.K.; Zhukov, S.A. Fragment size distribution and its probability distribution in blasted rock mass in a natural stone quarry. *Gorn. Z.* **2022**, *4*, 52–56. [CrossRef]
5. Djoudi, M.; Bensehamdi, S.; Fredj, M. Study of blasting effect on bench stability. *IOP Conf. Ser. Earth Environ. Sci.* **2021**, *833*, 012196. [CrossRef]
6. Senyur, M.G. A Statistical Analysis of Fragmentation After Single Hole Bench Blasting. *Rock Mech. Rock Eng.* **1998**, *31*, 181–196. [CrossRef]
7. Nefis, M.; Talhi, K. A model study to measure fragmentation by blasting. *Min. Sci.* **2016**, *23*, 91–104. [CrossRef]
8. Park, D.; Jeon, B.; Jeon, S. A Numerical Study on the Screening of Blast-Induced Waves for Reducing Ground Vibration. *Rock Mech. Rock Eng.* **2009**, *42*, 449–473. [CrossRef]
9. Abdulazeez, S.; Kudirat Oziohu, Y.; Hareya, Z.; Onimisi, A.; Hashim, M. Blasting efficiency in granite aggregate quarry based on the combined effects of fragmentation and weighted environmental hazards. *Min. Miner. Depos.* **2023**, *17*, 120–128. [CrossRef]
10. Adesida, P. Powder factor prediction in blasting operation using rock geo-mechanical properties and geometric parameters. *Int. J. Min. Geo-Eng.* **2022**, *56*, 25–32. [CrossRef]
11. Mancini, R.; Pelizza, S. Previsione dei consumi di esplosivo e di lavoro di perforazione nello scavo di gallerie. *Proc. I Convegno Internazionale Probl. Tec. Costr. Gall. Turin* **1969**, *III–IV*, 1075–1087.
12. Cardu, M.; Clerici, C.; Frisa Morandini, A.; Mancini, R.; Occella, E. An experimental research on the comminution law and work index concepts in jaw crushers. In Proceedings of the XVIII International Mineral Processing Congress, Sidney, Australia, 23–28 May 1993; pp. 97–102.
13. Hu, Y.; Lu, W.; Chen, M.; Yan, P.; Yang, J. Comparison of Blast-Induced Damage Between Presplit and Smooth Blasting of High Rock Slope. *Rock Mech. Rock Eng.* **2014**, *47*, 1307–1320. [CrossRef]
14. Mohamed, F.; Hafsaoui, A.; Talhi, K.; Menacer, K. Study of the Powder Factor in Surface Bench Blasting. *Procedia Earth Planet. Sci.* **2015**, *15*, 892–899. [CrossRef]

15. Del Greco, O.; Oggeri, C. Reinforcement design and control of rock slopes above tunnel portals in Northern Italy. *Int. J. Rock Mech. Min. Sci.* **2004**, *41* (Suppl. 1), 786–791. [CrossRef]
16. Berta, G. *L'esplosivo Strumento di Lavoro*; Italesplosivi: Milano, Italy, 1985.
17. Nova, R.; Zaninetti, A. An investigation into the tensile behaviour of a schistose rock. *Int. J. Rock Mech. Min. Sci. Geomech.* **1990**, *27*, 231–242. [CrossRef]
18. Rotzetter, G. Beseitigung der Bergsturzgefahr am Kirchberg in Meiringen (Schweiz) durch Sprengarbeit. *Nobel Hefte* **1977**, *43*, 101–106.
19. Dorval, P. Abattage d'un Diedre Rocheux Instable. Route 155, Grandes Piles. 1993. Available online: https://seeq.qc.ca/wp-content/uploads/2016/06/Article-2-Dynamitage-dun-di%C3%A8dre-rocheux-instable-Route-155-Grandes-Piles-P-Dorval.pdf (accessed on 9 January 2025).
20. Cardu, M.; Oggeri, C.; Peila, D.; Casale, M. Bonifiche con esplosivo. In Proceedings of the Conference: "Bonifica di Versanti Rocciosi per la Protezione del Territorio", Trento, Italy, 11–12 March 2004; pp. 251–278.
21. Casale, M.; Oggeri, C.; Peila, D. Improvements of safety conditions of unstable rock slopes through the use of explosives. *Nat. Hazards Earth Syst. Sci.* **2008**, *8*, 473–481. [CrossRef]
22. Prasad, S.; Choudhary, B.S.; Mishra, A.K. Effect of Stemming to Burden Ratio and Powder Factor on Blast Induced Rock Fragmentation– A Case Study. *IOP Conf. Ser. Mater. Sci. Eng.* **2017**, *225*, 012191. [CrossRef]
23. Hosseini, M.; Khandelwal, M.; Lotfi, R.; Eslahi, M. Sensitivity analysis on blast design parameters to improve bench blasting outcomes using the Taguchi method. *Geomech. Geophys. Geo-Energy Geo-Resour.* **2023**, *9*, 9. [CrossRef]
24. Zhang, Z.X.; Gong, F.; Kozlovskaya, E.; Aladejare, A. Characteristic Impedance and Its Applications to Rock and Mining Engineering. *Rock Mech. Rock Eng.* **2023**, *56*, 3139–3158. [CrossRef]
25. Jiyanov, A.; Buriyev, S.; Sunnatulloev, S. Technological schemes of blasting operations for ensuring stability of Amantaytau quarry slopes. *E3S Web Conf.* **2024**, *54812*, 05013. [CrossRef]
26. Ramulu, M.; Sinha, A.; Jha, A.K. Blast optimization with impedance matching using surface-wave tomography at an opencast coal mine. In *Lisbon Conference Proceedings*; Holmberg, R., Ed.; European Federation of Explosives Engineers: Vienna, Austria, 2011; ISBN 978-0-9550290-3-5.
27. Jha, A.K. Impedance Matching Algorithm for Selection of Suitable Explosives for Any Rock Mass—A Case Study. *J. Geol. Resour. Eng.* **2020**, *8*, 55–65. [CrossRef]
28. Imashev, A.; Mussin, A.; Adoko, A.C. Investigating an Enhanced Contour Blasting Technique Considering Rock Mass Structural Properties. *Appl. Sci.* **2024**, *14*, 11461. [CrossRef]
29. Engin, I.C. *A Practical Method of Bench Blasting Design for Desired Fragmentation Based on Digital Image Processing Technique and Kuz-Ram Model*; Rock Fragmentation by Blasting—Sanchidrián Edition; Taylor & Francis Group: London, UK, 2010; ISBN 978-0-415-48296-7.
30. Reccord, T. Rock Scaling contractor's risk. Managing Highway Rock Slope Scaling. In Proceedings of the Transportation Research Board 97th Annual Meeting, Washington, DC, USA, 7–11 January 2018; Transportation Research Board Circular E-C260; 2020; ISSN 0097-8515. Available online: https://trid.trb.org/View/1688103 (accessed on 9 January 2025).
31. Jelušič, P.; Vlastelica, G.; Žlender, B. Sustainable Retaining Wall Solution as a Mitigation Strategy on Steep Slopes in Soft Rock Mass. *Geosciences* **2024**, *14*, 90. [CrossRef]
32. Thomson, S. Controlling fragmentation. In *Fragmentation, International Mining*; Team Publ.: Berkhamsted, UK, 2012; pp. 80–84.
33. Gao, P.; Pan, C.; Zong, Q.; Dong, C. Rock fragmentation size distribution control in blasting: A case study of blasting mining in Changjiu Shenshan limestone mine. *Front. Mater.* **2023**, *10*, 1330354. [CrossRef]

*applied sciences*

MDPI

*Review*

# A Comprehensive Review of the Influence of Sensitizers on the Detonation Properties of Emulsion Explosives

**Andrzej Maranda [1], Dorota Markowska [2], Bożena Kukfisz [3,*] and Weronika Jakubczak [4]**

1 Łukasiewicz Research Network, Institute of Industrial Organic Chemistry, Annopol 6 Street, 03-236 Warsaw, Poland; andrzej.maranda@ipo.lukasiewicz.gov.pl
2 Faculty of Process and Environmental Engineering, Lodz University of Technology, Wolczanska Street 213, 90-924 Lodz, Poland; dorota.siuta@p.lodz.pl
3 Institute of Safety Engineering, Fire University, Slowackiego Street 52/54, 01-629 Warsaw, Poland
4 Internal Security Institute, Fire University, Slowackiego Street 52/54, 01-629 Warsaw, Poland; wjakubczak@apoz.edu.pl
* Correspondence: bkukfisz@apoz.edu.pl

**Abstract:** Emulsion explosives are extensively utilized in the global mining industry due to their superior water resistance, high safety standards, cost-efficiency, and robust performance. The basic component of these explosives is a water-in-oil emulsion matrix, which, in its initial state, lacks the capacity for detonation. The sensitization process, achieved through either physical or chemical means, is a critical step that enhances the emulsion's sensitivity to detonation, thereby improving its operational efficiency in blasting applications. This review presents a comprehensive and systematic analysis of the current scientific literature and experimental investigations concerning the impact of key sensitizing methods and agents on the detonation characteristics of emulsion explosives. Particular emphasis is placed on the classification of sensitizers, their physicochemical properties, and their interactions with the emulsion matrix. By examining various sensitization mechanisms, this study provides insights into the role and efficacy of both established and emerging sensitizing agents. The findings of this review highlight the pivotal role of sensitizer selection in defining the detonation performance of emulsion explosives, with implications for enhancing safety standards and ensuring the protection of both industrial operations and public safety. The most optimal sensitization method is chemical, utilizing cost-effective components that generate gas bubbles within the matrix. A key advantage is the in situ production of emulsion explosives, which eliminates the need for their transport on public roads, thereby enhancing safety and reducing the risk of terrorist threats.

**Keywords:** emulsion explosives; sensitizers; detonation parameters; matrix; safety

## 1. Introduction

Emulsion explosives (EEs) have become the most widely used blasting agents in the mining industry and for various other industrial applications, primarily due to their exceptional water resistance, insensitivity to mechanical stimuli facilitating the mechanical loading of blast holes, and remarkable chemical and physical stability, particularly in cartridge form. The basic component of EEs is a water-in-oil emulsion, referred to as the matrix, with densities typically ranging from 1.3 to 1.4 g/cm$^3$. This phase acts as the medium within which water droplets are suspended, forming a theoretically homogeneous material with continuous mechanical properties. However, such mixtures, in the absence of chemical compounds classified as explosives, exhibit only minimal detonation capability.

Following the relatively short era of slurry explosives, emulsion explosives have become one of the primary blasting agents used in the mining industry. A notable example is the Polish mining sector, where in 2022, bulk emulsion explosives (chemically sensitized) accounted for 75% of all blasting agents used in open-pit mines, with ANFO being the second most common choice. The observed decline in ANFO consumption is attributed, among other factors, to the rising cost of ammonium nitrate and the transition to water-bearing deposits. Similarly, in underground mining, both cartridge and bulk forms of emulsion explosives have dominated as the primary blasting agent (90%), gradually replacing dynamite. The widespread adoption of emulsion explosives is driven not only by their previously mentioned advantages, such as water resistance and minimal sensitivity to mechanical stimuli, but also by their detonation parameters, which position them slightly below dynamite yet significantly above ANFO.

EEs are recognized for their comparatively lower environmental impact relative to conventional explosives. Their detonation produces fewer toxic fumes, which benefits both the environment and the health of mine workers. By adjusting their formulation, it is possible to further reduce the emission of harmful gases, such as nitrogen oxides ($NO_x$), enhancing their environmental friendliness [1]. However, despite their environmental advantages, the production and use of emulsion explosives still contribute to various environmental impacts. Maranda et al. [2] stated that the greatest contribution to marine aquatic ecotoxicity potential occurs during the raw material extraction and production stages of emulsion explosives, while their detonation has the most significant impact on acidification and particulate matter emissions with inorganic compounds. Process contribution analysis indicates that ammonium and sodium nitrates are the primary contributors to the overall environmental burden. Bulk emulsion explosives have a 65% lower environmental impact compared to cartridge emulsion explosives.

The primary method of initiating a high-energy process in explosives is to act on it with suitably intense shock waves generated by the detonation of a blasting agent. The initiation of a rapid chemical reaction is associated with the high temperature generated by the shock wave's rapid adiabatic compression of the explosive. However, the impact of even a very intense shock wave causes a small average increase in the temperature of the loaded homogeneous material. Therefore, it is necessary to make structural changes to the explosive that will cause local foci characterized by increased energy to form behind the front of the initiating shock wave. These points (areas) are called hot spots. If the energy at such a point is high enough, then it can become the focus of a chemical reaction. If it develops according to the thermal mechanism, the point is an active hot spot, and if it disappears, it is an inactive hot spot. The phenomenon of hot spot formation in energetic materials remains incompletely understood, with multiple mechanisms proposed to explain their generation. These include the formation of matter streams during collisions between crystals and grains, and medium heating due to pore reduction, which is influenced by the viscosity or viscoelastic properties of the deformed material near pore surfaces or by the development of microfluidic flows during pore surface deformation and their interaction with explosive material. Other mechanisms involve local heating of the medium from the shock compression of entrapped gas bubbles, shock wave interactions near inclusions with high wave impedance, frictional heating from relative movement between explosive crystals and grains, and internal friction generated on slip planes within explosive crystals [3–7]. The dominance of any particular mechanism in the hot spot formation process may vary depending on the specific conditions, or multiple phenomena may act simultaneously. Based on these theoretical premises and the outlined mechanisms, various sensitization methods have been developed to enhance the explosive properties of the emulsion matrix. These methods include the following:

- The reduction in matrix density through chemical means;
- The introduction of low-bulk-density substances to decrease overall density;
- The incorporation of solid inhomogeneities to promote localized energy release;
- The addition of high-energy explosives, such as 2,4,6-trinitrotoluene (TNT), pentrite, nitrocellulose, smokeless powder, complex rocket fuels, hexogen, octogen, and their mixtures.

A co-occurrence analysis of the keywords "emulsion explosives", "sensitizers", and "detonation parameters" was conducted to elucidate the relationships and thematic connections between key research areas. The sizes of the circles and labels represent the relative weights of keyword co-occurrences. The network map was constructed using bibliometric data from 30 articles indexed in the Scopus database, published between 2004 and 2024. The visualization of the co-occurrence network was generated using VOSviewer software (version 1.6.20). The analysis identified 26 key elements, which were categorized into three distinct clusters (Figure 1). Our findings indicate a growing research emphasis on the production of emulsion explosives (EEs) through the chemical sensitization of various matrices. This trend is driven by the pursuit of a deeper understanding of how different sensitizing additives influence critical detonation parameters, such as detonation velocity and detonation pressure. The clustering of keywords highlights the interconnected nature of studies focusing on matrix composition modification, the development of novel sensitizers, and their subsequent impact on the energetic and physical properties of EEs.

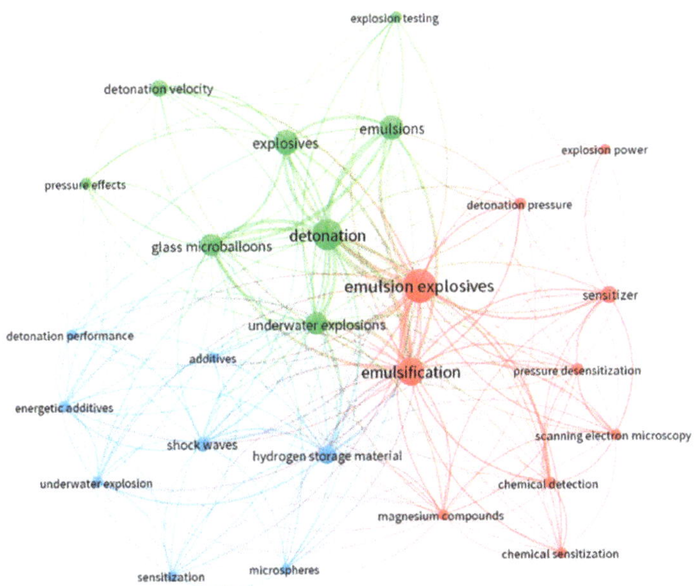


**Figure 1.** Co-occurrence analysis of keywords (emulsion explosives, sensitizers, and detonation parameters) in articles published between 2004 and 2024 (keyword co-occurrence threshold of 4).

The purpose of this review is to present a comprehensive synthesis of theoretical, practical, and application-oriented research on the sensitization of emulsion explosives. This review not only enhances our understanding of the underlying mechanisms of sensitization, but also lays the groundwork for the development of more effective sensitization strategies.

## 2. Review of Influence of Sensitizers on Detonation Properties of Emulsion Explosives

### 2.1. Chemical Density Reduction

This process involves the addition of substances into the matrix that decompose to release nitrogen or other gases in the form of microbubbles. These substances include N N,N-dinitrosopentaethyl-ethylenetetramine; hydrogen peroxide; azodicarbonamide; sodium, potassium, and barium inorganic peroxides; bicarbonates of alkali metals and alkaline earth metals; nitrates (III) of alkali metals or chemical systems of sodium nitrate–urea; alkali hydroborides–urea [8,9]; or a mixture of potassium manganate (VII) and hydrogen peroxide [10]. The encapsulated gas microbubbles act as "hot spots", facilitating the initiation and propagation of the EE detonation process.

The aeration of the emulsion using the sodium nitrate (III)–urea chemical system occurs due to the release of gases (carbon dioxide and nitrogen) according to the following reaction (Equation (1)):

$$3NaNO_2 + 3HNO_3 + 2(NH_2)_2CO \rightarrow 3NaNO_3 + 2CO_2 + 4H_2O + 3.5N_2\uparrow \quad (1)$$

The nitric acid (V) involved in the reaction acts as a pH-regulating agent. If $NaNO_2$ alone is used to sensitize the matrix, it will react with ammonium nitrate (V) according to Equations (2) and (3):

$$NaNO_2 + NH_4NO_3 \rightarrow NH_4NO_2 + NaNO_3 \quad (2)$$

$$NH_4NO_2 \rightarrow 2H_2O + N_2\uparrow \quad (3)$$

In total,

$$NaNO_2 + NH_4NO_3 \rightarrow NaNO_3 + 2H_2O + N_2\uparrow \quad (4)$$

In practice, if $NaNO_2$ is used as a stand-alone gassing agent, compounds containing thiocyanate anions are additionally dosed to increase the decomposition efficiency of the sodium derivative. The decomposition of $NaNO_2$ in the presence of ionic thiocyanate compounds follows Equation (5):

$$H^+ + SCN^- + HONO \rightarrow NOSCN + H_2O \quad (5)$$

NOSCN is formed, which is more active than nitric (III) or nitric (V) acids, and reacts in solution with amines or inorganic ammonium salts.

Another aerating substance is sodium bicarbonate. It already decomposes easily at 65 °C, and at 270 °C, carbon dioxide is released completely. The decomposition reaction is more intense under humid conditions. The aqueous solution of sodium bicarbonate has a slightly alkaline pH, and when acetic acid is added, decomposition occurs with the release of carbon dioxide. Therefore, in order to gasify the EE matrix, sodium bicarbonate and acetic acid are dosed simultaneously.

The effect of the degree of gasification, achieved through the application of sodium nitrate (III), on the density and detonation velocity of EEs was determined by Mishra et al. [10]. Experiments were performed for 30 min at different temperatures (31 °C, 47 °C, and 70 °C) and at five concentrations of gassing agent solution (2%, 4%, 6%, 8%, and 10%). At 31 °C, there was a monotonic decrease in density for 25 min, followed by stabilization for another 25 min. At temperatures of 47 and 70 °C, density stabilization occurred after 20 min. The minimum densities for temperatures of 3 °C 1, 47 °C, and 70 °C, reached values of 1.14, 0.92, and 0.94 g/cm$^3$, respectively. The density of the EE (1.15 g/cm$^3$) at which the maximum detonation velocity of 4381 m/s was achieved was determined. The maximum density at which the process disappeared was 1.27 g/cm$^3$.

An original gassing agent was proposed by Kramarczyk et al. [11,12], through the addition of sodium perchlorate to an aqueous solution of ammonium nitrate (sensitizers BK-1 and BK-2, Table 1). They were used to sensitize the commercial EE Emulinit 8L. Compared to the standardly aerated Emulinit 8L (using the EE sensitizer without the addition of sodium perchlorate), the resulting EE exhibited higher detonation velocity, brisance, and air blast parameters, lower content of harmful detonation products, and a shorter density reduction time (Table 2).

**Table 1.** Sensitizer's components [11].

| Component | Sensitizer | | |
|---|---|---|---|
| | EE | BK-1 | BK-2 |
| | Component Content [wt. %] | | |
| Ammonium nitrate | - | 30.0 | 47.1 |
| Water | 95.45 | 61.6 | 41.0 |
| Sodium perchlorate | - | 4.0 | 8.0 |
| Sodium nitrite | 4.50 | 3.0 | 3.3 |
| pH modifier | 0.05 | 1.5 | 0.7 |

**Table 2.** Sensitized EEs parameters when sensitized according to [11,12].

| Sensitizer | Detonation Velocity [m/s] | Peak Over-pressure [1] [kPa] | Impulse ABV [1] [Pa·s] | Explosive Gas Content [kg/dm$^3$] | |
|---|---|---|---|---|---|
| | | | | $NO_x$ | CO |
| EE | 4233 | 123.73 | 57.20 | 0.55 | 4.11 |
| BK-1 | 4647 | 127.13 | 57.53 | 0.33 | 2.51 |
| BK-2 | 5033 | 131.20 | 58.60 | 0.31 | 3.45 |

[1] 2 m from axis of EE sample.

Between 2013 and 2016, five studies by Cheng et al. [13–17] were published, in which the authors proposed the use of magnesium dihydride ($MgH_2$) as a sensitizing agent for EEs. The sensitizing agent was dihydrogen microbubbles formed during $MgH_2$ decomposition. Publications [17,18] present the results of studies on the parameters of emulsion explosives (EEs) containing matrices with a density of 1.31 g/cm$^3$. These matrices consisted of ammonium nitrate (75%), sodium nitrate (10%), paraffin wax (4%), diesel oil (1%), emulsifier (2%), and water (8%). The performance of EEs sensitized with magnesium dihydride ($MgH_2$) (98% purity, average particle size of 3 μm, and bulk density of 1.45 g/cm$^3$) was evaluated and compared to that of EEs containing glass microspheres (GMs) (average particle size of 55 μm, bulk density of 0.25 g/cm$^3$), a GM/Al mixture, and sodium nitrite ($NaNO_2$). The compositions and parameters of the EEs are shown in Table 3.

The high detonation capability of EEs is due to their structure consisting of the maximum homogenization of oxidants and combustible components, ensuring the maximum degree of rearrangement in the chemical reaction zone of the detonation wave, as well as gassing additives forming "hot spots" in the non-explosive matrix. The shock wave propagation in the EE, which does not initiate the detonation process in the loaded material, can cause its partial or complete desensitization, as a result of emulsion reversal (crystallization of oxidants [18]) or the deformation (destruction) of hot spots. The aforementioned phenomena may occur during blasting operations as a result of the "channel effect".

**Table 3.** EE compositions and parameters [13].

| Component | Sensitizer | | | |
|---|---|---|---|---|
| | GMs | GMs/Al | $MgH_2$ | $NaNO_2$ |
| | [wt.%] | | | |
| Matrix | 96.0 | 92.0 | 98.0 | 99.8 |
| GMs | 4.0 | 4.0 | - | - |
| Aluminum dust | - | 4.0 | - | - |
| $MgH_2$ | - | - | 2 | - |
| $NaNO_2$ | - | - | - | 0.2 |
| Parameter | | | | |
| Density [g/cm$^3$] | 1.21 | 1.24 | 1.29 | 1.24 |
| Underwater test maximum overpressure [MPa] | 10.89 | 10.72 | 13.12 | - |
| Energy [kJ/kg] | 2871 | 3187 | 3762 | - |
| Detonation energy [kJ/kg] | | | | |
| Theoretical | 3297 | 3684 | 3530 | 3297 |
| Experimental | 2728 | 3028 | 3574 | 2835 |
| Detonation velocity [m/s] | 4434 | 4389 | 5552 | - |
| Brisance [mm] | 16.1 | 16.2 | 19.1 | 16.85 |

Therefore, Cheng et al. conducted studies of the effects of shock waves of different intensities on EEs sensitized with $MgH_2$ and, comparatively, with GMs [14,15] or sodium nitrite [15–17]. They placed 30 g EE samples in a container filled with water at various distances from a 10 g hexogen booster (hexogen/wax 95/5, density-1.45 g/cm$^3$). The EE samples were subjected to the shock wave generated by the detonation of the booster. The unloaded and loaded samples were then initiated with the detonator and the overpressure course was determined using the underwater test method. Starting from the obtained magnitudes of maximum overpressures, the authors of papers [18–20] estimated the degree of desensitization ($S_d$) (Table 4) of loaded EEs based on Equation (6):

$$S_d = (P_0 - P_1)/(P_0 - P_d) \quad (6)$$

where $P_0$—peak overpressure of unloaded EE, $P_1$—peak overpressure of loaded EE, and $P_d$—peak overpressure of the booster.

**Table 4.** Compositions and degrees of desensitization of EEs sensitized with various additives [13–17].

| Component | Sensitizer | | | |
|---|---|---|---|---|
| | GMs | $NaNO_2$ | $MgH_2$ | |
| | [wt.%] | | | |
| Matrix | 96.0 | 99.8 | 98.0 | 99.0 |
| GMs | 4.0 | - | - | - |
| $NaNO_2$ | - | 0.2 | - | - |
| $MgH_2$ | - | - | 2.0 | 1.0 |
| Distance [mm] | Sensitization degree [%] | | | |
| 25 | 100 | 88.12 | 38.97 | 38.97 |
| 40 | 86.41 | 84.47 | 18.89 | 18.89 |
| 50 | 79.82 | 71.63 | 12.11 | 12.11 |
| 60 | 73.67 | 53.47 | 10.45 | 10.45 |
| 75 | 63.34 | 15.59 | 11.76 | 11.76 |

The data summarized in Table 3 clearly demonstrate that EEs sensitized with $MgH_2$ exhibit the highest energy parameters. This result is attributed to the contribution of hydrogen, generated during the decomposition of $MgH_2$, to high-energy reactions within the detonation wave zone. Additionally, when assessing the degree of desensitization of EEs with different sensitizers, Cheng et al. [13–17] (Table 4) observed that EEs containing $MgH_2$ displayed the highest resistance to non-initiation shock waves. However, it is puzzling that the authors of studies [13–17] reported identical desensitization values for EEs despite using two different $MgH_2$ contents.

Based on the observed relationship between the degree of desensitization and the type of EE, Cheng et al. [13–17] concluded that the process affecting the value of this parameter is not the phenomenon of matrix demulsification. Instead, it is attributed to changes occurring in the "hot spots" formed by individual sensitizers. Under the influence of the shock wave, some of the GMs are crushed and lose their sensitizing properties. In contrast, in the case of $MgH_2$, the hydrogen bubbles are flexible and can return to their original form when squeezed. The authors also believed that when added to the matrix, only part of the $MgH_2$ reacts. The remaining $MgH_2$ can react only as a result of exposure to a strong shock wave (temperature increase), in which case, we would be dealing with so-called "dynamic sensitization".

The gasification method is now increasingly used in the production of EEs, due to the cumbersome handling of either glass or polymer microspheres and their high prices. However, this yields EEs with lower physical stability, due to the greater possibility of gas bubbles migrating from the explosive mixture. Chemical aeration in particular should be used for in situ EE fabrication.

### 2.2. Addition of Substances with Very Low Bulk Density

Dosing substances containing occluded air or other gases is another method of reducing the density of the matrix, causing an increase in its detonation ability. Several materials may be used, including glass or plastic microspheres, cenospheres (CSs), glass or polystyrene beads, glass capillaries, and various grades of perlites.

The addition of GMs as sensitizing agents for EEs was already proposed in the first patents for EEs, such as those of EE inventors Bluhm [19] and Wade [20]. Detailed results on the effect of the content and dimensions of GMs on the detonation velocity of EEs have been reported by Lee and Persson [21] and in Xuguang's monograph [22]. The EEs tested in the paper [21] contained ammonium nitrate (66.91%), calcium nitrate (14.59%), water (12.0%), light mineral oil (5.0%), and the emulsifier SPAN 80 (1.5%). They were sensitized with QCel GMs made by the PQ Corporation. The cracked GMs were separated with isopropyl alcohol. The results of detonation velocity tests in 23.6 mm diameter charges are summarized in Table 5. The study [21] was conducted on EEs containing glass microspheres with diameters of 33 µm, 54 µm, and 125 µm. It was found that as the size of the glass microspheres increased, the detonation velocity decreased. The maximum detonation velocities recorded were 4620 m/s (125 µm), 5260 m/s (54 µm), and 5500 m/s (33 µm). Additionally, with decreasing microsphere size, these maximum velocities were observed at progressively higher densities—1.15 $g/cm^3$, 1.24 $g/cm^3$, and 1.27 $g/cm^3$, respectively.

From the data presented in Table 5, it can be seen that the dimensions of GMs have a significant effect on the detonation speed of EEs. Moreover, the sensitizer content, by adjusting the density, is the main factor determining the detonation velocity of EEs. The density dependence of the detonation velocity is of the same nature for EE as for other ammonium–salt explosive mixtures not sensitized with individual explosives. With increasing density, the detonation velocity initially increases until it reaches a maximum, and then decreases until the detonation process disappears. A similar characteristic of dependence of the detonation velocity on GM content was observed by Deribas et al. [23]. According to the breakdown proposed by Price [24] under experimental conditions, EEs behave like secondary explosives (e.g., ammonium nitrate (V), ammonium chlorate (VII), dinitrotoluene), defined as "non-ideal". The non-ideality of EEs sensitized with microspheres is further supported by the relationship between the critical diameter and GM content [25]. As the GM content increases, the critical diameter initially decreases, reaching a minimum, after which it begins to increase. The lowest critical diameter was observed at a GM content of 3%.

**Table 5.** EEs' detonation velocities as a function of GM content and dimensions [21].

| GMs Size [μm] | GMs Content [%] | Density [g/cm³] | Detonation Velocity [m/s] |
|---|---|---|---|
| 153 | 5.359 | 0.909 | 3891 |
| | 3.874 | 1.003 | 4218 |
| | 2.135 | 1.100 | 4214 |
| | 1.724 | 1.210 | 3754 |
| 108 | 8.985 | 0.900 | 3837 |
| | 6.645 | 0.998 | 4165 |
| | 4.586 | 1.103 | 4374 |
| | 3.014 | 1.202 | 4120 |
| 82 | 9.916 | 0.895 | 3779 |
| | 8.289 | 0.992 | 4203 |
| | 5.870 | 1.097 | 4551 |
| | 3.876 | 1.205 | 4615 |
| 64 | 11.972 | 0.891 | 3781 |
| | 9.030 | 0.999 | 4244 |
| | 6.544 | 1.112 | 4695 |
| | 4.321 | 1.204 | 4711 |

Other studies have shown that the size of GMs affects the properties of EEs, including detonation ability and generated pressure, after prior dynamic loading [26,27]. Additionally, it influences the degree of crystallization [28] and the density, which is crucial for producing EEs with low detonation velocity for high-energy cladding applications [29].

Further sensitizers that have ben used in EEs are binary systems of GMs with titanium dioxide or titanium [30]. The choice of $TiO_2$ as a system component was dictated by its resistance to oxidation and active water interaction [30,31], ensuring the stability of $T_iH_2$ in the EE matrix. Cheng et al. [30] studied the properties of EEs containing matrices with a density of 1.31 g/cm$^3$ consisting of ammonium nitrate (75%), sodium nitrate (10%), paraffin wax (4%), diesel oil (1%), emulsifier (2%), and water (8%). The matrix was sensitized with GMs and their mixtures with $TiH_2$ and titanium. GMs had a bulk density of 0.25 g/cm$^3$ with an average grain size of 55 μm. In contrast, the other additives had significantly higher bulk densities of 3.91 g/cm$^3$ ($TiH_2$) and 4.5 g/cm$^3$ (Ti), and average grain sizes of 48 μm. Underwater energy, detonation velocity, and aggregability were determined for the EEs tested.

In the underwater test, Cheng et al. [30] detonated EE charges of 10 g. In the first phase of their tests, they determined the pressure changes ($p$) at time ($t$) for mixtures containing different amounts of GMs (Figure 2). They obtained a maximum pressure value of 20.58 MPa at a GM content of 4%, which was used in the subsequent EEs tested. They then determined $p$ = f(t) for EEs containing different amounts of $TiO_2$ (Figure 3, Table 6). The estimation of shock wave energy, gas bubble energy, and total energy (Figure 4) was estimated according to the methodology presented in the paper [32,33]. For the two selected

EEs, the authors determined the velocity of detonation in charges placed in PVC pipes with a diameter of 40 mm, and the brisance using the Hess method (Table 7).

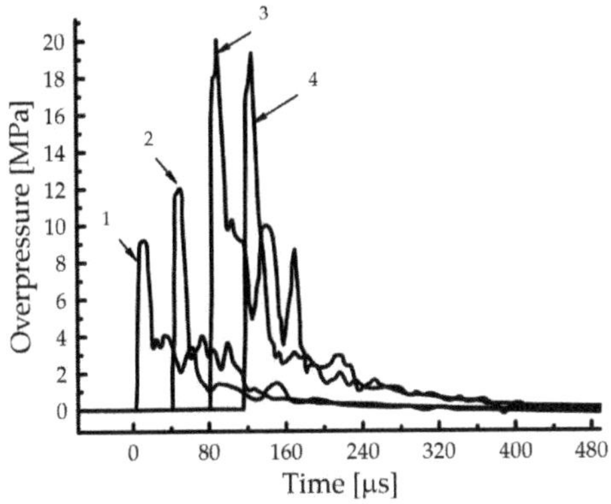


**Figure 2.** Pressure–time dependence for EE-GMs-$TiH_2$ containing different amounts of GMs [30]: 1—0%, 2—2%, 3—4%, 4—6%.

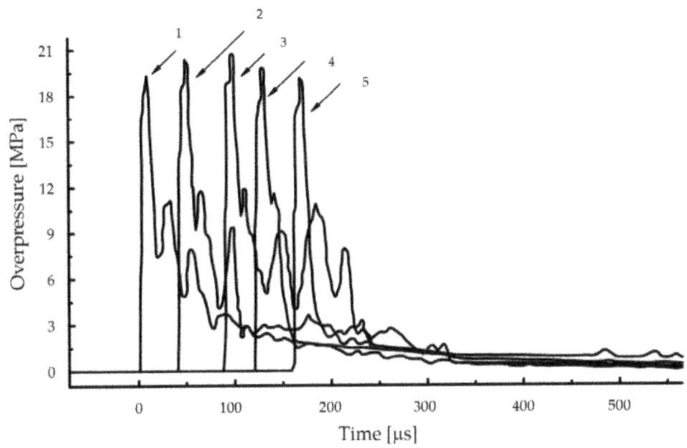


**Figure 3.** Pressure–time dependence for EEs-GMs-$TiH_2$ containing different amounts of $TiO_2$ [30]: 1—0%; 2—2%; 3—4%; 4—6%; 5—8%.

**Table 6.** Compositions and parameters of EEs-GMs-$TiH_2$ containing different amounts of $TiO_2$ [30].

| No | Component [%] | | | Maximum Overpressure [MPa] | Energy [MJ/kg] | | |
|---|---|---|---|---|---|---|---|
| | Matrix | GM | $TiH_2$ | | SW * | GB ** | Total |
| 1 | 95 | 4 | 1 | 19.01 | 0.658 | 1.63 | 2.69 |
| 2 | 94 | 4 | 2 | 20.58 | 0.712 | 1.68 | 2.83 |
| 3 | 92 | 4 | 4 | 20.72 | 0.695 | 1.66 | 2.79 |
| 4 | 90 | 4 | 6 | 19.92 | 0.681 | 1.64 | 2.74 |
| 5 | 88 | 4 | 8 | 19.46 | 0.679 | 1.65 | 2.75 |
| 6 | 96 | 4 | - | 21.77 | 0.704 | 1.50 | 2.58 |
| 7 | 94 | 4 | 2 *** | 19.16 | 0.665 | 1.58 | 2.63 |

* Shock wave, ** gas bubble, *** titanium.

**Table 7.** Parameters of EE with different $TiH_2$ contents [30].

| No | Component [%] | | | Density [g/cm$^3$] | Detonation Velocity [m/s] | Brisance [mm] |
|---|---|---|---|---|---|---|
| | Matrix | GM | $TiH_2$ | | | |
| 1 | 96 | 4 | - | 1.18 | 4534 | 16.1 |
| 2 | 94 | 4 | 2 | 1.11 | 2659 | 23.8 |

These results show that at a certain content of $T_iH_2$, the EEs have higher parameters (maximum shock wave pressure, explosion energy, detonation velocity, and brisance) than EEs sensitized with GMs alone. In contrast, the addition of titanium powder causes a decrease in maximum shock wave pressure and a slight increase in explosion energy compared to EEs sensitized with GMs alone. These results may suggest that hydrogen generated by the decomposition of $T_iH_2$ is involved in the chemical reactions of the detonation wave.

CSs formed during the combustion of lignite coal are another additive studied in terms of EE sensitization. They have a higher bulk density (0.3 g/cm$^3$) than typical GMs. In the work of Maranda et al. [34], the same qualitative dependence of the critical diameter of detonation on CS content was obtained as in the work [30] for GMs. However, the determined critical detonation diameter was much higher (16 mm) and occurred at a higher sensitizer content (18%). The results of studies of the effect of the content and dimension of CSs on the detonation parameters of EEs are presented in works [35,36]. Anshits et al. [35] tested EEs with matrices containing ammonium nitrate (76.9%), industrial oil (6.9%), emulsifier (1%), and water (15.3%), sensitized with CSs of several grain sizes and differing in bulk density. The dependence of the detonation velocity on the grain size of CSs is illustrated in Figure 4 [35].

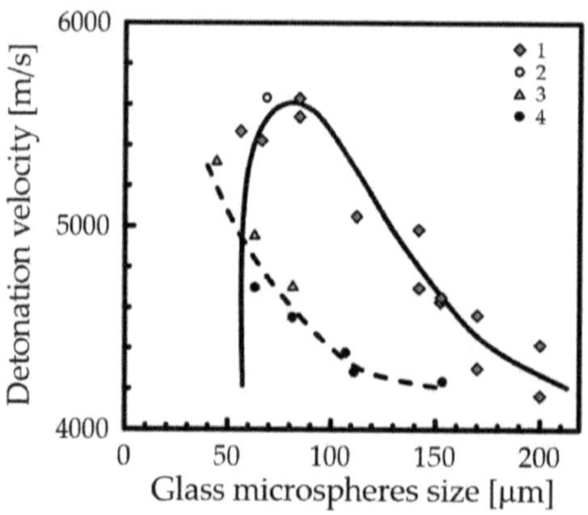


**Figure 4.** Dependence of EE detonation velocity on grain size of microspheres in charges of a given diameter: 1—CSs (d = 55 μm); 2—GMs (d = 55 μm); 3—GMs (d = 36 μm); 4—GMs (d = 23.6 μm) [35].

A maximum value of the velocity of detonation of 5.5–5.6 km/s was obtained, in charges of 55 mm in diameter, for a narrow fraction of CSs with sizes in a range from 70 to 100 μm (bulk density of 0.36 g/cm$^3$) and an EE density of 1.18–1.20 g/cm$^3$. The mass content of CSs was 8–10%. The critical diameter was 35–40 mm. Increasing the size of CSs to 200 μm resulted in a decrease in the velocity of detonation by 1.0–1.3 km/s. EEs containing an unseparated mixture of CSs had a detonation velocity of 4.2 m/s and a critical diameter of 45 mm.

Fang et al. [36] presented the content of the main components of the CSs in comparison to GMs. They were as follows: $S_iO_2$ (52.32% and 74.96%), $Al_2O_3$ (30.09% and 0%), and CaO (9.42% and 0%). For EEs containing a matrix composed of ammonium nitrate (75.0%), sodium nitrate (10.0%), paraffin wax (4.0%), diesel fuel (1.0%), emulsifier (2.0%), and water (8.0%), they conducted tests of detonation velocity (in 32 mm diameter charges) and crushability, among others. The variables were the content and grain size of CSs and GMs. The test results are summarized in Tables 8 and 9.

**Table 8.** Dependence of detonation velocity and brisance on CS (d50 = 58 μm) and GM (d50 = 57 μm) content, according to [36].

| Content [%] | | Density [g/cm$^3$] | Brisance [mm] | Detonation Velocity [m/s] |
|---|---|---|---|---|
| CSs | GMs | | | |
| 2 | - | 1.24 | 9.9 | ND |
| 4 | - | 1.21 | 10.7 | ND |
| 6 | - | 1.18 | 13.0 | 3027 |
| 8 | - | 1.17 | 13.3 | 3367 |
| 10 | - | 1.16 | 13.4 | 4138 |
| 12 | - | 1.15 | 14.7 | 4616 |
| 14 | - | 1.13 | 14.4 | 4137 |
| 16 | - | 1.11 | 13.9 | 4053 |
| - | 2 | 1.20 | 18.7 | 3653 |
| - | 3 | 1.18 | 19.7 | 4950 |
| - | 4 | 1.12 | 20.6 | 5176 |
| - | 5 | 1.10 | 20.2 | 4789 |

ND—no detonation.

**Table 9.** Detonation velocity and brisance of EEs as a function of grain size (d50) at optimum CS and GM content, according to [36].

| Content [%] | | Grain Size [μm] | | Density [g/cm$^3$] | Brisance [mm] | Detonation Velocity [m/s] |
|---|---|---|---|---|---|---|
| CSs | GMs | | | | | |
| 12 | - | 26 | - | 1.23 | 11.2 | ND |
| 12 | - | 58 | - | 1.15 | 14.7 | 4616 |
| 12 | - | 83 | - | 1.18 | 18.3 | 4970 |
| 12 | - | 142 | - | 1.20 | 13.8 | 3862 |
| - | 4 | - | 18 | 1.16 | 4.8 | ND |
| - | 4 | - | 35 | 1.14 | 14.5 | 4541 |
| - | 4 | - | 57 | 1.12 | 20.6 | 5176 |
| - | 4 | - | 102 | 1.09 | 14.7 | 4563 |

ND—no detonation.

The results of Fang et al. [36] showed that the optimum content of CSs compared to GMs is much higher, with values of 12% and 4%, respectively. Moreover, EEs containing GMs exhibited higher detonation parameters.

Another type of EE sensitizer is microballoons made from plastics [37–45]. A detailed study of the effect of microballoons made of acrylonitrile/vinylidene chloride with a bulk density of 0.027 g/cm$^3$ on the detonation velocity of EEs is presented in [37]. The variables were the degree of aeration (the matrix had a density of 1.39 g/cm$^3$) and the dimensions of the microballoons. Detonation velocity measurements were made in charges of 50 mm in diameter. With increasing aeration (decreasing density), the detonation velocity initially increased, followed by a subsequent decrease. Similarly to glass microspheres, the smaller the microballoons, the higher the maximum detonation velocity, which shifted towards a lower degree of aeration.

Similar results were reported by Mendes et al. [41]. Meanwhile, Yunoschev et al. [44] observed a critical diameter–density relationship of a similar nature to that described for GMs in [25]. It is worth noting that plastic microballoons, in addition to their sensitizing effect, also serve as combustible components, which should be considered when formulating EE compositions. The effect of perlite on the properties of EEs can be found in papers [22,28]. Yan et al. [28] studied EEs containing 2, 3, and 4% perlite. The degree of crystallization of the matrix caused by the influence of shock waves generated in water increased with increasing perlite content and was higher than for the same content of GMs. On the other hand, the dependence of aggregability on perlite content, shown in the monograph in [27], was of the same nature as the effect of GMs or microballoon content on the velocity of detonation. The maximum measured value of this parameter was 16 mm at a perlite content of 6%.

### 2.3. Introduction of Constant Inhomogeneities

The sensitization (reduction in the critical diameter) of liquid explosives using solid additives, such as aluminum or $Al_2O_3$ grains, was first described by Kurbangalina in 1969 [46]. A similar solution in the case of EEs concerns substances characterized by relatively high hardness: grains of ferrosilicon, ferrophosphorus, ferromanganese, or $Al_2O_3$ with a grain size of less than 180 μm [47]. The content of such additives, according to patent data, can be up to 50%. However, it seems that their optimal amount should be a few percent. Higher contents of them significantly reduce the detonation and thermochemical characteristics of EEs, due to their inertness in chemical terms, while at the same time taking heat away from the chemical reaction zone, due to the relaxation of thermodynamic parameters between the explosion products described by the additives.

### 2.4. Sensitization of EE Matrix with Explosives

The use of high-energy substances that exhibit explosive properties (e.g., TNT, pentrite, hexogen, octogen, nitrostarch, and smokeless powder) as ingredients that impart upon the EE matrix the ability to detonate from a standard No. 6 primer was first proposed by Tomic [48]. However, interest in incorporating such additives into EEs grew significantly during efforts to dispose of high explosives and propellants recovered from the decommissioning of munitions [48–60]. The purpose of their addition to EEs was to get rid of unnecessary explosive stockpiles and, at the same time, to obtain EEs with improved energy parameters. For example, increasing the 50/50 B composition from 30% to 50% resulted in an increase in detonation velocity from 5559 m/s to 6806 m/s [54,55]. Also, the addition of nitrocellulose powder caused an increase in EE detonation velocity. Charges initiated with an HC-14 booster at nitrocellulose powder contents of 15, 20, and 25% had detonation velocities of 4059, 4282, and 4920 m/s, respectively [56].

In addition to studies of the effect of de-elaborated explosives on the detonation parameters of EEs, analogous experiments were performed with several individual explosives [51–56]. Renick et al. [56] determined the effect of octogen on the critical diameter and detonation velocity of EEs with a composition of ammonium nitrate (71.5%), sodium nitrate (10.0%), water (12.0%), and oil phase (6.5%). Increasing the octogen content from 20% to 60% resulted in a decrease in critical diameter from 52 mm to 15 mm, and an increase in detonation velocity in a 51.5 mm diameter charge, from ~5300 m/s to ~8000 m/s.

Kűnzel et al. [59] used an EE matrix consisting of ammonium nitrate (V) (62.5%), sodium nitrate (V) (13,5%), water (15%), urea (3%), oil phase (6.5%), and custom explosive-erythritol tetranitrate (V) (ETN) for sensitization. The detonation velocities of EEs, measured in a 37 mm diameter plastic tube containing 10, 15, and 20% ETN, initiated with a booster made of Semtex, were 1900, 3490, and 4960 m/s, respectively. An EE containing 20% erythritol tetranitrate (V) was detonated from a fuse containing 0.72 g of pentrite.

Yang et al. [60] studied the effect of the addition of hexogen, with an average grain size of 90 μm, on the detonation velocity, brisance, and critical thickness of EEs with a matrix composition consisting of ammonium nitrate (75.0%), sodium nitrate (10.0%), and water oil phase-70, containing GMs (25%). An increase in hexogen content, from 0% to 20%, resulted in an increase in the detonation velocity at a layer thickness of 20 mm from 3120 to 3523 m/s, an aggregate from 11.9 to 17.6 mm, and a decrease in the critical thickness from 6.0 to 3.3 mm.

The method of the sensitization of emulsion explosives with explosive crushing and blasting is very rarely used. It uses raw materials that are sensitive to mechanical stimuli, which creates an explosive hazard, and the final product is much less resistant to impact and friction than typical EEs. This causes, among other things, a great limitation and difficulty in the mechanization of blasting work.

## 3. Conclusions

The presented literature review on the sensitization of emulsion explosive matrices demonstrates that several sensitization methods exist, each differing in its underlying mechanism. These methods are typically classified into two primary categories: chemical and physical. In chemical sensitization, specific chemical compounds are added to the EE matrix. The decomposition of these compounds produces gases that form bubbles, thereby reducing the density of the matrix. These gas bubbles are adiabatically compressed by the propagating shock wave generated by an initiator, leading to the formation of "hot spots" that are critical for the initiation of detonation. In contrast, physical sensitization encompasses several distinct mechanisms. The incorporation of low-bulk-density substances into the matrix similarly results in the formation of "hot spots" analogous to those produced by chemical gas generation. Furthermore, the introduction of solid heterogeneities—particularly materials with high hardness—represents a conventional approach to sensitizing explosives. The interaction of the shock wave with high-impedance inclusions results in localized temperature increases, which, in turn, trigger the exothermic decomposition of active molecules in their immediate vicinity. Additionally, the inclusion of high-detonation-capable explosive materials into the EE matrix leads to an exothermic reaction upon the passage of a shock wave, thereby creating centers that facilitate the propagation of detonation throughout the entire matrix.

An analysis of the described EE sensitization methods suggests that chemical sensitization is the most optimal approach, particularly when inexpensive components are employed to generate gas bubbles within the matrix. Its principal advantages include low cost and the ability to produce EE in situ. Sensitizing the matrix directly in the blasthole obviates the need for transporting explosive materials via public roads, a factor that is critical for ensuring transport safety and mitigating potential terrorist threats.

The broader field of explosive materials represents a highly specialized domain in both scientific research and industrial applications. Examples include the diverse EE formulations currently in use, as well as various explosive mixtures. Black powder, developed in the 9th century by the Chinese, remains one of the most effective ignition agents; nitrocellulose, synthesized in 1846 by Christian Friedrich Schönbein, continues to serve as a fundamental component in gunpowder formulations; and TNT, invented in 1863 by Julius Wilbrand, is the most commonly used low-sensitivity explosive in military mixtures. A similar scenario applies to emulsion explosives. Despite 56 years having elapsed since their initial patenting by Harold Bluhm, EEs are still regarded as the most advanced blasting agents in mining. Compared to the original patents, significant changes in composition have not been observed, with only modifications such as the incorporation of polyisobutylene succinic anhydride as an emulsifier [61,62] or the development of Low-Water-Composition EEs [63], which include additives like ammonium nitrate and aluminum powder.

Based on the current state of knowledge, it is challenging to delineate new research directions for EEs. A potential innovation in this field may be the development of explosive mixtures in which hydrogen peroxide acts as the oxidizer [64], building on the findings of previous studies by Araos et al. [65–68]. Notably, in the explosives investigated by Araos et al., a matrix based on a concentrated aqueous solution of hydrogen peroxide was sensitized using either glass microspheres or chemical methods.

**Author Contributions:** Conceptualization, A.M.; methodology, A.M.; data curation, B.K. and D.M.; writing—original draft preparation, A.M., B.K. and D.M.; writing—review and editing, A.M., B.K., D.M. and W.J.; funding acquisition, W.J. All authors have read and agreed to the published version of the manuscript.

**Funding:** This research was funded by the National Centre for Research and Development (Poland) under the project "Development of integrated knowledge management system for rescue phases: preparedness, prevention, response, and recovery, for fire protection and population safety", based on agreement GOSPOSTRATEG9/001G/2022 and the work was financed from funds granted to the Fire University, Warsaw, Poland by the Minister of the Interior and Administration, Poland for the maintenance and development of research potential under application RN-1.601.2.2025.

**Data Availability Statement:** Data are contained within the article.

**Conflicts of Interest:** The authors declare no conflicts of interest. The funders had no role in the design of the study; in the collection, analyses, or interpretation of the data; in the writing of the manuscript; or in the decision to publish the results. The authors declare that they have no involvement in any projects related to military applications or sensitive military research.

## References

1. Behera, R.; Biswal, T.; Panda, R.B. Recent Progress in Explosives: A Brief Review. In *Current Advances in Mechanical Engineering*; Acharya, S.K., Mishra, D.P., Eds.; Lecture Notes in Mechanical Engineering; Springer: Singapore, 2021. [CrossRef]
2. Maranda, A.; Wachowski, L.; Kukfisz, B.; Markowska, D.; Paszula, J. Valorization of Energetic Materials from Obsolete Military Ammunition Through Life Cycle Assessment (LCA): A Circular Economy Approach to Environmental Impact Reduction. *Sustainability* **2025**, *17*, 346. [CrossRef]
3. Włodarczyk, E. *Introduction to Explosion Mechanics*; Wyd. Naukowe PWN: Warszawa, Poland, 1994; ISBN 83-01-11594-1. (In Polish)
4. Maranda, A.; Gołąbek, B.; Kasperski, J. *Emulsion Explosives*; Wyd. Naukowo-Techniczne: Warszawa, Poland, 2008; ISBN 978-83-204-1427-9. (In Polish)
5. Chaudri, M.; Field, J. The Role of Rapidly Compressed Gas Pockets in the Initiation of Explosives. *Proc. R. Soc. A* **1974**, *340*, 113–128.
6. Bowden, A.D.; Yoffe, F.P. *Initiation and Growth of Explosion in Liquids and Solids*; Cambridge University Press: Cambridge, UK, 1952; ISBN 978-521-31233-2.
7. Bowden, A.D.; Yoffe, F.P. *Fast Reaction in Solids*; Butterwoths Scientific Publications: London, UK, 1958; ISBN -10 1114784273.
8. Marlow, J.M.; Bush, J.H. Thickened Emulsion Composition for Use a Propellants and Explosives. U.S. Patent 5,936,194, 10 August 1999.
9. Chrisp, D.J. Formation of Foamed Emulsion-Type Blasting Agents. U.S. Patent 4,008,108, 15 February 1977.
10. Mishra, A.K.; Rout, M.; Singh, D.R.; Jana, S.P. Influence of gassing Agent and Density on Detonation Velocity of Bulk Emulsion Explosives. *Geotech. Geol. Eng.* **2018**, *36*, 89–94. Available online: https://link.springer.com/article/10.1007/s10706-017-0308-7 (accessed on 7 January 2025). [CrossRef]
11. Kramarczyk, B.; Pytlik, M.; Mertuszka, P.; Jaszcz, K.; Jarosz, T. Novel Sensitizing Agent Formulation for Bulk Emulsion Explosives with Improved Energetic Parameters. *Materials* **2022**, *75*, 900. [CrossRef] [PubMed]
12. Kramarczyk, B.; Suda, K.; Kowalik, P.; Świątek, K.; Jaszcz, K.; Jarosz, T. Emulsion Explosives: A Tutorial Review and Highlight of Recent Progress. *Materials* **2022**, *15*, 4952. [CrossRef]

13. Cheng, Y.F.; Ma, H.H.; Shen, Z.W. Detonation Characteristic of Emulsion Explosives Sensitized by $MgH_2$. *Combust. Explos. Shock Waves* **2013**, *49*, 614–619. Available online: https://link.springer.com/article/10.1134/S0010508213050134 (accessed on 7 January 2025). [CrossRef]
14. Cheng, Y.F.; Ma, H.H.; Liu, R.; Shen, Z.W. Explosion Power and Pressure Desensitization Resisting Property of Emulsion Explosives Sensitized by $MgH_2$. *J. Energetic Mater.* **2014**, *32*, 207–218. [CrossRef]
15. Cheng, Y.F.; Ma, H.H.; Liu, R.; Shen, Z.W. Pressure Desensitization Influential Factors and Mechanism of Magnesium Hydride Sensitized Emulsion Explosives. *Propellants Explos. Pyrotech.* **2014**, *39*, 267–274. [CrossRef]
16. Cheng, Y.F.; Wang, Q.; Liu, F.; Ma, H.H.; Shen, Z.W.; Guo, Z.R.; Liu, R. The Effect of Energetic Additive Coated $MgH_2$ on the Power of Emulsion Explosives Sensitized by Glass Microballoons. *Cent. Eur. J. Energetic Mater.* **2016**, *73*, 707–713. [CrossRef] [PubMed]
17. Cheng, Y.F.; Yan, S.L.; Ma, H.H.; Shen, Z.W.; Liu, R. A New Type of Functional Chemical Sensitizer $MgH_2$ for Improving Pressure Desensitization Resistance for Emulsion Explosives. *Shock Waves* **2016**, *26*, 213–219. [CrossRef]
18. Huang, W.; Wu, H.; Yan, Y. Relationship Research between Crystallization Quantity of Emulsion Explosive and Desensitization Degree under Dynamic Pressure. *Adv. Mater. Res.* **2012**, *393–395*, 1389–1393. [CrossRef]
19. Bluhm, H.F. Ammonium Nitrate Emulsion Blasting Agents and Method of Preparing Same. U.S. Patent 3,447,978, 3 June 1969.
20. Wade, C. Water-In-Oil Emulssion Explosive Containing Entrapped Gas. U.S. Patent 3,715,247, 3 September 1970.
21. Lee, J.; Persson, P.A. Detonation Behavior Emulsion Explosives. *Propellants Explos. Pyrotech.* **1990**, *75*, 208–216. [CrossRef]
22. Xuguang, W. *Emulsion Explosives*; Metallurgical Industry Press: Beijing, China, 1993; ISBN 75-02-41574-2.
23. Deribas, A.A.; Medvedev, A.E.; Reshetnyak, A.Y.; Fomin, V.M. Detonation of Emulsion Explosives Containing Hollow Microspheres. *Dokl. Phys.* **2003**, *48*, 163–165. [CrossRef]
24. Price, D. Contrasting Pattems in Behavior of High Explosives. In *Symposium (International) on Combustion*; Elsevier: Amsterdam, The Netherlands, 1967; Volume 77, pp. 693–702. [CrossRef]
25. Yoshida, M.; Lida, M.; Tanaka, K.; Fujiwara, S.; Kusakabe, M.; Shiino, K. Detonation Behavior of Emulsion Explosives Containng Glass Microballoons. In Proceedings of the 8th Symposium (International) on Detonation, Albuquerque, NM, USA, 15–19 July 1985.
26. Sumiya, F.; Hirosaki, Y.; Kato, Y. Detonability of Emulsion Explosives Precompressed by Dynamie Pressure. *Sci. Tech. Energetic Mater.* **2004**, *65*, 88–93.
27. Sumiya, F.; Hirosaki, Y.; Kato, Y. Influence of Pressure Wave Propagating in Compressed Emulsion Explosives on Detonator. *Sci. Tech. Energetic Mater.* **2005**, *66*, 266–273.
28. Yan, S.H.; Wu, H.; Liu, F. Influence of Sensitizing Method on Crystallization of Emulsion Explosives under Dynamic Pressure. In Proceedings of the International Autum Semina on Propellants, Explosives and Pyrotechnics. Theory and Practice of Energetic Materials, Kunming, China, 22–25 September 2009.
29. Sil'vestrov, V.V.; Plastinin, A.V. Investigation of Low Detonation Velocity Emulsion Explosives. *Combust. Expl. Shock Waves* **2009**, *45*, 618–626. [CrossRef]
30. Cheng, F.; Meng, X.R.; Feng, C.T.; Wang, Q.; Wu, S.S.; Ma, H.H.; Shen, Z.W. The Effect of the Hydrogen Containing Material $TiH_2$ on the Detonation Characteristics of Emulssion Explosives. *Propellants Explos. Pyrotech.* **2017**, *42*, 585–591. [CrossRef]
31. Morsi, K.; Daoush, W.M. Al-TiH2 Composite Particles as Foaming Precursors for Metallic Foams. *Scr. Mater.* **2015**, *705*, 6–9. [CrossRef]
32. Rajak, D.K.; Kumaraswamidhas, L.A.; Das, S. Investigation and Characterization of Aluminium Alloy Foams with $TiH_2$ as a Foaming Agent. *Mater. Sci. Technol.* **2016**, *32*, 1338–1345. [CrossRef]
33. Bjarholt, G. Expansion Works in Underwater Detonation. In Proceedings of the 6th Symposium (International) on Detonation, San Diego, CA, USA, 24–27 August 1976.
34. Maranda, A.; Włodarczyk, E.; Serafmowicz, J. Analysis of Detonation Parameters of Emulsion Explosives Sensitized with Glass Microspheres Enveloping Air. *Biul. WAT* **1986**, *35*, 25–38. (In Polish)
35. Anshits, A.G.; Anshits, N.N.; Deribas, A.A.; Kasatkina, N.S.; Plastinin, A.V.; Reshetnyak, A.Y.; Sil'vestrov, V.V. Detonation Velocity of Emulsion Explosives Containing Cenospheres. *Combust. Expl. Shock Waves* **2005**, *47*, 591–598. [CrossRef]
36. Fang, H.; Cheng, Y.-F.; Tao, C.; Su, H.; Gong, Y.; Chen, Y.; Shen, Z.-W. Effects of Content and Particle Size of Cenospheres on the Detonation Characteristics of Emulsion Explosive. *J. Energetic Mater.* **2021**, *39*, 197–214. [CrossRef]
37. Chaudhri, M.M.; Almgren, L.A.; Persson, A. Detonation Behaviour of a „Water-In-Oil" Type Emulsion Explosives Containing Glass Microballoons of Selcted Sizes. In Proceedings of the 10th Symposium (International) on Detonation, Boston, MA, USA, 12–16 July 1993.
38. Vattipalli, M.R.; Ghosh, P.K. Elastomeric Polymers as Potent Additives for Emulsion Explosives. In Proceedings of the 27th International Annual Conference of ICT. Energetic Materials-Technology, Manufacturing and Processing, Karlsruhe, Germany, 25–28 June 1996.

39. Hirosaki, Y.; Murata, K.; Kato, Y.; Itoh, S. Detonation Behavior of Emulsion Explosives. In Proceedings of the 32nd International Annual Conference of ICT. Ignition, Combustion and Detonation, Karsluhe, Germany, 3–6 July 2001.
40. Hirosaki, Y.; Murata, K.; Kato, Y.; Itoh, S. Detonation Characteristics of Emulsion Explosives as Functions of Void Size and Volume. In Proceedings of the 12th Symposium (International) on Detonation, San Diego, CA, USA, 1 July 2002.
41. Mendes, R.; Ribeiro, J.; Plaksin, I.; Campos, J.; Tavaras, B. Differences between the Detonation Behavior of Emulsion Explosives Sensitized with Glass or with Polymeric Microballoons. *J. Phys. Conf. Series* **2014**, *500*, 052030. [CrossRef]
42. Bednarczyk, E.; Maranda, A.; Paszula, J.; Papliński, A. Studies of Effect of Aluminium Powder on Selected Parameters of Emulsion Explosives Sensitized with Microballoons. *Chemik* **2016**, *70*, 41–50.
43. Bordzilovski, S.A.; Karakhanov, S.M.; Plastinin, A.V.; Rafeichik, S.I.; Yunoshev, A.S. Detonation Temperature of Emulsion Explosive with a Polymer Sentizier. *Combust. Expl. Shock Waves* **2017**, *53*, 730–737. [CrossRef]
44. Yunoshev, A.S.; Plastinin, A.V.; Rafeichik, S.I. Detonation Velocity of an Emulsion Explosive Sensitized with Polymer Microballoons. *Combust. Expl. Shock Waves* **2017**, *53*, 738–743. [CrossRef]
45. Yunoshev, A.S.; Bordzilovski, S.A.; Voronin, M.S.; Karakhanov, S.M.; Makarov, S.N.; Plastinin, A.V. Detonation Pressure of an Emulsion Explosive Sensitized by Polymer Microballoons. *Combust. Expl. Shock Waves* **2019**, *55*, 426–433. [CrossRef]
46. Kurbangalina, R.K.H. Dependence of the Critical Diameter of Liquid Explosives on the Powder Content. *Prik. Mekh. Tekh. Fiz.* **1969**, *4*, 133–138. (In Russian)
47. Mullay, J. Solid Sensitizer for Water-In-Oil Emulsion Explosives. U.S. Patent 4,453,989, 12 June 1984.
48. Tomic, E.A. Emulsion Type Explosives Composition Containing Ammonium Stearate or Alkali Metal Stearate. U.S. Patent 3,770,522, 6 November 1973.
49. Machacek, O.; Eck, G.R. Waste Propellants and Smokeless Powders as Ingredients in Commercial Watergel Explosives. In Proceedings of the 23rd International Annual Conference of ICT, Waste Management of Energetic Materials and Polymers, Karlsruhe, Germany, 30 June–3 July 1992.
50. Rosa, P.C.; Walter, B.G.; Machacek, O. Solid Sensitizer for Beneficial of Containing Wastes. U.S. Patent 5,612,507, 1997.
51. Munson, W.O. Demilitarization of Large Rocket Motors and Propellant Utilization. In *Application of Demilitarized Gun and Rocket Propellants in Commercial Explosives*; Machacek, O., Ed.; Kluwer Academic Publishers: Dortrecht, The Netherlands; Boston, MA, USA; London, UK, 1999; ISBN 0-793-6697-2.
52. Matseevich, B.V.; Mokhova, N.V.; Malygin, N.K. Used of Converted High Energy Value Explosive Materials as Industrial Energetic Materias. In *Application of Demilitarized Gun and Rocket Propellants in Commercial Explosives*; Machacek, O., Ed.; Kluwer Academic Publishers: Dortrecht, The Netherlands; Boston, MA, USA; London, UK, 1999; ISBN 0-793-6697-2.
53. Glinskiy, V.P.; Mardasov, O.F.; Mochova, M.V.; Shalygin, N.K.; Obrazcova, B.F. Creation of Safe on Manipulation Industrial Explosives and Products for Mining Industry on the Basis of Gunpowder. In *Application of Demilitarized Gun and Rocket Propellants in Commercial Explosives*; Machacek, O., Ed.; Kluwer Academic Publishers: Dortrecht, The Netherlands; Boston, MA, USA; London, UK, 1999; ISBN 0-793-6697-2.
54. Nemec, O.; Novotny, M.; Yungova, M.; Zeman, S. Preliminary Verification of Fortification of W/O Type Emulsion with Demilitarized Explosives Based on TNT. In Proceedings of the 14th Seminar on New Tends in Research of Energetic Materials, Part II, Pardubice, Czech Republic, 13–15 April 2011.
55. Nemec, O.; Yungova, M.; Zeman, S. Modification of W/O Emulsions by Demilitarized Composition B. *Propellants Explos. Pyrotech.* **2013**, *38*, 142–146. [CrossRef]
56. Renick, J.D.; Persson, P.A.; Sanchez, J.A. Detonation Properties of Mixtures of HMX and Emulsion Explosives. In Proceedings of the 9th Symposium (International) on Detonation, Portland, OR, USA, 28 August–1 September 1989.
57. Lipińska, K.; Lipiński, M.; Maranda, A. Demilitarized Double Base Propellants as Ingredients of Commercial Explosives. *Cent. Eur. J. Energetic Mater.* **2005**, *2*, 69–78.
58. Biegańska, J. Using Nitrocellulose Powder in Emulsion Explosives. *Combust. Expl. Shock Waves* **2011**, *47*, 366–368. [CrossRef]
59. Kunzel, M.; Nemec, U.; Matys, R. Erythritol Tetranitrate as a Sensitizer in Ammonium Nitratebased Explosives. *Cent. Eur. J. Energetic Mater.* **2013**, *70*, 351–358.
60. Yang, M.; Ma, H.; Shen, Z.H. Effects of RDX Powders on Detonation Characteristics of Emulsion Explosives. *J. Energetic Mater.* **2019**, *37*, 459–474. [CrossRef]
61. Zhang, K.M.; Zhao, H.R. Historical Perspective. Perspectives in the Stability of Emulsion Explosives. *Adv. Colloid Interface Sci.* **2022**, *307*, 1027. [CrossRef] [PubMed]
62. Zhang, K.M.; Xu, M.X.; Hao, X.; Zhao, H.R. Peculiarities of Rheological Behavior of Highly Concentrated Water-In-Oil Emulsion: The Role of Droplet Size, Surfactant, Oil and Ammonium Nitrate Content. *J. Mol. Liq.* **2018**, *272*, 539. [CrossRef]
63. Gołąbek, B.; Kasperski, J. The Influence of the Composition and Structure of Emulsion Explosives on Their Detonation Parameters. Ph.D. Thesis, Central Institute of Mining, Katowice, Poland, 2006.
64. Hypex Bio. Sustainable Mining Explosive. Available online: https://www.swedishmininginnovation.se/wp-content/uploads/2023/06/HypexBio.pdf (accessed on 28 January 2025).

65. Araos, M. Improved Explosive Composition. Patent WO 2013/0132272, 31 January 2013.
66. Araos, M.; Onederra, I. Detonation Characteristics of Alternative Mining Explosives Based on Hydrogen Peroxide as the Oxidizing Agent. In Proceedings of the 7th EFEE World Conference on Explosives and Blasting, Moscow, Russia, 15–17 September 2013; pp. 182–188.
67. Araos, M.; Onederra, I. Development of a Novel Mining Explosive Formulation to Eliminate Nitrogen Oxide Fumes. *Min. Technol.* **2015**, *124*, 16–23. [CrossRef]
68. Araos, M.; Onederra, I. Detonation Characteristics of a NOx-Free Mining Explosive Based on a Sensitized Mixture of Low-Concentration Hydrogen Peroxide and Fuel. *Cent. Eur. J. Energetic Mater.* **2017**, *14*, 759–774. [CrossRef]

MDPI AG
Grosspeteranlage 5
4052 Basel
Switzerland
Tel.: +41 61 683 77 34

*Applied Sciences* Editorial Office
E-mail: applsci@mdpi.com
www.mdpi.com/journal/applsci

www.ingramcontent.com/pod-product-compliance
Lightning Source LLC
Chambersburg PA
CBHW061843210926
51850CB00006B/957